POWER
OVER PEOPLES

THE PRINCETON ECONOMIC HISTORY
OF THE WESTERN WORLD

Joel Mokyr, Editor

POWER
OVER PEOPLES

TECHNOLOGY, ENVIRONMENTS,

AND WESTERN IMPERIALISM,

1400 TO THE PRESENT

DANIEL R. HEADRICK

PRINCETON UNIVERSITY PRESS   PRINCETON AND OXFORD

❋ CONTENTS ❋

❈ ACKNOWLEDGMENTS ❈

Many people and institutions have helped me bring this book to fruition. To all of them I owe a deep debt of gratitude.

First, I wish to thank my friends Joel Mokyr, Alex Roland, Brad Hunt, Michael Bryson, and Suzanne Moon for their encouragement and helpful advice. I am sincerely grateful to Giancarlo Casale for allowing me to read and learn from his dissertation, "The Ottoman Age of Exploration," before it was available to the public. And I thank Brigitta van Rheinberg, Clara Platter, Jill Harris, Heath Renfroe, and Jennifer Backer of Princeton University Press for the enthusiasm with which they guided my manuscript through the editorial process.

I am most grateful to Roosevelt University for granting me a research leave, as well as to the helpful staff at the Murray Green Library of Roosevelt University for locating unusual books. I also thank the Regenstein and Crerar libraries of the University of Chicago for providing me with more information than I dreamed existed and a quiet space to read a small part of it.

A number of institutions and scholarly societies invited me to present parts of this work: the Great Lakes History Conference at Grand Valley State University in Grand Rapids, Michigan; the Society for Military History Conference in Manhattan, Kansas; the Technology in the Colony Conference at Harvey Mudd College in Claremont, California; Embry-Riddle Aeronautical University in Daytona Beach, Florida; the University of Illinois in Chicago; the Newberry Library in Chicago; and the Media and Imperialism Conference at the University of Amsterdam. To them and to the audiences that asked intelligent and challenging questions, I express my gratitude.

I thank my wife, Kate, for her unstinting support and encouragement.

But most of all, I would like to thank my friend and mentor, William H. McNeill, for four decades of inspiration, advice, and guidance in understanding and writing world history. I dedicate this book to him.

Daniel Headrick
New Haven, Connecticut, 2008

POWER
OVER PEOPLES

❊ Introduction ❊

Imperialism and Technology

For over five centuries, Europeans and their overseas descendants dominated the oceans of the world and much of its land and its peoples as well. This domination has been challenged many times, as it is once again in our times. Now that imperialism has returned to the forefront of world events, it is time to revisit its history and learn its lessons.

On Imperialism

Western imperialism is but the most recent example of a phenomenon going back to ancient times and culminating in the conquests of Genghis Khan. The first phase in the expansion of Europe, often called the Old Empires, began in the early sixteenth century with the Spanish conquest of Mexico and Peru and the Portuguese domination of the Indian Ocean; by the beginning of the nineteenth century, however, Western efforts in China, Central Asia, Africa, and the Americas were running into diminishing returns. Then in the mid-nineteenth century came a renewed spurt of empire-building—the New Imperialism—that lasted until the outbreak of World War II. Since World War II we have been in a third phase, one in which the Western powers (including Russia) have tried to hold on to their colonies and dependencies and even expand their spheres of influence, but in vain.

Historians have written in great detail about Western imperialism, often under the name "expansion of Europe." The second phase, or New Imperialism, has long been the object of controversy among historians because of its extraordinary speed and scope; by one account, the land area of the world controlled by Europeans increased from 35 percent in 1800 to 84.4 percent in 1914.[1] In trying to explain this phenomenal expansion, historians have focused on the motives of the explorers, missionaries, merchants, military men, diplomats, and political leaders involved. These motives were as diverse as the protagonists. Some wanted to spread Christianity or Western ethics, laws, and culture around the world. Others sought valuable commodities or markets for their goods or investment opportunities. Yet others saw imperial expansion as a means of achieving personal glory, national prestige, or strategic advantage. And, of course, many had several concurrent motives.[2] But in their fascination with the motivations of the imperialists, most historians took for granted that the European powers and the United States had the technical and financial means to turn their ambitions into reality. A few referred to these means as "prerequisites" or "a disequilibrium" or "a power advantage" without further investigation.[3] Others considered the issue too trivial to mention.

Yet events require not only motives and opportunities but also means. By what means did imperialists carry out their ambitions? This is the question to which I turned twenty-five years ago in a book called *The Tools of Empire: Technology and European Imperialism in the Nineteenth Century.*[4] In it, I described the relationship between technological innovations and the European colonial conquests in Africa and Asia during the New Imperialism. Among the factors that explain this dramatic expansion, certain technological innovations—in particular steam engines, better firearms, and medical advances—played a major role. Technology is now widely recognized as a necessary, if not sufficient, explanation for the New Imperialism in Africa and Asia.

As an author, it is gratifying to realize that my book has reached a large audience and that the conclusions I drew about the role of technology in nineteenth-century European imperialism have been cited and disseminated in numerous other books. However, in drawing conclusions from the events of one time and place, there is the temptation to generalize from that specific situation to others—in short, to turn a contingent conclusion into a *law* of history. If one accepts the idea that technological innovations were essential to the European conquests of the nineteenth century, does it then follow that technological factors explain other conquests at other times in the past? Does it follow that the key to successful conquests in our own times lies in possessing more powerful technologies than one's opponents? Or was the case of nineteenth-century European imperialism a fluke, an aberration? These are the questions that prompted me to write this book.

On Technology

Before I attempt to address these questions, let me propose a definition of technology. Simply put, by technology I mean all the ways in which humans use the materials and energy in the environment for their own ends, beyond what they can do with their bodies. Under the rubric of technology I would include not only artifacts and domesticated plants and animals but also the skills needed to use them and the systems in which they are embedded: playing the violin, for instance, but not singing; riding a horse, but not running; writing a letter or talking on the telephone, but not speaking to someone within earshot; using drugs, rather than prayers, to combat disease. In their creation as well as in their use, all technologies are the results of human ingenuity. The history of technology is the story of humans' increasing ability to manipulate nature, from Stone Age hand axes to nuclear bombs, from dugout canoes to supertankers, from gardening to genetic engineering.

As technologies change, new ones are often deemed "superior" to the old ones they replace. What we mean by a "superior" technology is one that gives its owners more power over nature—for instance, the ability to travel faster, to communicate further, to live longer, healthier lives, or to kill more efficiently—than to those who lack it. But such superiority is instrumental, that is, it allows people to do more. This is not the same as moral superiority. We must be careful not to confuse the two.

We associate technological innovations with Western civilization. Seen over the course of human history, however, the technological advantage of the West over other cultures is a recent phenomenon. Until the fifteenth century, the Chinese and the Arabs were at the forefront of technology. Not until the mid-fifteenth century did western Europe begin to forge ahead. The innovativeness of the West came from two sources. One is a culture that encouraged the domination of nature through experimentation, scientific research, and the rewards of capitalism. The other is the competitive nature of the Western world, in which states powerful enough to challenge one another—Spain, France, Britain, Germany, Russia, and the United States—at one time or another vied for dominance over Europe. Nor were states the only competitive elements in Europe civilization. Bankers and traders competed with one another and encouraged competition among kings and states. Individuals, some of them rootless adventurers, sought glory, wealth, and honors through heroic deeds.[5]

Technologies are never evenly distributed. Their skewed distribution allows—though it does not oblige—those who possess a given technology to share it with others, to withhold it from them, or to use it against them. Thanks to the technologies they command, some people enjoy longer, healthier, more comfortable and exciting lives than others. Weapons, means of surveillance, and systems of organization can be used to coerce or intimidate. This disparity gives some people an advantage over others; in the words of philosopher Leon Kass, "What we really mean by 'Man's Power over Nature' is a power exercised by some men over other men, with a knowledge of nature as their instrument."[6] It is the disparities in knowledge of nature—and in the institutions such as universities, governments, and corporations that transform that knowledge into practical applications—that have fueled the disparities of power associated with technological change.

Although no technology forces people to use it, any new form of power over nature creates a powerful temptation to do so. There have been cases in which nations and their leaders have turned away from using a well-known technology; since 1945, for example, the nuclear powers have avoided using their atomic weapons. But all too often nations, like individuals, succumb to temptation. Once it becomes possible, for instance, to put men on the moon or to keep a body alive after its brain has died, it is tempting to do so. Likewise, it is tempting to use one's technological advantage to coerce other people into doing one's bidding. In most societies, this disparity is evident in the police powers of the state. Between states, it is expressed in unequal economic or military might, and sometimes in war. When a powerful state uses force or the threat of force to impose its will on a weaker society, especially when the weaker society belongs to another culture, we call that imperialism.

What are the connections between technological innovations and Western imperialism? One of the connections, as historian Michael Adas has argued, is Western arrogance, a belief that technological superiority demonstrates religious, cultural, and even biological superiority over non-white peoples.[7] Another connection is the desire to conquer and control other peoples; a technological advantage is itself a motive for imperialism. The temptation to coerce is especially compelling when weaker societies fail to live up to the expectations of the powerful, for instance by practicing a different religion, treating their people in ways that offend, threatening their neighbors, or withholding valuable resources.

From the fifteenth to the eighteenth centuries, as historians Carlo Cipolla and Geoffrey Parker have shown, European technological superiority manifested itself outside Europe in the form of ships and guns.[8] During the nineteenth century, as I argued in *The Tools of Empire*, the key technologies were steamboats, steamships, rifles, quinine prophylaxis, and the telegraph, all of them products of the Industrial Revolution. The twentieth century saw a great many technological advances, the most striking being airplanes. Not surprisingly, these periods of technological creativity correspond to eras of Western expansion. The competitive nature of Western society has been the engine of both technological innovations and imperialism.

Yet our study of the relations between technology and imperialism is not complete without considering two other factors. One is the environments in which imperial expeditions took place. Nature is extremely varied, and so is its influence on historical events. Western imperialism was subject to environmental forces as much as by the encounters between peoples. In some situations, environmental factors greatly aided the conquerors; the diseases imported by Spaniards to the Americas are the most famous case. In others, such as diseases in Africa, the environment was an obstacle to conquest. But saying that environments influence events does not mean that geography determines history, as Jared Diamond and others have asserted, but only that it poses challenges to the technical ingenuity of the protagonists.[9]

Although most technological advances originated in the West in the period we are studying, it does not follow that other cultures were passive victims. Some societies, it is true, submitted to Western conquest and domination. Others attempted to emulate the technology of the West, sometimes successfully, as in the case of Japan, but more often unsuccessfully, like early nineteenth-century Egypt. Yet others—and these are the most interesting cases—found alternative ways of resisting the Western pressure, relying on indigenous or simple technologies; we call the resulting conflicts asymmetrical. Power over nature may be permanent, but power over people is often ephemeral.

The Goal and Organization of This Book

The goal of this book is to analyze the role of technology in the global expansion of Western societies from the fifteenth century to the present. To explain this role, we must take into account three factors. One is the use of technology to master particular natural environments, in other words, its power over nature. The second is the technological innovations that permitted the Western powers to conquer or coerce non-Western societies. And the third is the responses of non-Western societies, technological and otherwise, to the Western pressures. In short, this book aims to be a technological, environmental, and political history of Western imperialism in the past six hundred years.

The first phase of European expansion began in the fifteenth century, at a time when Christian Europe, though dynamic and growing in population, was hemmed in on the south and east by powerful and hostile Muslim states. To escape the confinement of their continent and adjacent seas, a few adventurous Europeans turned toward the ocean. But the ocean was a dangerous environment. Chapter 1,

"The Discovery of the Oceans, to 1779," therefore recounts how Europeans mastered the Atlantic, Indian, and Pacific oceans through the technologies of shipbuilding and navigation.

Their goals were not exploration but military, commercial, and religious domination. Chapter 2, "Eastern Ocean Empires, 1497–1700," discusses the use of new naval technologies in establishing naval empires, beginning with that of the Portuguese in the Indian Ocean. On the oceans, Europeans encountered little resistance and often no other ships of any sort. In coastal waters and on narrow seas, however, they encountered barriers that were both environmental and technological. There, they faced stiff opposition and sometimes suffered serious reverses at the hands of the Ottomans, the Chinese, and the Gulf Arab states.

At the same time the Portuguese were attempting to dominate the Indian Ocean and its approaches, Spaniards were undertaking the conquest of a land empire in the Americas. This story has received a great deal of scholarly attention, of course, so chapter 3, "Horses, Diseases, and the Conquest of the Americas, 1492–1849," concentrates on the roles that technologies new to the Western Hemisphere, especially horses and steel weapons, played in the conquest. It also highlights the importance of the diseases that Spaniards introduced into the New World. But it contrasts this story of triumph with another, less well-known one, namely the resistance of certain Indian tribes and the resulting inability of the Europeans to extend their conquests into the grasslands of South and North America.

Chapter 4, "The Limits of the Old Imperialism: Africa and Asia to 1859," takes the story to the Eastern Hemisphere where it considers two anomalies. One was the Portuguese failure to emulate in Africa the victories of the Spaniards in the Americas; the other is the equally surprising conquest of India by the British. In both cases, the role of technology is analyzed and found wanting. Instead, the better explanation for the African experience is the disease environment, and for India, it is tactics and organization rather than weapons. By the early nineteenth century, however, the European advantage based on weapons, tactics, and organization had reached its limits, as shown by the British defeat in Afghanistan, the very costly

French conquest of Algeria, and the many failed attempts at penetrating sub-Saharan Africa.

Just when Europeans seemed to have reached the limit of their
ability to conquer other peoples, they acquired new means of advancing where their predecessors had been stymied. In the early
nineteenth century, we enter the second phase, an era marked by
the technological innovations of the Industrial Revolution and the
advances in science since the Enlightenment. Western industrialization had two kinds of impacts on the rest of the world: the demand
for its products and the means of conquest and colonization. On
the demand side, Western industrialization stimulated a ravenous
appetite for raw materials and exotic stimulants. At the same time,
industrialization provided Western nations with the means to expand their spheres of influence and impose their will on non-Western peoples for the purpose of obtaining these necessities and
achieving the other goals of the empire-builders. The next three
chapters examine the three technical and scientific breakthroughs
that had the most impact on the New Imperialism of the nineteenth
century: steamboats, medical technologies, and firearms. They will
cover much of the ground covered in *The Tools of Empire* but in
more detail and extend the scope of its argument to the Americas.

On water, the breakthrough occurred when steam power was applied to boats and ships. This allowed steam-powered gunboats to
enter the shallow seas and rivers that had been closed to the sailing
ships of earlier times. The development of steam navigation and its
consequences for the relations between the West and the non-West
are the subject of chapter 5, "Steamboat Imperialism, 1807–1898."

For four centuries, Africa had been closed to Europeans because
of the diseases it harbored, especially malaria. In chapter 6, "Health,
Medicine, and the New Imperialism, 1830–1914," we look at the
medical advances in the nineteenth century that opened Africa to
European penetration, as well as the role that disease, medicine, and
public health played in other parts of the world.

The nineteenth century witnessed the most rapid expansion of
Europeans and their descendants into regions previously off-limits,
an expansion variously known as the Scramble for Africa, the Winning of the West, or *la Conquista del Desierto*. What made this expan-

sion not only possible but cheap, easy, and rapid was the revolution in firearms, a by-product of the Industrial Revolution and of the rivalries and wars among Europeans and Americans. This revolution and its consequences for the non-Western world are the subject of chapter 7, "Weapons and Colonial Wars, 1830–1914."

By the turn of the nineteenth century global relations entered a new phase, as non-Western peoples began to acquire weapons and to learn tactics similar to those that Europeans had successfully employed for a generation. Just as the old methods were no longer as effective as they had once been, Americans and Europeans created an entirely new technology—aviation—that promised to give them back the advantage they were losing on the ground. Chapter 8, "The Age of Air Control, 1911–1936," looks at the impact of airplanes on various imperial ventures by Italy, the United States, Great Britain, and Spain before World War II.

After World War II, despite incredible technological advances, those who possessed the most advanced technologies were no longer able to dominate those who did not. Although France, the Soviet Union, and the United States achieved dominance of the air over insurgents with weak air defenses, these insurgents defeated them on the ground, in mountains, forests, wetlands, and cities. This will be the theme of chapter 9, "The Decline of Air Control, 1946–2007."

Thus we will end with a paradox: the greater power over nature that superior technology provides does not necessarily confer power over people with less advanced technologies. Yet the pursuit of ever more advanced technologies continues unabated and, with it, the temptation to use them against other peoples.

Notes

1. D. K. Fieldhouse, *Economics and Empire* (Ithaca: Cornell University Press, 1973), p. 3.

2. Among the many scholars who have analyzed the motives of the imperialists, we might mention J. A. Hobson, Ronald Robinson and John Gallagher, Vladimir Lenin, Henri Brunschwig, Hans-Ulrich Wehler, David Landes, Hannah Arendt, Carleton J. H. Hayes, William Langer, Joseph Schumpeter, Geoffrey Barraclough, and D. K. Fieldhouse.

3. This is the case, for example, of Fieldhouse (*Economics and Empire*, pp. 460–61), who wondered, "Why did the critical period of imperialism happen to occur in these thirty years after 1880?"

4. Daniel R. Headrick, *The Tools of Empire: Technology and European Imperialism in the Nineteenth Century* (New York: Oxford University Press, 1981).

5. This is how Felipe Fernández-Armesto explains the sudden expansion of Europe in the fifteenth and sixteenth centuries; see *Pathfinders: A Global History of Exploration* (New York: Norton, 2006), pp. 144–45.

6. Leon Kass, "The New Biology: What Price Relieving Man's Estate?" *Science* 174 (November 19, 1971), p. 782.

7. Michael Adas, *Machines as the Measure of Men: Science, Technology, and Ideologies of Western Dominance* (Ithaca: Cornell University Press, 1989).

8. Carlo Cipolla, *Guns, Sails, and Empires: Technological Innovation and the Early Phases of European Expansion, 1400–1700* (New York: Random House, 1965); Geoffrey Parker, *The Military Revolution: Military Innovation and the Rise of the West, 1500–1800*, 2nd ed. (Cambridge: Cambridge University Press, 1996).

9. For a brilliant popular example of geographical determinism, see Jared M. Diamond, *Guns, Germs, and Steel: The Fates of Human Societies* (New York: Norton, 1997).

Chapter I

The Discovery of the Oceans, to 1779

On September 8, 1522, a small ship, the *Victoria*, berthed at the Spanish port of Seville. The next day, its eighteen weakened and bedraggled crew members walked barefoot, clad only in their shirts and carrying long candles, to the church of Santa Maria de la Victoria to give thanks for their safe return. Theirs was the first ship to sail around the world. Their arrival marked a milestone in the long struggle to master the seas and oceans of the world.

For centuries, ambitious and adventurous Europeans had sought ways to escape the narrow confines of their subcontinent. Rumors of mythical lands—Guinea, the land of gold, the fabled isle of Antilia, the kingdom of Prester John, and Marco Polo's empire of the Great Khan—enticed them with dreams of conquest, fame, and riches. Yet they were blocked to the south by Arabs and to the east by Turks, hostile Muslim states that barred the way to the world beyond. Earlier attempts to break out of Europe by way of the eastern Mediterranean—the Crusades—had failed. That left only one way out: the North Atlantic Ocean beyond the coastal waters of Europe. In this chapter we will recount the story of the navigators and the ships, as well as the knowledge of navigation and geography with which they mastered this vast and dangerous new environment, and leave their encounters with other peoples to later chapters.

Five Seafaring Traditions

People had navigated the oceans long before the sixteenth century. Coastal residents everywhere had built boats and devised techniques of navigation suited to the specific maritime environment on whose shores they lived, some bold mariners venturing out of sight of land for days or weeks at a time. During those centuries, five great maritime traditions evolved.[1]

Among the great seafaring traditions, that of the Polynesians surely ranks first. Captain James Cook, the first European to thoroughly explore the Pacific Ocean in the eighteenth century, remarked of them: "It is extraordinary that the same Nation could have spread themselves over all the isles of this Vast Ocean from New Zealand to this [Easter] Island which is almost a fourth part of the circumference of the Globe."[2] Thirty-five hundred years before Cook, the Lapita people, ancestors of the Polynesians, had sailed from the Bismarck Archipelago to Vanuatu and New Caledonia, a distance of over a thousand miles. By 1300 BCE they had reached Fiji and two hundred years later, Tonga and Samoa. In the first centuries CE they reached the Marquesas and Society Islands; by the fifth century, they were in Easter Island, six thousand miles from the Bismarcks; by the eighth century, they had settled Hawaii and New Zealand.

Pacific islanders conquered the ocean in open, double-hulled dugout canoes. Again, Captain Cook: "In these Proes or Pahee's as they call them from all accounts we can learn, these people sail in those seas from Island to Island for several hundred Leagues, the Sun serving them for a compass by day and the Moon and Stars by night."[3] Their craft may have been simple, but their skills were nothing short of astounding. They sailed for weeks out of sight of land without instruments, navigating by the position of the sun and stars and by the feel of long ocean swells. Long before they spotted land, they could sense its presence over the horizon by watching the flight of birds that roosted on land and flew out to sea during the day to catch fish. They could also read the color of the clouds; if the underside of a cloud appeared green, it meant there was an island below it.[4]

Though Polynesians reached a third of the way around the world, their craft and skills were limited to the tropics. Near the equator, the stars travel directly overhead rather than at an angle to the horizon as they do in the temperate zones, making star-sighting more reliable.[5] In the Pacific, the trade winds blow from east to west during most of the year, but unlike in the Atlantic, they reverse themselves briefly in December and January. By carefully choosing the date of their departure, mariners could rely on winds to bring them home, should they fail to find land.[6] Their techniques would not have served them in the temperate zone, nor could their sailors have survived cold climates in open boats. Finally, their craft, though suitable for long exploration voyages, were too small to carry much cargo. Once they had discovered and settled all the inhabitable islands from New Zealand to Hawaii, they lost the motivation to continue long-distance voyages. By the fifteenth century, their travels were almost entirely local, and the outlying islands—Hawaii, Easter, New Zealand, and the Chathams—became isolated from the rest of Polynesia.

As a highway of trade, the Indian Ocean was far more important than the Pacific. North of the equator, the monsoon winds follow a regular pattern. From November through April, when the Asian landmass is cold, cool, dry air flows south and west toward the Indian Ocean. Between June and November, the continent warms up, causing air to rise, sucking in winds from the ocean that bring "monsoon" rains. This alternation made sailing predictable and relatively safe. South of the equator, the situation is very different. In this region, the trade winds blow from east to west, but with many storms, and sailing ships avoided it.

Seafaring between Arabia and India dates back thousands of years. Before the fifteenth century, mariners from Persia, Arabia, and Gujarat in northwestern India regularly sailed throughout the Indian Ocean and as far as China and Korea. Their ships, called dhows, were of a construction unique to the Indian Ocean.[7] Their planks, made of teak from southern India, were laid edge to edge and sewn together with coconut fibers, with ribs inserted into the finished hull to strengthen it. There was no deck, so cargo was protected with palm-leaf thatch or leather. This form of construction was light, inexpensive, and flexible and could be repaired at sea. It served well, and still serves, for small ships up to two hundred tons, but was not strong enough for larger ships.[8]

Dhows were equipped with triangular lateen sails. These sails, held up by a long boom attached to a mast with a pronounced forward rake, could be trimmed to suit different winds and could sail much closer to the wind than square sails. Because the boom was longer than the mast that supported it, however, it had to be taken down in order to be shifted from one side to the other. Hence sailing against the wind by tacking, or zigzagging, was difficult and dangerous.[9] It was also unnecessary, for Indian, Persian, and Arabian mariners, well acquainted with this seasonal pattern since ancient times, knew to wait patiently for a favorable wind.[10] Once at sea, they navigated by the stars like the Polynesians; as the Qu'ran says: "He it is who hath appointed for you the stars, that ye guide yourself thereby in the darkness of land and sea. We have made the signs distinct for the people that have knowledge."[11] Versatile as dhows were north of the equator, they were ill equipped to sail south of Zanzibar on the East African coast, where winds were much stronger and more erratic than in the monsoon region, and where the North Star was no longer visible above the horizon. Sailing beyond Mozambique (fifteen degrees south) was both dangerous and commercially unattractive.

In the early thirteenth century, mariners adopted the magnetic compass.[12] They also used an instrument called *kamal*, a string tied to the middle of a piece of wood. The navigator held the stick in such a way that one end seemed to touch the horizon and the other the North Star, while gripping the other end of the string with his teeth. By the length of string between the stick and his face, he could estimate the angle of the star above the horizon, hence his latitude.[13] Sailing manuals helped captains and pilots navigate safely; among the best known were the Greek *Periplus of the Erythrean Sea*, written in the first century CE, and the *Kitab al-Fawa'id fi Usul 'Ilm al-Bahr wa'l-Qawa'id* (Book of Useful Information on the Principles and Rules of Navigation) of Ahmad ibn Majid, written in 1489–90, which served sailors well into the nineteenth century.[14]

In the age of sailing ships, the winds largely determined the patterns of trade. Ships arriving from China and Southeast Asia, on the one hand, or from East Africa and the Middle East, on the other, unloaded their merchandise in South Indian ports, loaded up with Indian goods, and sailed home when the monsoon changed. Rather than keeping ships at anchor for months awaiting a change in the wind, merchants shipping pepper and spices between Southeast Asia and the Middle East preferred to store their goods in warehouses. As goods traveled in stages in different bottoms rather than all the way in the same ship, the ports of the Malabar coast of South India, especially Calicut, became the great entrepôts of the Indian Ocean.

The same was true of the trade between China and India. Since the early fifteenth century, the main entrepôt on the Malay coast was Melaka, where Chinese junks exchanged cargoes with Indian or Arab dhows. At the entrance to the Persian Gulf, Hormuz served the same function. Hormuz, Calicut, Malacca, and several lesser ports were the nodes of the great Indian Ocean trading network. From the eighth century on, trade was largely in the hands of Muslims, for most Hindus shunned ocean travel. It is through largely peaceful trade and the persuasive efforts of traders that Islam spread throughout the lands bordering the Indian Ocean.[15] Land warfare was common throughout the region, but warships and naval battles were rare in the Indian Ocean before 1497, though piracy was endemic in Malay waters.

The third great seafaring tradition was that of China. For several thousand years the Chinese had navigated their coastal waters and nearby seas, but until the Song dynasty (960–1279), trade with Southeast Asia and the Indian Ocean was carried in foreign bottoms.[16] After 1127, when northern China was conquered by warriors from Central Asia, Chinese merchants living south of the Yangzi turned to long-distance trade, regularly sending ships to Southeast Asia, the East Indies, and India; some may have ventured as far as the Persian Gulf and the Red Sea. Once the Chinese had entered long-distance trade, even foreign merchants preferred to travel in Chinese ships.

Under the Song, seafaring was left to private enterprise, for the rulers were more concerned with their northern land frontiers than with the sea. This changed with the Mongols, who defeated the Song and founded the Yuan dynasty in 1266. The emperor Kublai Khan is said to have built 4,400 ships with which he hoped to conquer Japan.[17] The Ming, who overthrew the Mongols in 1368, built several thousand warships and merchant vessels. In the early fifteenth century the shipyards of Longjiang, near Nanjing, employed between twenty and thirty thousand workers.[18] Throughout this period, China had by far the largest navy and merchant marine in the world.

Figure 1.1. A fourteenth-century Chinese junk, the *Fengzhou*, with bamboo sails.
Illustration by Lilana Wofsey Dohnert.

Oceangoing Chinese ships, called junks, were not only numerous, they were also much bigger and more stoutly constructed than the modest vessels of the Indian Ocean. They were built with a flat bottom for shallow waters, a keel that could be lowered at sea, and a stern-post rudder. Planks were attached with iron nails to watertight bulkheads and caulked with resin and vegetable fibers to form a watertight hull. Their three to twelve masts carried sails made of canvas or bamboo matting able to withstand gale-force winds. They could carry up to one thousand crew members and passengers and over a thousand tons of cargo.[19]

Until the late fifteenth century, Chinese navigation was far ahead of the rest of the world. Chinese ships, like those of the Indian Ocean, sailed with the monsoons, south in the winter and north in the summer. Their captains had maps and star charts and could calculate their latitude from the angle of the stars above the horizon.

They used the magnetic compass as early as the late eleventh century, a century before it reached the Middle East or Europe.[20] As a Chinese text written in the early twelfth century explained: "The ship's pilots are acquainted with the configuration of the coasts; at night they steer by the stars, and in the daytime by the sun. In dark weather they look at the south-pointing needle."[21]

In 1405 Zhu Di, the Yongle emperor, sent a fleet of 317 ships carrying twenty-seven thousand men under the command of Admiral Zheng He to Southeast Asia and into the Indian Ocean. The largest ships in the fleet were approximately four hundred feet long, almost five times the length of Christopher Columbus's Santa María. These "treasure ships" carried silk, porcelain, and ceremonial and household goods as trade items and as gifts to foreign rulers. They were accompanied by warships, troop transports, horse transports, supply ships, and water tankers.

This was only the first of seven great fleets that China sent out in the early fifteenth century. These expeditions visited Vietnam and Indonesia, India and Ceylon, southern Arabia, and several ports in East Africa. They concluded treaties of trade and tribute with the places they visited and, in some cases, overthrew or captured recalcitrant rulers. Along with the usual trade goods, they brought back objects previously unknown in China, such as magnifying glasses and a giraffe. After Zhu Di's death in 1424, the expeditions ceased for several years, but his successor, Zhu Zhanji, the Xuande emperor, ordered a seventh and last voyage in 1432–33 to repatriate foreign dignitaries who had been brought to China by earlier expeditions. Thereafter, foreign trade was severely limited, and Chinese ships were destroyed or allowed to rot.[22]

Why these extraordinary expeditions? When Zhu Di reached the throne, China clearly had the means, both technical and economic, to build the largest and most seaworthy fleet in the world. But the motivation was personal; it was Zhu Di's decision to use that fleet to enhance his prestige by establishing trade and tribute relations with the rulers of Southeast Asia and the Indian Ocean, not to acquire new lands. In the early fifteenth century, the Chinese could have sailed to America and around the world, had they wanted to. This has led some writers of historical fiction to imagine that they did so.[23] But uncharted oceans and undiscovered lands were of little interest to an emperor who wanted, more than anything, the respect of the known world. The Chinese, therefore, never left the western Pacific and Indian oceans.

Why then did the expeditions end so abruptly? Although the personal decision by Zhu Zhanji was the most important factor, there were political, economic, and strategic reasons for it. At the imperial court, Confucian scholars replaced the coalition of eunuchs, merchants, Buddhists, and Muslims who had been prominent under Zhu Di. Furthermore, the Grand Canal that linked the Yangzi Valley to the Yellow River and to Beijing was finally completed in 1411. Thereafter, thousands of riverboats transported grain from central China to the capital and bound China into a single trade zone. No longer was it necessary to use seagoing transports, always vulnerable to storms and pirates. At the same time, China was once again threatened by nomadic warriors from the northern steppe, and the government of Zhu Zhanji devoted its resources to rebuilding the Great Wall and defending its northern frontier. The Treasure Fleets, which were extraordinarily costly, must have seemed an extravagance that could never produce enough benefits to justify their continuation.[24]

The fourth seafaring tradition was that of the Mediterranean, a sea that had been sailed upon since very ancient times. In the Middle Ages, two distinct kinds of ships sailed its waters. The first kind was the galley, descendant of the trireme and quadrireme of Greco-Roman times. Galleys were built for speed and maneuverability. Long and narrow, they were propelled by rows of oarsmen with the assistance, in a following wind, of a square sail. Some were warships designed for ramming and boarding enemy ships; others carried passengers and costly freight such as spices. They needed a crew of 75 to 150 rowers and could only carry provisions for a few days. Because their light construction made them poorly suited to heavy seas, they generally sailed only during the summer months.

By the beginning of the fourteenth century, Italian shipbuilders were constructing great galleys, larger than traditional ones, to transport pilgrims to the Holy Land and passengers and precious cargo throughout the Mediterranean. In the summer months Venetian and Genoese galleys sailed as far as Southampton in England or Bruges in the Low Countries. With three masts and sails, they could sail with the wind, but they needed a crew of two hundred for additional speed and maneuverability and to sail against the wind. Galleys were displaced by sailing ships in the sixteenth century but survived into the eighteenth century as warships in the Mediterranean Sea.

The other kind of ship was the large, slow, unwieldy round ship, built to carry heavy cargo. Like galleys, round ships were carvel-built, that is, a frame was built first, to which planks were then nailed edge to edge. To the single mast and square sail of Roman transports, medieval Mediterranean shipbuilders added a second and sometimes a third mast. With a small crew, these ships were cheap to maintain but spent a lot of time in harbor, waiting for a following wind.[25]

The fifth seafaring tradition arose late, off of western Europe. Unlike the Mediterranean, the North Atlantic Ocean and the North and Baltic seas are cold and dangerously stormy, even in summer. The most daring navigators to sail those seas were the Vikings who raided the coasts and rivers of Europe between the ninth and eleventh centuries. For their raiding parties, they built longboats that were even more maneuverable than the war galleys of the Mediterranean: from five to eleven times longer than they were wide, they were built with a shallow draft to be rowed up rivers as well as at sea. They carried a short mast and a small square sail, but mostly relied on the muscles of their crew members, all battle-hardened warriors. For cargo, the Vikings built *knarrs*, wider boats with an open hold and a square sail and oars for use when sailing in contrary winds.[26] A few Norsemen sailed across the Atlantic in such boats; we do not know how many died trying.

Both kinds of Viking ships disappeared by the twelfth century, replaced by cogs built for the merchants of the growing ports of northern Europe. These were large tubby vessels with a length-to-beam ratio of three to one and a capacity of up to three hundred tons. They carried no oars but relied entirely on a single mast with a large square sail, making them hard to maneuver when the wind was not directly astern. Like longboats, they were clinker-built, a method of construction (also known as lapstrake) in which overlapping planks were built up from a massive keel and internal bracing was inserted later. This made them stronger but used more wood than the carvel-built ships of the Mediterranean and the Indian Ocean. Cogs rode so high in the water that they were invulnerable to the boarding tactics of the Vikings, while archers on their high castles fore and aft could rain arrows down on any would-be attacker. For all their clumsiness, cogs served as both merchantmen and warships.[27] Two Chinese inventions, the stern-post rudder and the compass, reached Europe in the late twelfth or thirteenth century and allowed cogs to keep their bearing even on cloudy days and in heavy seas.[28]

The Portuguese and the Ocean

New methods of navigation and new kinds of ships adaptable to all oceans resulted from the marriage of several nautical traditions. The place where this happened was Portugal, at the intersection of the Mediterranean and North Atlantic seafaring systems.

At first sight, Portugal would seem to be a most unlikely nation to found the first empire of the sea. It was a small kingdom, with little more than a million inhabitants in the fifteenth century. With a population of farmers and fishermen and few natural resources, it was very poor compared to the wealthy city-states of Italy or such large kingdoms as France or England. Furthermore, it was frequently at war, either with its large neighbor Castile or with the Muslim states of North Africa. Yet this minor kingdom at the southwestern corner of Europe succeeded in becoming a world power for a century, paving the way for all the empires that followed.

The Portuguese did not discover India, Arabia, or East Asia but only confirmed what they knew from travelers' reports. What they really discovered was not land but the ocean itself, and a means of crossing it. This discovery, as momentous as that of the Americas, was not the result of luck but of systematic trial and error.

Several factors came together fortuitously in this small kingdom in the mid-fifteenth century. The first was the Portuguese Christians' age-old hatred of Muslims and the urge to continue fighting them long after they had been driven out of Portugal itself. Another equally powerful motive was the desire to reach the sources of gold and spices that Europeans craved. But motivations alone do not lead to actions; also needed are the means to carry out these motivations. The means in this case were ships and navigation, two fields in which the Portuguese led the world for half a century. What brought means and motivations together was the extraordinary personality of Henry the Navigator.

Henry (1394–1460) was the son of King John I of Portugal. In 1415, he participated in the capture of Ceuta in northern Morocco. He later became governor of the Order of Christ, a knightly religious order that helped fund his projects. He seems to have been motivated by a desire not only to fight Muslims but also to discover a route to the kingdom of Ghana, depicted on a Catalan atlas of 1375 as a land of gold, and the realm of Prester John, a mythical Christian kingdom rumored to exist somewhere beyond the lands of Islam.[29] At Sagres, at the very southwestern tip of Europe, Henry founded a center devoted to oceanic exploration and surrounded himself with geographers and astronomers, as well as adventurers and impoverished hidalgos.[30] This center engaged in three related activities: training captains and pilots, sending expeditions down the coast of Africa, and gathering astronomic and oceanographic information that would assist the expeditions.

The voyages began in 1419. The first attempts were very unproductive, for neither the ships nor the sailors' skills and knowledge were up to the task. People remembered that the Vivaldi brothers had sailed out into the Atlantic with a fleet of galleys in 1291 and never returned.[31] The first Portuguese to venture down the coast of Africa sailed in *barcas* or *barineles*, open or partially covered boats of twenty-five or thirty tons, about seventy-five feet long by sixteen wide, of a sort used for fishing and coastal trading, not long-distance voyages. They carried a crew of eight to fourteen men and were equipped with one or two masts with square sails and oars in case they encountered contrary winds. Like ancient Greek mariners, they sailed along the coast by day and anchored at night.

After passing Morocco, the land of their enemies, the Portuguese mariners reached the coast of the Sahara Desert, a place with no human beings, no water, and no food. Sailing south was easy, for the trade winds carried the boats along. Returning, however, meant inching their way up the coast, either rowing against the wind or taking advantage of the breezes that blew out to sea at night and toward land during the heat of the day. Even after Henry had sent out a ship almost every year for fifteen years, his sailors had not gone beyond Cape Bojador, less than a thousand miles from Portugal.[32] The chronicler of these events, Gomes Eanes de Zuzara, explained:

> sailors said, that after that cape there are no people or any town; the land is no less sandy than the deserts of Libya where there is no water, nor trees, nor green grass; and the sea is so shallow that a league from the coast the bottom is less than a fathom [c. 5 1/2 feet] deep. The currents are so strong that any ship that passes it will never never return. That is why their predecessors never dared to pass it.[33]

Meanwhile, other explorers had sailed west and discovered the Azores in 1427–31. At that latitude, close to forty degrees north, the winds are called westerlies, for they blow toward Europe, bringing rain to the lands north of Lisbon. Unlike the smooth and reliable trade winds, the westerlies are variable and often stormy. Yet they provided an easier and faster way home than beating along the African coast. The discovery of the *volta do mar largo*, or "long ocean return," was a turning point in the history of navigation.[34]

Finally in 1434, Gil Eannes passed Cape Bojador in a *barca*. After that, Henry's sailors sailed much faster and more boldly down the coast of Africa. In 1445 Dinís Dias rounded Cape Verde, the westernmost tip of Africa; ten years later Alvise da Cadamosto, a Venetian in Henry's employ, discovered the Cape Verde Islands and the Gambia River. By the time Henry died in 1460, Portuguese ships were trading with Guinea, exchanging cloth and iron goods for pepper, gold, and slaves. Exploration had become a profitable business.

Barcas and *barineles* were simply not suited, however, for voyages to the Guinea Coast that lasted months. Nor was it possible to sail those waters in Nordic cogs or Mediterranean round ships, which could not maneuver along unknown coasts or sail in unfavorable winds. What opened the oceans to exploration was another kind of ship, the caravel.

The caravel was a hybrid, combining the best features of both the Mediterranean and the North Atlantic ships that visited the harbors of Portugal. By the 1430s, Portuguese shipbuilders had for decades been building ships using the carvel method, which produced a strong hull with less wood than the clinker method of northern Europe. Their ships had a length-to-width ratio between three to one and four to one, intermediate between Mediterranean galleys and Atlantic cogs. Like cogs but unlike galleys, caravels were pure sailing ships, never meant to be rowed, and had a full deck, an aft castle, and a stern-post rudder. The first caravels were fifty to seventy tons; later ones were as large as 150 tons, with an average of one hundred tons.[35]

Caravels derived their rigging from both the Mediterranean and Atlantic traditions. The first caravels had two masts carrying lateen sails. This allowed them to sail close to the wind and maneuver near coasts. Later in the fifteenth century, shipbuilders added a third mast. A captain now had a choice. He could use only lateen sails, creating a *caravela latina* for coastal exploration, or a combination of square sails on the mainmast and foremast and a lateen sail on the mizzenmast, an arrangement known as *caravela redonda*, for greater speed at sea in a following wind. This combination created a ship that was fast and maneuverable, could sail close to the wind, and could carry enough water for a crew of twenty to twenty-five men for a month at sea and enough food for four months. From the 1440s to the end of the century, only caravels were seaworthy enough to explore the Atlantic safely. Cadamosto called them "the best sailing ships afloat."[36]

For all their virtues, caravels had severe limitations. On long voyages, they were crowded and uncomfortable. The crew slept on the deck. Daily rations consisted of biscuits, cheese, rice, or beans, and salted meat or fish. Each crew member got a quart of water and a pint and a half of wine a day. Food spoiled rapidly. Fruit and vegetables quickly ran out, leaving the crew vulnerable to scurvy after a few weeks at sea. Carrying enough water for extended voyages at sea or along the barren Saharan coast was difficult. Improvements in casks and adding vinegar to inhibit microorganisms contributed to

Figure 1.2. A fifteenth-century *caravela latina*, with triangular lateen sails
for tacking against the wind. Illustration by Lilana Wofsey Dohnert.

the safety of long voyages.[37] More serious, from the captain's point
of view, was the fact that caravels were too small to carry any appre-
ciable quantity of cargo, and hence were not suited for very long
voyages or trade with distant lands. Their lateen sails, while well
suited to sailing at an angle to the wind, made them more difficult to
handle than square sails when sailing with the wind. As navigators
learned to sail further out to sea and to locate advantageous winds,
they began to prefer larger ships with square sails.

By the late fifteenth century, therefore, shipbuilders had designed
one-hundred- to four-hundred-ton ships with a broader beam and a
deeper hull than a caravel. They went by the name *nau* (*nao* in
Spanish) or carrack.[38] Their hulls resembled cogs more than caravels; they were basically cargo ships with a wide beam and a deep

Figure 1.3. A fifteenth-century *caravela redonda*, with two square and one lateen
sail, for sailing with the wind. Illustration by Lilana Wofsey Dohnert.

hold. However, they were full-rigged, with three masts, the main
and foremasts carrying square sails (and sometimes topsails as well)
and the mizzenmast carrying a lateen sail to make them more maneuverable; some added a small spritsail on the bowsprit, a short
mast canted forward at the bow. Such a ship was Columbus's *Santa
María*. The other ocean explorers of the sixteenth century generally
had mixed fleets of caravels and carracks.[39]

By 1470, Portuguese ships had reached Fernando Po and São
Tomé, islands in the Gulf of Guinea just north of the equator. In
the equatorial region, the navigators encountered the low-pressure
zone that brings rain to the land but something far more dangerous
at sea: the doldrums that can leave a ship stranded, baking in the
sun with sails hanging limp, for weeks at a time.

The expeditions continued to push forward. In 1482–84 Diogo
Cão explored the mouth of the Congo River. Then, in 1487, Bartholomeu Dias proceeded down the African coast with two small caravels and a supply ship. Past the Congo estuary, he found himself
battling against both the northward-flowing Benguela Current and
the southeast trade winds, making the voyage long and arduous. Past

Lüderitz Bay, on the barren coast of Namibia, he left his supply ship behind and headed out to sea seeking a more favorable wind. And find it he did: the South Atlantic westerly, also known as the Roaring Forties, carried his two caravels eastward, past the southern tip of Africa. Turning north, he landed at Mossel Bay, on the Indian Ocean coast of South Africa. There, he turned back west and headed home, stopping at the Cape of Storms (later renamed the Cape of Good Hope), 250 miles west of Mossel Bay.[40]

When Dias returned to Portugal, he brought with him two vital pieces of information. One was that he had found a way around Africa and into the Indian Ocean, the long-cherished dream of Henry the Navigator. The other was that in the South Atlantic westerlies blew at the latitude of the Cape. The road to India now lay open.[41]

Armed with that knowledge, Vasco da Gama sailed for India on July 8, 1497, with a fleet of four ships. Two of them, the *São Gabriel* and the *São Rafael*, were specially built under the supervision of Bartholomeu Dias. They were square-rigged, three-masted *naus* of 100 to 120 tons, built larger and taller than the caravels that Dias had used in order to carry more cargo and better withstand high waves. A caravel named *Berrio* and a nameless store ship accompanied them.[42]

Ten years had passed since Dias had returned from his voyage, and historians have wondered about the delay. Perhaps King John II hesitated to incur the expense, and further expeditions had to await the decision of his successor, the enthusiastic twenty-six-year-old Manuel, nicknamed "the Fortunate," who reigned from 1495 to 1521. It has also been suggested that John, learning of Columbus's plan to reach India by sailing west, sent out expeditions to investigate the still-secret route around Africa. As historian J. H. Parry put it: "it is not unreasonable to surmise—though positive evidence is lacking—that the ten years' interval had been used to collect information about the wind system of the central and south Atlantic, information which could only have come from voyages of which no record survives."[43]

Whether from prior knowledge or lucky guesswork, da Gama chose not to follow in Dias's footsteps. Like all sailors heading toward the Gulf of Guinea, he followed the northeast trade winds to Cape Verde. There, he left the familiar path and sailed south across the equator, away from Africa, then southwest at an angle to the southeast trade winds, thereby avoiding the winds and currents that had retarded Dias's progress along the African coast. Around thirty degrees south, he turned east and reached southern Africa at Saint Helena Bay, 130 miles north of the Cape. The voyage from Cape Verde to South Africa had taken almost three months, the longest any ship had ever been out of sight of land. By swinging west, da Gama had discovered another *volta* in the South Atlantic, a mirror image of the one north of the equator.[44]

Da Gama's small fleet had difficulty sailing around southern Africa and north into the Indian Ocean against the Mozambique Current. In March 1498 he arrived at Malindi, where he secured the services of a pilot who guided his fleet across the Arabian Sea to India in just twenty-seven days. Some historians have identified this pilot as Ahmad ibn Majid, the famed Arab navigator; this is a myth, however, for Portuguese sources do not name the pilot but identify him as a Gujarati, not an Arab.[45] In May 1498, da Gama's fleet arrived in Calicut, the major entrepôt in southern India.

After three months in Calicut and nearby towns, the fleet left India. The return to Portugal was difficult. Leaving too late in the season to catch the monsoon, da Gama's ships battled against the wind for three months to reach East Africa. During that voyage, so many of his crew died of scurvy that he had to burn the *São Rafael*. Sailing beyond the Cape of Good Hope was easier, for he understood the winds. What was left of his little fleet reached Lisbon in July and August 1499.

Navigation

Ships alone, no matter how seaworthy, could not open the oceans to reconnaissance. Sailors also had to know where they were and how to get back, a more daunting task than building seaworthy ships. While caravels and carracks were designed and built by craftsmen familiar with the sea, the development of navigation could come only from systematic data-gathering, experimentation, and calculation. This was the contribution of the Portuguese reconnaissance of the ocean to the Scientific Revolution that was beginning to transform the world.

The first requirement of sailors heading into unknown waters was a set of instruments to tell them in which direction they were sailing. The compass allowed ships to sail in the winter when the sky was cloudy. By the fourteenth century, instrument makers had developed the modern compass: a round box containing a magnetized iron wire glued to a circular card pivoting on its center and divided into 360 degrees.[46]

To determine their latitude, or how far north or south they had sailed, mariners estimated the angle of the North Star over the horizon. In 1454, Cadamosto, off the mouth of the Gambia, noted that the North Star was "about the third of a lance above the horizon." The following year, a few navigators began using the quadrant, a quarter-circle with a scale of zero to ninety degrees engraved on the curved edge, two pinholes on one straight edge, and a plumb line attached to the intersection of the two straight edges. By sighting the North Star through the pinholes and reading the degree to which the free end of the plumb line pointed, a navigator could determine the angle of the star with the horizon. By assuming that a degree of latitude measured 16 2/3 leagues—later revised to 17 leagues—he could estimate how far south of Lisbon he had sailed.[47]

The quadrant, though an improvement over pure guesswork, had serious flaws. It could not be used at sea where the rolling of the ship made the plumb line swing. And as one approached the equator, the North Star sank ever lower, until at five degrees north it disappeared below the horizon, rendering the quadrant useless. As they entered the southern hemisphere, explorers discovered stars no European had ever seen before. Among the new constellations, the Southern Cross was near but not on the axis of the Earth's rotation, and therefore could not play the same role as the North Star did in the northern hemisphere.

A new instrument, the seaman's astrolabe, helped. It consisted of a circular dial suspended by a ring. A metal rod called an alidade pivoted around the center of the disk; pinholes at each end could be aligned with the sun so that a ray of light shining through the upper sight fell on the lower sight, and the angle of the sun above the horizontal could be read off the dial. Nonetheless, mariners could not determine their latitude at sea with any accuracy until well into the sixteenth century.

In the course of the sixteenth century, the seaman's astrolabe was displaced by the cross-staff. This simple instrument, based on the *kamal* used by Indian Ocean navigators, consisted of a long piece of wood that could slide at a right angle to another, shorter one. The navigator held one end of the long stick near his eye while sliding the short one back and forth until one end seemed to touch the horizon and the other end the sun. He could then estimate the sun's altitude by the position of the sliding stick. Sighting was done at noon when the sun reached its zenith.

That alone, however, was not sufficient to determine one's latitude, since the sun's angle above the horizon varied not only with latitude but also with the seasons. What was needed was a table of the sun's declination, or angle above the celestial equator, at noon for every day of the year. This was a problem for mathematicians and astronomers, not craftsmen.[48]

As Portuguese ships sailed ever further into the South Atlantic, the problem became acute. In 1484, King John II convened a committee of astronomers and cosmographers to find the best method of determining latitude by observing the sun. Among the experts assigned to this task was Abraham Zacuto, a Jew from Salamanca who had written the *Almanach Perpetuum* (1473–78) with tables of the sun's declination based on the work of Arab astronomers of the Umayyad dynasty. The commission had a simplified table of declinations, based on Zacuto's work and that of Johannes Müller of Königsberg, drawn up in Latin for the benefit of literate seamen. In 1485, Joseph Vizinho, another Jewish astronomer, sailed as far as the Gulf of Guinea to test these tables. This work remained a state secret until 1509, when it was printed under the title *Regimento do astrolabio e do quadrante*.[49] If seamen knew, with any degree of accuracy, their latitude in either hemisphere, it was thanks to techniques introduced by these astronomers in the late fifteenth century. However, longitude, or one's east-west position, was impossible to measure from a ship until the eighteenth century.

Despite these advances, most mariners of the age of reconnaissance possessed little or no knowledge of celestial navigation. The Portuguese mathematician and cosmographer Pedro Nunes wrote in the mid-sixteenth century,

> Why do we put up with these pilots, with their bad language and barbarous manners; they know neither sun, moon, nor stars, nor their courses, movements or declinations; or how they rise, how they set and to what part of the horizon they are inclined; neither latitude nor longitude of the places on the globe, nor astrolabes, quadrants, cross staffs or watches, nor years common or bisextile, equinoxes or solstices?[50]

Instead of trusting astronomy, instruments, and manuals of navigation, most mariners—even Christopher Columbus—used a more traditional method, known as dead reckoning. To estimate his position at sea, a captain needed to know three things: the direction in which his ship was sailing, as shown by a compass; its speed, which he estimated by timing how long it took for a piece of flotsam to drift past his ship; and the time, measured by an hourglass. Such estimations were easily thrown off in foggy or cloudy weather or when the ship was carried by a current. When a ship was tacking upwind, the calculations were even more complex and prone to errors. Indeed, so uncertain was dead reckoning that most captains practiced latitude sailing, that is, sailing north or south to the latitude of their destination, then turning east or west until they reached it, a method that Polynesians would have found primitive.[51] Merchant ships that practiced latitude sailing along well-established routes were easy prey to pirates.

Until the mid-fifteenth century, European knowledge of the world was based on Ptolemy's *Geography*, a work that enshrined ancient Greek misinformation, such as placing Asia just across the Atlantic Ocean and describing the Indian Ocean as a closed lake. Ptolemy was the first to introduce the concepts of latitude and longitude and the meridians and parallels, imaginary lines that represented these concepts on the map. Most seamen cared very little about Ptolemy, however, for their voyages were short and generally close to shore. An entirely different geography informed their view of the sea, one that was based on experience and had been passed from master to apprentice since ancient times.

In the Mediterranean seamen used *portolani*, or charts, to guide ships sailing from place to place. In western European waters mariners used the rutters (*routiers* in French), or pilot-books, with information on tides and the depth of the sea near coasts. These were accompanied by charts giving bearings and distances, but without parallels or meridians to indicate latitude or longitude.

Such guides and charts were useful in known waters but prone to cumulative errors on the ocean. To record the findings of oceanic explorers and enable ships to return to a given place, geographers working for Henry the Navigator and his successors drew up *portolans* of the West African coast. When sailing out of sight of land, however, navigators needed charts with rhumb lines that they could follow by keeping a constant compass bearing. Even after Mercator published his world map in 1569, navigators preferred plain charts that did not require any knowledge of mathematics.[52]

After da Gama's return in 1498, the Portuguese crown sent fleets to the Indian Ocean on a regular basis. Throughout the Indian Ocean and beyond, to the Indonesian archipelago and the South China Sea, the Portuguese always found local pilots willing to serve as guides to the weather, the seas, and the islands.

In the process, the private research institute that Prince Henry had established at Sagres was transformed into a branch of the government, the Casa da India e da Mina (House of India and of Mines), a combined ministry of the navy, of colonies, and of overseas trade. It operated a school for cartographers and another for pilots and navigators that taught celestial navigation. Upon their return, captains and pilots reported on their voyages and submitted their ships' logs and instruments. The Casa da India also gathered intelligence from travelers and agents. With that information, the chief cartographer updated the *padrão*, a standard map of the world upon which charts for subsequent voyages were based. The Casa da India also produced rutters or sailing directions and descriptions of the coasts of the Indian Ocean, the Red Sea, Southeast Asia, and much of the Indonesian archipelago (but not Australia). Among the most important was Tomé Pires's *Suma Oriental*, describing in detail the eastern seas from the Red Sea to Japan.[53] The Portuguese government kept these documents secret from potential European competitors throughout much of the sixteenth century.[54]

The Spanish Voyages

In medieval Europe, myths were more abundant than truths, and the credulous outnumbered the skeptics. Many believed in the kingdom of Prester John, the fabled islands of Antilia and Saint Brendan's, and the existence of a Terra Australis in the southern hemisphere large enough to balance Eurasia in the northern hemisphere. Few learned men doubted these fables, and even those who believed were not prepared to investigate their veracity. One man was certain he knew an easier way to reach the fabled land of India than the long and arduous route the Portuguese had found.

No man has had so many nations, provinces, districts, cities, towns, streets, squares, and colleges and universities named after him as Christopher Columbus, and all because he made the biggest mistake in history. No story demonstrates so well the importance of contingency—also known as blind luck—as Columbus's voyage across the Atlantic.

Among the many myths circulating in his day, Columbus seized upon those that confirmed his vision and added a few of his own, all intended to prove, as he wrote in the margins of his books: "India is near Spain," "the end of Spain and the beginning of India are not far distant but close, and it is evident that this sea is navigable in a few days with a fair wind," and other statements to that effect.[55]

Columbus got his ideas from a variety of sources. One was Ptolemy's *Geography*, recently translated into Latin, which described the Indian Ocean as a closed sea and gave an estimate of the size of the earth that was one-fifth too small. Another was Pierre d'Ailly's book *Imago Mundi*, citing classical and Muslim scholars to exaggerate the size of Asia and underestimate the extent of the oceans, and concluding that a voyage west to Asia was eminently feasible.[56] In the margin of his copy of d'Ailly's book, Columbus estimated the circumference of the earth to be nineteen thousand statute miles, one-quarter less than its actual size.

Columbus lived in Portugal from 1476 to 1485. While there, he heard of a letter from the Florentine Paolo dal Pozzo Toscanelli to the king of Portugal, explaining how to reach China by sailing west for five thousand miles, with stops in the mythical island of Antilia and in Cipangu (Marco Polo's term for Japan). Columbus wrote to Toscanelli, who sent him a copy of the original letter and some maps. From d'Ailly, Toscanelli, and Marco Polo (and his own wishful thinking), Columbus concluded that Asia was thirty degrees wider than it actually is, and that Japan was fifteen hundred miles east of Asia; hence he concluded that Japan was only twenty-four hundred nautical miles from Europe, one-fifth the actual distance.[57] Even Admiral Samuel Eliot Morison, Columbus's biographer and admirer, had to admit it was a "colossal miscalculation . . . an extraordinary perversion of the truth."[58]

In 1484–85, Columbus presented his ideas to King John II, hoping for royal support and subsidies for an expedition. The king, after consulting his *Junta dos Mathematicos*, rejected Columbus's scheme as "vain, simply founded on imagination, or things like that Isle Cypango of Marco Polo."[59] Disappointed in Portugal, Columbus left for Spain in 1485. Queen Isabella and King Ferdinand granted him an audience in January 1486, then called a special junta of scholars, astronomers, geographers, and mariners to study his plans. Columbus showed the committee a map of the world and promised to find India less than three thousand miles west of Spain. This junta, like the Portuguese, rejected his arguments.

Turned away by the Spanish monarchs, Columbus went back to Lisbon, arriving just in time to witness the return of Bartholomeu Dias from Africa. Though King John II was well disposed toward Columbus, the knowledge that Dias had brought back, namely that the Atlantic and Indian oceans were connected, meant that the Portuguese had no further use for Columbus. Ever hopeful, Columbus returned to Spain. Once again, Isabella and Ferdinand called a council of experts to review his proposal and, once again, the council confirmed the judgment of the first junta, saying that his project "rested on weak foundations" and judging its achievement "uncertain and impossible to any educated person."[60] In short, it is not just with hindsight that we can ridicule Columbus's ideas; even his contemporaries thought his scheme was a "colossal miscalculation."

Rejected by the Spanish monarchs a second time, Columbus departed for France. No sooner had he left the court, however, than King Ferdinand had second thoughts and decided to fund Columbus's expedition after all and sent his guards out to bring him back. It was a gamble, a small price to pay to keep an unlikely but just possible discovery out of the hands of a rival monarch.[61] Nothing could contrast more with the rational and systematic Portuguese method of exploring than Columbus's dream and Ferdinand and Isabella's gamble.

No voyage in history is as celebrated as Columbus's first expedition. His fleet consisted of one eighty-five-foot *nao*, the *Santa María*, and two caravels, the *Niña* and the *Pinta*. From the small Spanish port of Palos de Moguer, they sailed to the Canary Islands, where Columbus had the *Niña* re-rigged into a *caravela redonda*, better for running before the wind.[62] From the Canaries, the voyage to the Bahamas, propelled by the northeast trade winds, was as smooth as anyone could have wished and only took five weeks, a remarkably short time. The return was more difficult. Columbus first tried to sail back the way he came but was beaten back by the trades. He then tacked toward the northeast until he reached the latitude of the Azores, where he found the westerlies that carried him home.[63]

There is some disagreement regarding his use of instruments. He seems to have brought along quadrants and astrolabes, but they were probably too imprecise to be used at sea. He may or may not have used the new Portuguese method of determining latitude by solar declination.[64] What is certain, however, is that he was a master at dead reckoning with a keen instinct for navigating uncharted waters.

Columbus's second voyage in 1493–95 was a major colonizing expedition, the start of the Spanish Empire in the New World. Knowledge of the sea was gained more slowly. Of the seventeen ships that made the voyage out, Columbus sent twelve back early. These ships, under Antonio de Torres, made the trip from Santo Domingo to Cadiz in thirty-five days; Admiral Morison called it "a record unbroken for centuries." Columbus himself, however, tried to sail directly east from Guadeloupe, fighting the trades almost all the way.[65] Not until many such experiences did Spanish navigators discover the best route to take: northeast from Cuba along the coast of North America, aided by the Gulf Stream, then east from the Carolinas to Europe, carried by the westerlies.

The two decades after Columbus's first voyage were a time of intense investigation of the oceans. Spanish expeditions explored the Caribbean and the coasts of the Americas from the Amazon to the Carolinas. John Cabot, an Italian working for England, explored the coast of Newfoundland. The Portuguese discovered the coast of Brazil and explored the Indian Ocean and the Indonesian archipelago. By then, most Europeans (except Columbus) were convinced that they had found a "New World," one that the Spanish referred to as *las Indias* but that the German geographer Martin Waldseemüller named America after the Italian explorer Amerigo Vespucci. Though the new continent offered tantalizing possibilities, it also barred the way to Asia, the original goal of Columbus. Then in 1513, Vasco Nuñez de Balboa crossed the Isthmus of Panama and saw an ocean stretching out before him, one that surely would lead to the Spice Islands that Portugal monopolized.

Meanwhile, Ferdinand Magellan, a Portuguese soldier of fortune who had served eight years in the East, sought support for an expedition to the Spice Islands by going west instead of east. This was Columbus's goal, but Magellan's was based on a much more complete and realistic knowledge of world geography than Columbus ever had. The Portuguese court had no interest in Magellan's plan, however. In fifteenth-century China, every expedition had obeyed the wishes of the emperor, but in Europe, explorers and adventurers made good use of the rivalry between states, playing one kingdom off against another. And so, like Columbus, Magellan turned to Spain. In 1517 he received the endorsement of the Casa de contratación, the official body in charge of navigation and external trade, the Spanish equivalent of Portugal's Casa da India.[66] Two years later, he sailed from Spain on one of the most spectacular, and harrowing, voyages of exploration ever undertaken.[67]

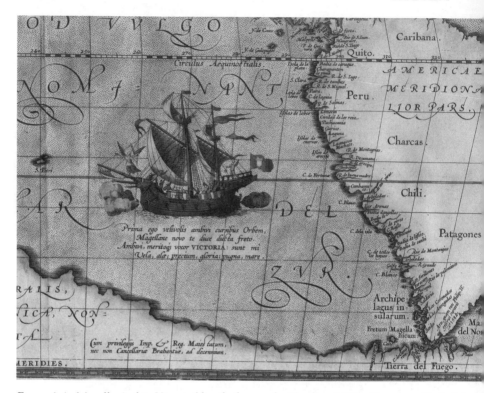

Figure 1.4. Magellan's ship *Victoria* (detail of a map by Ortelius, 1590). Courtesy of helmink.co

Magellan started out with two medium-sized *naos* of about a hundred tons, the *San Antonio* and the *Trinidad*, and three smaller ones, including the *Victoria*. They were second-rate ships and the naval stores and other cargo they carried were also defective, Magellan having been cheated by the ships' chandlers who had provisioned his fleet. This, and the hostility between him and his officers and crew, almost wrecked the expedition before it even reached the Pacific Ocean.

Magellan and his officers were up-to-date navigators, however. Since the days of Columbus, the Portuguese and, to a lesser extent, the Spaniards had made great advances in celestial navigation. Magellan's expedition carried thirty-five compasses, eighteen half-hour glasses, seven astrolabes, and twenty-one mariner's quadrants, as well as tables of the sun's declination. They also had the latest available charts. Hence they were able to note fairly accurately the latitude of the places they visited. They were also well armed with cannon, muskets, pikes, swords, suits of armor, and other weapons and brought along trade goods such as bells, bracelets, and velvet cloth.[68]

Longitude was another matter entirely. Before leaving, Magellan had been accompanied by a mathematician and astronomer named Rui Falero who claimed to have a method of determining longitude. Not only was Falero wrong, he was insane and had to be locked up in a *casa de locos*.[69] Though Magellan and others knew that the world was quite a bit larger than Columbus had claimed, they did not know how much larger.

Magellan and his fleet spent five months crossing the Atlantic and exploring the east coast of South America. From the end of March to the end of August 1520 they wintered on the coast of Patagonia, waiting for warmer weather before seeking a passage to the Pacific. Finding and navigating through that passage—the infamous Straits of Magellan—took thirty-eight days in the face of contrary winds and terrible storms.

On November 18, 1520, Magellan's fleet, now reduced to three *naos* (one having returned to Spain and another having been wrecked) left the straits that bear his name. His crews had been decimated by cold and disease and had little food left. They sailed north, then northwest into the tropics where they picked up the trade winds, then west into the unknown. Magellan, underestimating the size of the earth like all other Europeans at the time, expected to reach Asia in a few weeks. He may also have expected that there would be islands along the way, as there were in the Atlantic. Ironically, the ocean they sailed through is studded with islands and archipelagos, but Magellan's fleet missed them. Instead, they sailed through a trackless ocean, with a constant wind to their backs and an endless succession of sunny days and starry nights, an ocean so calm that Magellan called it *pacific*. For four months they sailed on, starving and many dying of scurvy, before reaching Guam in the Marianas on March 6, 1521.[70] After a few days to pick up food and fresh water, they sailed on, reaching the Philippines on March 13. There, little over a month later, Magellan died in a skirmish with native warriors.

It is in the Philippines that the crew realized they had reached "the Indies" at last, for Magellan's personal slave, Enrique, whom he had purchased in Malacca many years before, could speak the language of the people they encountered. The question remained: how far west had they come or, more precisely, were they east or west of a line 180 degrees west (or east) of the Line of Demarcation set by the Treaty of Tordesillas of 1494 that granted the Indian Ocean and the Spice Islands to Portugal and most of the Western Hemisphere to Spain?

The question of longitude was of more than academic interest, since both Spain and Portugal claimed the Indonesian archipelago and its precious Spice Islands as being within their half of the globe. By the time they knew for certain, both nations were being challenged by the Dutch, whom no treaty or papal bull could restrain.

Yet the question is also of academic interest, as historians debate exactly *who* was the first person to circumnavigate the globe. Morison and others have long claimed Magellan was the first, as having sailed in 1511 with a Portuguese fleet to Ambon and Banda, two of the Spice Islands that lie four to six degrees east of where Magellan landed in the Philippines.[71] Others, however, deny that Magellan ever sailed east of Malacca or visited the Spice Islands. In that case, his slave Enrique, who probably came originally from Indonesia or the Philippines, would be the first. In any event, neither of them was so honored. Instead, it was one of Magellan's captains, Juan Sebastián del Cano (or Elcano), who reached Spain in September 1522 and was awarded a coat of arms with a globe emblazoned with the motto *Primum Circumdedesti me*, or "Thou first circumnavigated me."[72]

In doing so, Magellan and del Cano showed that the world was much larger than anyone had previously thought, that the Pacific Ocean was larger than the Atlantic and Indian oceans combined, and that they were all connected and open to daring navigators. It was upon del Cano's return that Spain officially designated the Casa de contratación as the official cartographic institute designated to keep and update the *padrón real*, or standard map of the world upon which captains' charts were based. Diogo Ribeiro's map of the world, drawn in 1529, was the first to show in their proper size the oceans and the coasts of most of the continents, except the northern Pacific, the west coast of America, and Australia.[73]

Spanish navigators since Columbus had learned, by trial and error, how to return from the Americas to Spain by using the westerlies north of the thirtieth parallel. Upon del Cano's return, they also realized how to sail west across the Pacific. Learning how to cross that ocean from west to east was a very much harder lesson, however.

After Magellan's death, one of his captains, Gonzalo Gómez de Espinosa, made his way with the flagship *Trinidad* to the Moluccas to stock up on spices. Believing that if he sailed home via the Indian Ocean he risked being captured by the Portuguese, he tried to sail across the Pacific to Panama. From the Moluccas he headed north to forty-two or forty-three degrees north, but starvation, scurvy, bitter cold, and twelve days of storms forced him back to the Moluccas, where he surrendered to the Portuguese.[74] The second Spanish expedition to follow Magellan's route, that of Francisco García Jofré de Loaisa in 1526–27, ended in the Moluccas in a skirmish with the Portuguese. A third expedition, this one sent out from Mexico under Alvaro de Saavedra in 1527–28, tried twice to sail back to Mexico but failed. A fourth expedition, led by Ruy López de Villalobos, sailed from Mexico to the Philippines in 1542–43 but was captured by the Portuguese. One of his ships, under Bernardo de la Torre, escaped and reached thirty degrees north but, facing contrary winds, had to return to the Philippines.[75]

By this time, carracks and caravels were giving way to a new kind of ship. Galleons were as large as carracks but with finer lines and a greater length-to-beam ratio that made them as seaworthy and maneuverable as caravels. Most galleons carried four masts with many sails: square main and topsails on the fore and main masts, often with smaller topgallant sails above them; lateen sails on the mizzenmast and bonaventure (or fourth) mast, and spritsails at the bow. Such a profusion of sails made it easier and safer for the crew to add or remove sails quickly in changing wind conditions. By the mid-sixteenth century, the design of ships had reached a plateau; as naval historian Björn Landström put it, "Within the period of a hundred years the sailing ship had undergone more profound development than during the 5,000 years of its history that had passed and more than was to occur during 400 years to come."[76]

In 1564 a fleet of six ships under Miguel de Legazpi sailed from Mexico to establish Spain's claim to the Philippines on a permanent basis. The following year one of Legazpi's captains, Alonso de Arellano, deserted with the small supply ship *San Lucas* and headed east across the Pacific. He was followed by *San Pablo* under Fray Andrés de Urdaneta, an expert navigator who had sailed on Loaisa's ill-fated expedition. Both tacked northeastward against the trade winds until they reached 40 to 43 degrees north, the latitude of northern Japan, then sailed east with the westerlies across the northern Pacific to California, then southeast along the coast to Mexico. Arellano's voyage took 111 days and Urdaneta's 114.[77] The path across the Pacific Ocean was now open.

Three years after Arellano and Urdaneta's historic voyages, the government of Mexico (then called New Spain) inaugurated a shipping line between Acapulco and Manila that lasted almost 250 years. Every year, several galleons left Acapulco laden with Mexican silver and returned from Manila with silk and other fine products of China. The trade was very lucrative for those engaged in it, but the Spanish government, in the grip of the mercantilist economics, viewed it with suspicion as diverting precious silver from its coffers. In 1593 King Philip II tried to limit the trade to two ships per year of no more than three hundred tons apiece in either direction. Such were the profits that this decree was often ignored, and galleons of seven hundred to two thousand tons sailed this route.[78] They were built in the Philippines of local teak under the direction of Spanish shipwrights. They were reputed to be the strongest and most seaworthy in the world, and needed to be because the route was the longest and most dangerous of all.

The Manila galleons usually left Acapulco in February or March. The voyage to Manila meant eight to ten weeks of smooth and easy sailing with the trade winds. The return, however, was horrendous. The galleons left Manila preferably between mid-June and mid-July in order to catch the southwest monsoon but before the late summer typhoon season. Once out of the Philippine archipelago, a passage that often took several weeks, the galleon headed northeast until the forties, then east.[79] Loaded to the gunwales with precious cargo, they seldom carried enough food and fresh water. The voyage across the northern Pacific took anywhere from four to seven months, during which the crew and passengers endured frequent storms and suffered from thirst, hunger, and scurvy. Ordinarily 30 to 40 percent died and the death rate often reached 60 to 70 percent.[80]

The captains who followed this route hewed to the well-known method of latitude sailing. Once Urdaneta had shown the way to return to Mexico, they dared not depart from it; besides, the government expressly forbade any deviation. By following the prescribed route, the Manila galleons missed the Hawaiian Islands, for they sailed too far south going west and too far north going east.

Completing the Map of the Oceans

The late sixteenth and seventeenth centuries witnessed the decline of Spain and Portugal, who were challenged and increasingly replaced by the Dutch and later by the English and French. It was an age of frequent wars among the European powers and of attacks by pirates and state-sponsored corsairs (the distinction was not always clear) upon the Spanish silver fleets and the few ships that Portugal still managed to send out. In the process, the newcomers not only acquired the technology and knowledge of the Iberians, they also expanded the Europeans' knowledge of the oceans.

After the mid-sixteenth century, ship design evolved slowly. The most important change was the growing differentiation between merchant ships and warships. The Dutch pioneered a new kind of merchant ship, the *fluyt* or "flyboat," a slow ungainly ship with a broad beam and a flat bottom that carried few or no guns and required a much smaller crew than carracks and galleons. Fluyts were cheaply built of Norwegian or Baltic timber using wind-powered sawmills and cranes. The Dutch also invented the gaff sail, the forerunner of the triangular fore-and-aft sails seen on modern yachts; it permitted sailing close to the wind but required much less labor to handle than lateen sails. Being cheap to build and to man, fluyts allowed the Dutch to dominate maritime trade in the North Atlantic during the entire seventeenth century.[81] They were not well suited, however, for very long voyages to the East Indies and the Pacific. There galleons, more stoutly built and heavily armed, predominated.

Meanwhile the English were building specialized warships for both nearby and distant seas. The "race-built" galleons designed and used by the privateer John Hawkins and his nephew and successor, Francis Drake, were faster and more maneuverable than the ponderous Iberian galleons, a contrast tellingly demonstrated in the destruction of the Spanish Armada in 1588. Thereafter, navies, privateers, and pirates alike used race-built galleons, which were faster and more heavily armed than merchant ships but had limited cargo space. When not engaged in fighting in the endless wars of the time between the European naval powers, they conducted raids on commerce, hoping to bring back silver, gold, and other valuable booty. Such were the ships with which Drake circumnavigated the globe in 1577–80, raiding Spanish settlements on the west coast of South America and capturing Spanish ships along the way. Such also were the ships with which the Dutch seized a Spanish convoy laden with silver in 1628 and with which the English captured Jamaica in 1655, Portobelo in Panama in 1739, and Havana and Manila in 1762. Though these events took place thousands of miles from Europe, they belong to European history as much as they do to the history of the seas on which they were played out.

Despite the entry of new contenders for naval hegemony and colonial possessions and the incessant wars that they provoked, large parts of the oceanic world remained unknown to Europeans until well into the eighteenth century.

The Dutch gained precious knowledge of the oceans by espionage. In 1592, the leading merchants of Amsterdam, eager to trade with India, obtained crucial information from a Dutch resident of Lisbon, Cornelius de Houtman, and especially from Jan Huygen Linschoten, who had spent seven years in the service of the archbishop of Goa and who published his *Navigatio ac Itinerarium* in 1595, revealing the secrets of Portuguese navigation in the Indian Ocean and beyond.[82] The Dutch gained a foothold in the Indonesian islands in

1604. In 1611, Henrik Brower, like Bartholomeu Dias 123 years earlier, was carried past the Cape of Good Hope by the westerly Roaring Forties. Rather than turn north toward India as the Portuguese did, he followed the wind until he encountered the trade winds that carried him north to the Sunda Straits between Sumatra and Java. On the return trip he used the trade winds south of the equator, beyond the reach of the monsoons; in effect, he had discovered another *volta do mar* in the southern Indian Ocean. In following this route too far to the east, a Dutch ship discovered the west coast of Australia in 1616. In 1642–44 Abel Janzoon Tasman explored New Zealand and the coast of Australia as far as Tasmania. As there seemed no possibility of finding spices or precious metals there, Europeans ignored these lands for another two centuries.[83]

Instruments also evolved slowly. The back-staff, invented by John Davis in 1595, allowed a navigator to determine the height of the sun above the horizon while facing away from it, unlike the cross-staff on which it was based. At the same time, English mariners began dropping overboard a log attached to a line with knots spaced at seven-fathom intervals. They could then measure how much of the line had played out in a given length of time, as measured by a sand-glass. The ship's speed, measured in knots, was then recorded in a book called a log.

Seventeenth-century advances in navigation and in knowledge of the oceans, while significant, pale compared to the second wave of exploration in the second half of the eighteenth century. During that second wave, British and French explorers surveyed the parts of the Pacific Ocean long avoided by the Manila galleons. They charted Australia, New Zealand, Hawaii, and other Pacific islands as well as the west coast of North America and the northeast coast of Asia, proving that there was neither a great southern continent nor a northwest passage between Europe and Asia.

The primary motivation for this second wave is most certainly the Enlightenment and the reawakening of interest in scientific matters that led the governments of France and Britain, in particular, to fund costly research expeditions. Yet motivations, in this case as in others, are not sufficient to explain the new wave of explorations. In earlier times, exploration of the Pacific had been hindered not by a lack of desire but by the inability of navigators to locate themselves in the vast and trackless ocean and by the terrible death toll of sailors on long voyages. What opened the Pacific to exploration in the eighteenth century was the development of methods of determining longitude at sea and of preventing scurvy.

For centuries, the greatest obstacle to exploration of the oceans was the problem of longitude.[84] Until the mid-eighteenth century, mariners had no way to determine how far east or west they had traveled. Attempts to determine longitude by magnetic declination, that is, by the angle between the magnetic and the true north, proved unreliable. Magellan's guess as to the location of the Philippines was off by fifty-three degrees, or about four thousand miles. Seventeenth-century geographers thought the Mediterranean Sea was fifteen degrees or five hundred miles longer than it really is. Islands were discovered, then disappeared and were not found again for decades.[85]

Ships furled their sails at night for fear of running aground; such fears were not baseless. In 1707 a British fleet, returning from the Mediterranean, was approaching the southwestern tip of Cornwall. The commander, Admiral Sir Clowdisley Shovell, and his officers had miscalculated their longitude and ran aground on Scilly Isles, with a loss of four ships and two thousand men.[86]

In 1598, King Philip III of Spain, believing that the loss of the Armada in 1588 was due to a miscalculation, offered a prize to anyone who could find a means of determining longitude at sea. The Dutch followed suit. In 1714 the British Parliament, reacting to the disaster of 1707, passed the Longitude Act, offering a reward of twenty thousand pounds (several million dollars in today's money) to anyone who could determine longitude at sea within half a degree, or thirty miles at the latitude of the Caribbean, at the end of a six-weeks' voyage. It also set up a Board of Longitude to scrutinize the proposals.[87]

In the late seventeenth and eighteenth centuries, cartography, like other branches of knowledge, was transformed from a description of the world into a scientific discipline expressed in mathematical terms. Astronomy, the mother of the Scientific Revolution, played a major role in this transformation.[88] Since the sixteenth century, it was known that the difference in longitude between two points on the planet could be determined if one knew the difference in time between those two points. Since the earth revolved 360 degrees every twenty-four hours, if noon was an hour later in one place than in another, then the two places were 360 degrees divided by twenty-four, or fifteen degrees apart. The problem was how to tell in one place what time it was at another place. Even the best clocks of the early eighteenth century were so erratic and unreliable they were of no use; as Isaac Newton told a parliamentary committee, "by reason of the Motion of a Ship, the Variation of Heat and Cold, Wet and Dry, and the Difference of Gravity in Different Latitudes, such a Watch hath not yet been made."[89]

Astronomers, however, had devised three methods of determining the time by the motions of the heavenly bodies. One method was that of eclipses first proposed by Hipparchus in 160 BCE. Since astronomers could predict the time of an eclipse in Europe quite exactly, the difference between the time in Europe and solar time somewhere else gave the difference in longitude between those two points. Unfortunately, eclipses were so rare as to be of little use to mariners.

The second method, discovered by Galileo, was that of the moons of Jupiter. Since that planet's moons passed behind or in front of it a thousand times a year, and those eclipses are seen at the same moment from every point on earth, a traveler equipped with a table (called an ephemeris) giving the time of the eclipses could thereby determine his longitude. In the late seventeenth century, using such a table prepared by the French astronomer Jean-Dominique Cassini, astronomers went out to places around the world and reported, for the first time, their exact longitude. With their reports, cartographers such as Guillaume Delisle and Jean-Baptiste Bourguignon d'Auville were able to draw globes and world maps that were vastly more accurate than anything known before the eighteenth century. On their maps, the Mediterranean Sea and the Atlantic and Indian oceans were much smaller, and the Pacific much larger, than previously believed. Unfortunately, this method required a long telescope, a pendulum clock, and complicated calculations, making it impossible to use onboard a ship.[90]

A third astronomical method, called lunar distances, was much more promising. It was based on the idea that the moon moves across the sky at a different rate than the stars. By observing the angle between the moon and a given star, and knowing the time at which the same angle between the same two bodies was seen at another meridian, a navigator could deduce his longitude. To do so, he needed an instrument capable of measuring with precision the angle between the moon and a star; a clock able to keep time within a minute a day between noon (as determined by the sun) and night-time (when the moon and the stars were visible); an ephemeris showing, at regular intervals, the lunar distances or angles between the moon and certain stars; and the skill to make the appropriate calculations.

The instrument to measure angles was the octant, presented to the Royal Society by John Hadley and tested by the Admiralty in 1731–32. It could not only measure the angle between two celestial objects within one minute of arc (equivalent to one nautical mile at the equator), it also had an artificial horizon, a necessity at night when the real horizon was invisible. Even more accurate and useful was the sextant, introduced in 1757, which incorporated a telescope.[91]

Because the moon's path across the sky is erratic, creating a table of lunar distances was far more complicated than devising the ephe-merides of the moons of Jupiter and occupied some of the best math-ematicians of the eighteenth century. In 1713 Newton compiled a table that yielded errors of up to three degrees (or two hundred miles at the equator); even so, he complained that doing so was "difficult, and it is the only problem that ever made my head ache."[92] In 1755 Tobias Mayer, professor of mathematics at the University of Göt-tingen, produced a set of tables that were accurate within 37 minutes of arc, sufficiently so that the Board of Longitude awarded his widow three thousand pounds in recognition of his achievement. In 1763 the British astronomer Nevil Maskelyne published *The British Mari-ner's Guide* based on Mayer's calculations. Two years later, now ap-pointed astronomer-royal, he began publishing annual editions of the *Nautical Almanac and Astronomical Ephemeris*, with lunar tables starting in 1767.

From then on, determining longitude at sea was within reach of navigators. However, doing so required not only instruments and tables but also considerable knowledge of mathematics. To get a reading within a degree of longitude required four observers to take four or five sets of observations within six to eight minutes, then average them out; the result then had to be adjusted for parallax (the angle between the ship's position and the center of the earth) and refraction (the distortion caused by the atmosphere close to the horizon). Maskelyne himself needed four hours to work out the longitude from his observations; later improvements in the table reduced that time to thirty minutes. While brilliant in conception, the lunar distances method was clearly beyond the reach of most ship's captains, who continued to rely on dead reckoning.[93]

Meanwhile, another method of determining longitude appeared to compete with the lunar distances: an accurate clock. This was the contribution of John Harrison, not a scientist but a craftsman, the son of a carpenter, self-taught and single-minded in his pursuit of the perfect timekeeper. Harrison spent the years from 1728 to 1735 building H-1, an ungainly seventy-two-pound clock, made mainly of wood, with an intricate mechanism that ran flawlessly on a trip from England to Portugal and back. His next clock, H-2, was made of brass and even heavier, with anti-friction rollers, a grasshopper escapement, and a bi-metallic pendulum rod that compensated for temperature; it was never sent to sea. His third clock, H-3, took seventeen years to build and was almost as large.

It was Harrison's fourth clock, completed in 1759, that met the standard set by the Longitude Act. Unlike its predecessors, H-4 was only 5.2 inches in diameter, the size of a large pocket watch, but its mechanism was more intricate than any ever created before. Unfortunately for Harrison, none other than Nevil Maskelyne sat on the Board of Longitude that was to judge its merits, and Maskelyne had personal reasons to deny Harrison the prize he sought. In 1762 the clock was put on a ship to Jamaica. Upon its return, it registered an error of less than five seconds. On a second trial in 1764, this time to Barbados and back, H-4 was found to run less than one-tenth of a second fast per day. Still dissatisfied, the Board of Longitude demanded that Harrison hand over all four of his clocks and withheld the prize money. It took the personal intervention of King George II and an act of Parliament for Harrison to be, at last, rewarded for his accomplishment.[94]

There can be no doubt that Harrison had met the most stringent standards of the Longitude Act. But it was a long way from there to giving seamen an accurate reading of their location at sea. To become a practical instrument for mariners, Harrison's H-4 had to be reproducible, which is why the board had demanded that he hand it over. Once in possession of H-4, the board handed it to a rival clockmaker, Larcum Kendall, who made a replica of it, called K-1, in 1770. In France, meanwhile, the Bureau of Longitude offered a prize for an accurate clock, as did the Academy of Sciences. Two clockmakers, Pierre Le Roy and Ferdinand Berthoud, competed for these prizes. Between 1767 and 1772, their clocks underwent a series of tests on several long sea voyages. Le Roy won the academy prize, while Berthoud was appointed clockmaker to the French navy.[95]

Until the mid-eighteenth century, navigation over long distances was among the most dangerous occupations in the world. Many expeditions lost half or more of their crew members. Some had to burn some of their ships for lack of seamen to sail them. Others had to turn back or failed to return. The cause was scurvy, a breakdown of the body's tissues due to a lack of vitamin C. Getting lost at sea could be deadly. In 1741 Commodore George Anson, commanding a small fleet off Cape Horn, was caught in a storm. After fifty-eight days during which he thought he had sailed two hundred miles west, he found himself close to where he started out. He then headed north to the thirty-fifth parallel, then east, seeking the Juan Fernández Islands. Instead, he found himself off the coast of Chile and had to turn west again. By the time he reached his goal, half his crew had died of scurvy.[96]

Navigators had known since the early sixteenth century that scurvy could be prevented, and sometimes cured, by eating fresh food, especially citrus fruits; but transporting fresh produce on long voyages was simply impossible. Substitutes, like the quince preserves that Magellan brought for his officers, were too costly for common sailors. In the calculus of costs and benefits, the health of sailors ranked lower than other cargo. The eighteenth century saw both an advance in knowledge and a shift in values. James Lind, a doctor in the British fleet, published his classic *Treatise of the Scurvy* in 1753, in which he advocated the use of lemon juice as a preventive and a cure.[97] It made long expeditions much less dangerous.

The development of two methods of determining longitude at sea and the possibility of preventing scurvy coincided with the second wave of maritime exploration, that of the Pacific Ocean. The best known and most productive of these expeditions were those led by James Cook.

Cook, a man of humble background, taught himself astronomy, celestial navigation, and marine cartography. As a young man he established his reputation by drawing accurate charts of the coast of Newfoundland and Labrador. In 1768 he was chosen to take the botanist Joseph Banks and a team of scientists to Tahiti to observe the transit of Venus, when the planet passed directly in front of the sun. Their vessel, the *HMS Endeavour*, was an old but sturdy ship of 368 tons, originally built to haul coal in the stormy seas around Great Britain. After visiting Tahiti, Cook spent six months charting the coasts of New Zealand and eastern Australia before returning home in 1771. On this voyage he used a sextant and Maskelyne's *Nautical Almanac* to determine his longitude.

He also tested Dr. Lind's ideas about scurvy. He rationed the consumption of salt beef and pork—the traditional food of sailors at sea—and banned cheese and butter. Instead, he fed his crew as much fresh food as he could obtain, especially oranges and lemons. For times when those items were not available, his ships carried supplies of sauerkraut, raisins, mustard, and vinegar. He also insisted on scrupulous cleanliness, ventilating the crew's quarters and washing them with fresh water and drying them with stoves whenever possible. When he arrived in Java after many months at sea, all his crew members were in good health. Within a few weeks in the unhealthy port of Batavia (now Djakarta), forty crew members fell ill and seven died of dysentery and malaria; another twenty-three died on the way home from Java to England.[98]

On his second voyage around the world in 1772–74, Cook circumnavigated the world at high latitudes between forty and seventy degrees south, proving there was no Terra Australis other than Australia, New Zealand, and Antarctica. His ship, *HMS Resolution*, and its sister ship, *HMS Adventure*, like *Endeavour* before them, were well-used cargo haulers. Cook took along Kendall's K-1, along with three clocks made by another clockmaker, John Arnold. Periodically, he verified the accuracy of the clocks by the lunar distance method. The Arnolds gave him trouble, but K-1 was unfailingly accurate. In the year from November 1773 to October 1774, it was off by only nineteen minutes and thirty-one seconds. Cook wrote: "Mr Kendall's watch . . . exceeded the expectations of the most zealous advocate and by being now and then corrected by lunar observations has been our faithful guide through all the vicissitudes of climates."[99]

On his third voyage, starting in 1776, Cook commanded the *Resolution* and a sister ship, *Discovery*. He again visited New Zealand and Tahiti, then headed north, where he discovered the Hawaiian Islands, previously unknown to Europeans. From there he sailed to Alaska and through the Bering Straits into the Arctic Ocean, then back to Hawaii, where he was killed in a skirmish with native Hawaiians in 1779. On this voyage he took with him Kendall's K-3, finished in 1774. As on his second voyage, he verified the accuracy of this clock by the method of lunar distances.[100]

Other explorers of the Pacific followed Cook's example: Jean-François de la Pérouse, William Bligh (of the *HMS Bounty*), George Vancouver, and Bruni d'Entrecasteaux took marine clocks with them. As the French maritime historian Frédéric Philippe Marguet wrote: "Between longitude and the discovery of the Pacific, there is such a close connection that one can say that the voyages that took place at the end of the eighteenth century and at the beginning of the nineteenth would have been neither as numerous nor as fruitful if dead reckoning had been the only possible method of navigation."[101]

Conclusion

The discovery of the oceans from the fifteenth to the eighteenth centuries ranks as one of the great epics of humankind. Before the Europeans, Polynesian, Arab, South Asian, and Chinese seafarers had accomplished remarkable feats of navigation, but always within the boundaries of a single maritime environment: Polynesians in the tropical Pacific, Arabs and Indians in the Indian Ocean, and Chinese and Malays in the waters of East, Southeast, and South Asia. Only the Europeans developed ships and methods of navigation that allowed them to sail on any ocean except the Arctic, to confront any weather from the equatorial doldrums to the Roaring Forties, and to survive for months out of sight of land. By data-gathering, experimentation, and the application of mathematics, they found ways of determining their location at sea. In the process, they explored all the world's oceans and filled in the map of the world.

By the early nineteenth century, all the oceans of the world that could be navigated by sailing ships were known to Europeans. Except for the Arctic Ocean and the coasts of Antarctica, almost every sea and coastline and island had been visited and charted. The winds and currents were also familiar to navigators. The introduction of steam power later in the century did little to increase mariners' knowledge of the seas; it simply made travel faster, safer, and more reliable. The discovery of the sea was drawing to a close, but the domination of its peoples was only beginning.

Notes

1. The best introduction to shipbuilding and navigation before the fifteenth century is J. H. Parry, *The Discovery of the Sea* (New York: Dial Press, 1974), chapters 1 and 2.

2. James Cook, *The Voyage of the Resolution and Adventure*, ed. John C. Beaglehole (Cambridge: Hakluyt Society, 1961), p. 354.

3. James Cook, *The Voyage of the Endeavour, 1768–1771*, ed. John C. Beaglehole (Cambridge: Hakluyt Society, 1955), p. 154.

4. On Polynesian navigation, see Ben R. Finney, *Voyage of Rediscovery: A Cultural Odyssey through Polynesia* (Berkeley: University of California Press, 1994); Will Kyselka, *An Ocean in Mind* (Honolulu: University of Hawaii Press, 1987); and David Lewis, *We the Navigators: The Ancient Art of Landfinding in the Pacific* (Honolulu: University Press of Hawaii, 1994).

5. Parry, *Discovery*, p. 33.

6. Finney, *Voyage*, p. 13.

7. Clifford W. Hawkins, *The Dhow: An Illustrated History of the Dhow and Its World* (Lymington: Nautical Publishing, 1977); Patricia Risso, *Oman and Muscat:*

An Early Modern History (New York: St. Martin's, 1986), p. 216. Under the generic term *dhow* are many varieties of ships: *baghlahs, booms, sambuqs,* etc.

8. K. N. Chaudhuri, *Trade and Civilization in the Indian Ocean* (Cambridge: Cambridge University Press, 1985), pp. 146–50; Simon Digby, "The Maritime Trade of India," in Tapan Raychaudhuri and Irfan Habib, eds., *The Cambridge Economic History of India* (Cambridge: Cambridge University Press, 1981), vol. 1, pp. 128; P. Y. Manguin, "Late Medieval Asian Shipbuilding in the Indian Ocean," *Moyen Orient et Océan Indien* 2, no. 2 (1985), pp. 3–7; W. H. Moreland, "The Ships of the Arabian Sea about A.D. 1500," *Journal of the Royal Asiatic Society* 1 (January 1939), p. 66; George F. Hourani, *Arab Seafaring in the Indian Ocean in Ancient and Early Medieval Times,* revised and expanded by John Carswell (Princeton: Princeton University Press, 1995), pp. 91–105. Iron nails were adopted in the fifteenth century but only for larger ships built for long voyages; see Ahsan Jan Qaisar, *The Indian Response to European Technology and Culture, AD 1498–1707* (Delhi and New York: Oxford University Press, 1982), pp. 23–27.

9. I. C. Campbell, "The Lateen Sail in World History," *Journal of World History* 6 (Spring 1995), pp. 1–24; Alan McGowan, *The Ship,* vol. 3: *Tiller and Whipstaff: The Development of the Sailing Ship, 1400–1700* (London: National Maritime Museum, 1981), p. 9; Parry, *Discovery,* pp. 17–20.

10. William D. Phillips, "Maritime Exploration in the Middle Ages," in Daniel Finamore, ed., *Maritime History as World History* (Salem, Mass.: Peabody Essex Museum, 2004), p. 51.

11. Quoted in E.G.R. Taylor, *The Haven Finding Art: A History of Navigation from Odysseus to Captain Cook* (London: Hollis and Carter, 1956), p. 126.

12. Amir D. Aczel, *The Riddle of the Compass: The Invention That Changed the World* (New York: Harcourt, 2001); Philip de Souza, *Seafaring and Civilization: Maritime Perspectives on World History* (London: Profile Books, 2001), p. 34; J. H. Parry, *The Establishment of the European Hegemony, 1415–1715: Trade and Exploration in the Age of the Renaissance* (New York: Harper and Row, 1961), p. 17; Moreland, "Ships," p. 178; Chaudhuri, *Trade and Civilization,* p. 127.

13. Taylor, *Haven,* pp. 123–29.

14. G. R. Tibbetts, *Arab Navigation in the Indian Ocean before the Coming of the Portuguese* (London: Royal Asiatic Society, 1971), pp. 1–8.

15. C. R. Boxer, *The Portuguese Seaborne Empire, 1415–1825* (New York: Knopf, 1969), pp. 45–46; Moreland, "Ships," pp. 64, 174.

16. Mark Elvin, *The Pattern of the Chinese Past* (Stanford: Stanford University Press, 1973), p. 137.

17. William H. McNeill, *The Pursuit of Power: Technology, Armed Force, and Society since A.D. 1000* (Chicago: University of Chicago Press, 1982), p. 43.

18. Louise Levathes, *When China Ruled the Seas: The Treasure Fleet of the Dragon Throne, 1405–1433* (New York: Oxford University Press, 1994), pp. 75–76.

19. Ibid., pp. 81–82; Digby, "Maritime Trade of India," pp. 132–33; Chaudhuri, *Trade and Civilization,* pp. 141–42, 154–56; Elvin, *Pattern,* p. 137.

20. Aczel, *Riddle*, pp. 78–86; Elvin, *Pattern*, p. 138; Parry, *Discovery*, p. 39.

21. Aczel, *Riddle*, p. 86.

22. The story of the Zheng He expeditions is the subject of Louise Levathes's *When China Ruled the Seas* and Edward L. Dreyer's *Zheng He: China and the Oceans in the Early Ming Dynasty, 1405–1433* (New York: Pearson, 2007). See also Francesca Bray, *Technology and Society in Ming China* (Washington, D.C.: AHA Publications, 2000), pp. 21–22; J. R. McNeill and William H. McNeill, *The Human Web: A Bird's Eye View of World History* (New York: Norton, 2003), pp. 125–26, 166–67; and Felipe Fernández-Armesto, *Pathfinders: A Global History of Exploration* (New York: Norton, 2006), pp. 109–17.

23. For an example of fiction disguised as history, see Gavin Menzies, *1421: The Year China Discovered America* (New York: Bantam, 2002).

24. The analogy with the U.S. expeditions to the moon in the 1960s and 1970s is striking. On the end of the expeditions, see Lo Jung-pang, "The Decline of the Early Ming Navy," *Oriens Extremus* 5 (1958), pp. 151–62.

25. J. H. Parry, *The Age of Reconnaissance: Discovery, Exploration, and Settlement, 1450–1650* (Cleveland: World Publishing, 1963), pp. 54–63; Pierre Chaunu, *L'expansion européenne du XIIIe au XVe siècle* (Paris: Presses Universitaires de France, 1969), pp. 274–78; Richard W. Unger, "Warships and Cargo Ships in Medieval Europe," *Technology and Culture* 22 (April 1981), pp. 233–52.

26. John R. Hale, "The Viking Longship," *Scientific American* (February 1998), pp. 56–62; Unger, "Warships and Cargo Ships," 241.

27. Hale, "Viking Longship," p. 62; Unger, "Warships and Cargo Ships," pp. 240–45.

28. Lynn White, Jr., "Technology in the Middle Ages," in Melvin Kranzberg and Carroll W. Pursell, eds., *Technology in Western Civilization* (New York: Oxford University Press, 1967), vol. 1, p. 76; Chaunu, *L'expansion européenne*, p. 279; Unger, "Warships and Cargo Ships," p. 244.

29. On Henry the Navigator, see Michel Vergé-Franceschi, *Un prince portugais du XVème siècle: Henri le Navigateur, 1394–1460* (Paris: Ed. Félin, 2000); and Peter Russell, *Prince Henry "the Navigator": A Life* (New Haven: Yale University Press, 2000). See also Fernández-Armesto, *Pathfinders*, p. 131.

30. Fernández-Armesto, *Pathfinders*, p. 148.

31. Phillips, "Maritime Exploration," p. 55.

32. Quirino da Fonseca, *Os navios do infante D. Henrique* (Lisbon: Comissão Executiva das Comemorações do Quinto Centenário da Morte do Infante D. Henrique, 1958), pp. 15–40; João Braz d'Oliveira, *Influencia do Infante D. Henrique no progresso da marinha portugueza: Navios e armamentos* (Lisbon: Imprensa Nacional, 1894), pp. 17–20; Henrique Lopes de Mendonça, *Estudios sobre navios portuguezes nos secolos XV e XVI* (Lisbon: Academia Real das Sciencias, 1892), pp. 15–17; Eila M. J. Campbell, "Discovery and the Technical Setting, 1420–1520," *Terrae Incognitae* 8 (1976), p. 12; Boies Penrose, *Travel and Discovery in the Renaissance,*

1420–1620 (Cambridge, Mass.: Harvard University Press, 1960), p. 269; Parry, *Discovery*, pp. 109–22.

33. Quoted in Luis de Albuquerque, *Introdução à história dos descubrimentos* (Coimbra: Atlantida, 1962), p. 249.

34. Roger Craig Smith, *Vanguard of Empire: Ships of Exploration in the Age of Columbus* (New York: Oxford University Press, 1993), p. 40.

35. For ships of this early period, tonnage measurements are very approximate and based on guesswork. Generally, they refer to displacement (i.e., weight) rather than, as later, to cargo capacity.

36. Quoted in Penrose, *Travel and Discovery*, p. 35.

37. Fernández-Armesto, *Pathfinders*, p. 143.

38. Clinton R. Edwards, "Design and Construction of Fifteenth-Century Iberian Ships: A Review," *Mariner's Mirror* 78 (November 1992), pp. 419–32, and McGowan (*The Ship*, vol. 3, p. 10) say that *nau* and carrack were the same. K. M. Mathew, *History of the Portuguese Navigation in India, 1497–1600* (Delhi: Mittal Publications, 1988), pp. 280–92; John H. Pryor, *Geography, Technology, and War: Studies in the History of the Mediterranean, 649–1571* (New York: Cambridge University Press, 1988), pp. 39–43; and Smith, *Vanguard*, pp. 31–32, distinguish between the two. In all likelihood, there were no fixed types but a great variety of ships later identified in an imprecise manner.

39. Clinton R. Edwards, "The Impact of European Overseas Discoveries on Ship Design and Construction during the Sixteenth Century," *GeoJournal* 26, no. 4 (1992), pp. 443–52; Parry, *Age of Reconnaissance*, pp. 53, 66.

40. On the Portuguese expeditions between 1460 and 1496, see Parry, *Discovery*, pp. 133–42.

41. Alfred Crosby, *Ecological Imperialism: The Biological Expansion of Europe, 900–1900* (Cambridge: Cambridge University Press, 1986), pp. 108–16; Parry, *Discovery*, pp. 130–31.

42. Sanjay Subrahmanyam, *The Career and Legend of Vasco da Gama* (Cambridge: Cambridge University Press, 1997), p. 79; Smith, *Vanguard*, pp. 32, 46–47.

43. Parry, *Age of Reconnaissance*, p. 139; see also Oliveira, *Influencia*, pp. 24–25.

44. Subrahmanyam, *Career*, pp. 83–85; Parry, *Discovery*, pp. 169–70; Crosby, *Ecological Imperialism*, p. 118; Penrose, *Travel and Discovery*, p. 50; McGowan, *The Ship*, vol. 3, p. 18.

45. Parry, *Age of Reconnaissance*, pp. 140–41; Parry, *Discovery*, p. 174; and A.J.R. Russell-Wood, *The Portuguese Empire, 1415–1808: A World on the Move* (Baltimore: Johns Hopkins University Press, 1998), p. 18, identify the pilot as ibn Majid. Tibbetts (*Arab Navigation*) refutes this idea. Subrahmanyam (*Career*, pp. 121–28) argues that the French orientalist Gabriel Ferrand concocted the ibn Majid story in the 1920s, using the flimsiest of evidence.

46. Aczel, *Riddle*, pp. 61, 103–4.

47. On navigation and astronomy in the late fifteenth century, see Parry, *Discovery*, pp. 155–62, and Taylor, *Haven*, pp. 158–60.

48. Albuquerque, Introdução, pp. 233–400; Mathew, Portuguese Navigation, pp. 6–34; Taylor, Haven, pp. 158–59; Parry, Age of Reconnaissance, p. 93; Parry, Discovery, p. 148; Penrose, Travel and Discovery, p. 264.

49. Parry, Age of Reconnaissance, pp. 94–96; Parry, Discovery, pp. 148–49; Penrose, Travel and Discovery, pp. 44–45, 265. On the contributions of Arabs and Jews, see Albuquerque, Introdução, pp. 255–63, and Mathew, Portuguese Navigation, pp. 34–38.

50. Samuel Eliot Morison, Admiral of the Ocean Sea: A Life of Christopher Columbus (Boston: Little Brown, 1942), pp. 186–87.

51. Finney, Voyage, pp. 266–67; Parry, Discovery, p. 147.

52. Aczel, Riddle, pp. 124–25; Parry, Age of Reconnaissance, pp. 84–114.

53. Tomé Pires, The Suma Oriental of Tomé Pires: An Account of the East, from the Red Sea to Japan, Written in Malacca and India in 1512–1515, ed. and trans. Armando Cortesão (London: Hakluyt Society, 1944).

54. Auguste Toussaint, History of the Indian Ocean, trans. June Guicharnaud (Chicago: University of Chicago Press, 1966), pp. 115–17; Parry, Age of Reconnaissance, p. 96.

55. Quoted in Morison, Admiral, pp. 93–94.

56. William D. Phillips and Carla Rahn Phillips, The Worlds of Christopher Columbus (Cambridge: Cambridge University Press, 1992), pp. 76–79. On Columbus's beliefs, see also Morison, Admiral, chapter 6: "The Enterprise of the Indies."

57. Phillips and Phillips, Christopher Columbus, pp. 108–10.

58. Samuel Eliot Morison, The European Discovery of America: The Southern Voyages (New York: Oxford University Press, 1974), p. 30.

59. Ibid., p. 31; see also Morison, Admiral, pp. 68–69, and Phillips and Phillips, Christopher Columbus, pp. 110–11.

60. Phillips and Phillips, Christopher Columbus, pp. 120–31; Morison, Admiral, p. 75; Morison, European Discovery, p. 40.

61. Phillips and Phillips, Christopher Columbus, p. 132.

62. On the construction and rigging of caravels, see Carla Rahn Phillips, "The Caravel and the Galleon," in Robert Gardiner, ed., Cogs, Caravels, and Galleons: The Sailing Ship, 1000–1650 (Annapolis, Md.: Naval Institute Press, 1994), pp. 91–114; Chaunu, L'expansion européenne, 283–88; Parry, Discovery, pp. 28–29, 140–43; Penrose, Travel and Discovery, 269–70; and Smith, Vanguard, pp. 34–41.

63. Phillips and Phillips, Christopher Columbus, 108; Morison, European Discovery, pp. 82–85.

64. Phillips and Phillips (Christopher Columbus, p. 75) say he carried up-to-date tables of solar declination. Parry (Discovery, pp. 202–3) says he knew very little celestial navigation and nothing of the new Portuguese method. According to Morison (European Discovery, p. 55), the instruments Columbus possessed were so crude they could only be used ashore.

65. Morison, European Discovery, pp. 119, 138.

66. The Casa de contratación, modeled on the Portuguese Casa da India, began informally in 1503 and became an official branch of the government in 1524; see Morison, *European Discovery*, p. 474.

67. There are many fine works on Magellan, the most recent being Tim Joyner's *Magellan* (Camden, Maine: International Marine, 1992), and Laurence Bergreen's *Over the Edge of the World: Magellan's Terrifying Circumnavigation of the Globe* (New York: Morrow, 2003).

68. Donald D. Brand, "Geographical Exploration by the Spaniards," in Herman R. Friis, ed., *The Pacific Basin: A History of Its Geographical Exploration* (New York: American Geographical Society, 1967), pp. 111–13; Morison, *European Discovery*, pp. 177, 343; Parry, *Discovery*, p. 270.

69. Parry, *Discovery*, pp. 265–66.

70. Brand, "Geographical Exploration," pp. 112–18; Morison, *European Discovery*, pp. 405–9; Parry, *Discovery*, p. 276.

71. Morison, *European Discovery*, pp. 316–17.

72. This is the view of Brand ("Geographical Exploration," p. 118); see also Simon Winchester, "After dire straits, an agonizing haul across the Pacific," *Smithsonian* 22, no. 1 (April 1991), pp. 92–95.

73. Parry, *Discovery*, p. 287; Brand, "Geographical Exploration," p. 112; Taylor, *Haven*, p. 174; Morison, *European Discovery*, pp. 474–75.

74. William Lytle Schurtz, *The Manila Galleon* (New York: Dutton, 1959), pp. 217–18; Brand, "Geographical Exploration," p. 119.

75. Pierre Chaunu, "Le Galion de Manille: Grandeur et décadence d'une route de la soie," *Annales ESC* 4 (October–December 1951), p. 450; Morison, *European Discovery*, pp. 477–93; Brand, "Geographical Exploration," pp. 119–21.

76. Björn Landström, *The Ship: An Illustrated History* (New York: Doubleday, 1961), p. 118. See also J. H. Parry, *The Spanish Seaborne Empire* (New York: Knopf, 1966), p. 134.

77. Chaunu, "Galion," pp. 451–52; Brand, "Geographical Exploration," pp. 129–30; Schurtz, *Manila Galleon*, pp. 219–20; Morison, *European Discovery*, pp. 493–94.

78. Parry, *Spanish Seaborne Empire*, p. 132; Chaunu, "Galion," p. 453; Schurtz, *Manila Galleon*, p. 193.

79. Brand, "Geographical Exploration," p. 130; Schurtz, *Manila Galleon*, pp. 217–21.

80. Parry, *Spanish Seaborne Empire*, p. 132; Chaunu, "Galion," p. 453.

81. Parry, *Age of Reconnaissance*, p. 67; Gardiner, *Cogs*, p. 9.

82. K. M. Panikkar, *Asia and Western Dominance* (New York: Macmillan Collier, 1969), p. 46.

83. Charles R. Boxer, *The Dutch Seaborne Empire, 1600–1800* (New York: Knopf, 1965), p. 197.

84. On the longitude question, see William J. H. Andrewes, *The Quest for Longitude: Proceedings of the Longitude Symposium, Harvard University, Cambridge, Massa-*

chusetts, November 4–6, 1993 (Cambridge, Mass.: Collection of Scientific Instruments, Harvard University, 1996); and a brief popular account, Dava Sobel, *Longitude* (New York: Penguin, 1996).

85. Numa Broc, *La géographie des philosophes: Géographes et voyageurs français au XVIIIᵉ siècle* (Paris: Editions Ophrys, 1975), pp. 16, 281; John Noble Wilford, *The Mapmakers* (New York: Knopf, 1981), p. 129.

86. Wilford, *Mapmakers*, p. 128; Sobel, *Longitude*, pp. 11–12.

87. Rupert Gould, *The Marine Chronometer: Its History and Development* (London: J. D. Potter, 1923), pp. 254–55; Gould, "John Harrison and His Timekeepers," *Mariner's Mirror* 21 (April 1935), p. 118; David Landes, *Revolution in Time: Clocks and the Making of the Modern World* (Cambridge, Mass.: Harvard University Press, 1983), pp. 112, 146; Lloyd A. Brown, *The Story of Maps* (New York: Dover, 1980), p. 227.

88. On cartography in this period, see Daniel Headrick, *When Information Came of Age: Technologies of Knowledge in the Age of Reason and Revolution, 1700–1850* (New York: Oxford University Press, 2000), chapter 4: "Displaying Information: Maps and Graphs."

89. Quoted in Wilford, *Mapmakers*, p. 131.

90. Charles H. Cotter, *A History of Nautical Astronomy* (New York: American Elsevier, 1968), pp. 184–86; J. B. Hewson, *A History of the Practice of Navigation* (Glasgow: Brown, Son and Ferguson, 1951), pp. 223–50; Frédéric Philippe Marguet, *Histoire générale de la navigation du XVᵉ au XXᵉ siècle* (Paris: Société d'éditions géographiques, maritimes et coloniales, 1931), pp. 127–31; Broc, *La géographie des philosophes*, pp. 16–33.

91. Landes, *Revolution in Time*, p. 152; Sobel, *Longitude*, pp. 89–91; Taylor, *Haven*, p. 256.

92. Quoted in Cotter, *History*, p. 195.

93. Derek Howse, *Greenwich Time and the Discovery of Longitude* (Oxford: Oxford University Press, 1980), pp. 62–69; Marguet, *Histoire générale*, pp. 185–94; Landes, *Revolution in Time*, pp. 151–55; Cotter, *History*, pp. 189–237; Wilford, *Mapmakers*, pp. 130–35.

94. Harrison's life, one of the most dramatic David-and-Goliath stories in the history of technology, has been told many times. See Landes, *Revolution in Time*, pp. 146–62; Gould, "John Harrison"; Wilford, *Mapmakers*, pp. 128–37; Sobel, *Longitude*, pp. 61–152; and Taylor, *Haven*, pp. 260–63.

95. Broc, *La géographie des philosophes*, pp. 282–84; Marguet, *Histoire générale*, pp. 148–84; Landes, *Revolution in Time*, chapter 10.

96. Sobel, *Longitude*, pp. 17–20; Richard I. Ruggles, "Geographical Exploration by the British," in Friis, *Pacific Basin*, p. 237.

97. James Lind, *A Treatise of the Scurvy: Containing an Inquiry into the Nature, Causes, and Cure, of That Disease* (Edinburgh, 1753). See also Christopher Lloyd and Jack L. S. Coulter, *Medicine and the Navy, 1200–1900*, vol. 3: *1714–1815* (Edin-

burgh and London: Livingstone, 1961), pp. 293–322; and Alfred F. Hess, *Scurvy, Past and Present* (Philadelphia: Lippincott, 1920), pp. 172–204.

98. J. C. Beaglehole, *The Exploration of the Pacific* (Stanford: Stanford University Press, 1966), pp. 256–57.

99. John C. Beaglehole, *The Life of Captain James Cook* (London: Hakluyt Society, 1974), pp. 410, 423, 438; Sobel, *Longitude*, pp. 149–50.

100. Beaglehole, *Exploration*, p. 311; Sobel, *Longitude*, pp. 144, 154–55; Gould, "Harrison," p. 126.

101. Frédéric Philippe Marguet, *Histoire de la longitude à la mer au XVIIIe siècle, en France* (Paris: Auguste Challamel), p. 217.

Chapter 2

Eastern Ocean Empires, 1497–1700

Mastery of the environment was a necessary, but far from sufficient, step toward achieving dominion over the sea. On the world's oceans, European mariners encountered other navigators. Some of them—Pacific islanders, Native Americans, and West Africans—sailed in open canoes; they were adept at navigating, some over long distances, but when hostilities erupted, they were no match for the larger European ships and their cannon. On the Indian Ocean and on the seas bordering East Asia, however, were ships as large as those of the Europeans and sometimes larger. Although piracy was common, naval warfare was almost unknown before 1498.

Only the Ottomans, who reached the Indian Ocean forty years after the Portuguese, had experience with naval warfare. In their struggle for mastery of the Indian Ocean, two types of ships and two ways of fighting confronted one another: Atlantic sailing ships and Mediterranean oared galleys. Each was designed for, and dominated, a particular marine environment. In the clash between two imperialisms—the Portuguese and the Ottoman—victory or defeat depended on leadership, motivation, and fighting qualities, but always within the narrow limits imposed by technologies and environments.

The Portuguese in the Indian Ocean

The first ships that Vasco da Gama encountered since leaving Portugal were in the harbor of Mozambique. Hostilities quickly broke out between the inhabitants, who were Muslims, and the Christian interlopers. Before departing, da Gama bombarded the town. The same happened at Mombasa, where da Gama had to use his guns to persuade the inhabitants to let him get fresh water. Not until he reached Malindi did he receive a friendly welcome and the offer of a pilot to guide his fleet across the Arabian Sea to Calicut in India.[1]

Calicut was a predominantly Hindu city under the protection of the Hindu kingdom of Vijayanagara, but it was also home to a large population of Muslim traders from Persia, Arabia, and Gujarat. There, and in ports throughout the Indian Ocean, religious tolerance was the norm and trade was largely free and peaceful, with low duties and taxes, good port and warehousing facilities, and well-developed banking and legal systems. The only limitations on trade were the products available and the capacity and cost of ships to transport them.[2]

Upon arrival, da Gama sent a man ashore. When a merchant from Tunis asked what the Portuguese were doing there, the man replied: "We came to seek Christians and spices."[3] In Calicut, they found an abundance of spices, especially pepper, a precious commodity to Europeans who needed to hide the taste of spoiled meat in an age before refrigeration. But da Gama was not prepared to trade, for he had brought only coarse cloth, cheap hardware, and beads, wares of the sort that were in demand on the Guinea coast but of no interest to the people of Calicut. He also failed to bring gifts for his hosts, a breach of etiquette in much of the world. The Zamorin or ruler of Calicut received the Portuguese in a civil manner, but their behavior was "abrupt and hectoring." The Muslim traders, some of whom were from North Africa and had first-hand experience with European Christians, became hostile. After three months of hard bargaining, da Gama obtained enough spices to make his voyage a financial success, but in the process he alienated the ruler and the people of Calicut.[4]

When Vasco da Gama returned to Lisbon in 1499, King Manuel celebrated the event by proclaiming himself "Lord of Guinea and of the Conquest of the navigation and commerce of Ethiopia, Arabia, Persia and India."[5] With that declaration, Portugal entered a new era; no longer was the goal of its expeditions discovery or trade but empire-building. This meant attacking a well-organized commercial network run by merchants who were predisposed, for religious reasons, to be hostile toward the Portuguese. As historian J. H. Parry explains:

> From the Portuguese point of view the physical destruction of Arab commercial shipping, besides being a pious duty, might become a competitive necessity. This would involve piracy and naval aggression on an enormous scale. . . . If the Portuguese seriously proposed to break into the trade of the Indian Ocean, by the route which da Gama had discovered, they would have to use their guns.[6]

The second fleet that left Lisbon in March 1500 under the command of Pedro Alvares Cabral was as much a military expedition as a commercial venture. It started out with thirteen ships and over a thousand men. Sailing far to the southwest from the Cape Verde Islands, Cabral discovered Brazil. Of the thirteen ships, only six reached Calicut in September 1500. At first, relations were tense but formal. Then fighting broke out in the town between the Portuguese and Muslim traders and fifty-four Portuguese were killed. In retaliation, Cabral bombarded Calicut, causing four or five hundred deaths and seizing and burning a number of Muslim ships along with their crews. He then sailed to the nearby cities of Cochin and Cannanore, where he was welcomed as an ally against Calicut. Although only four of his ships returned to Lisbon, Cabral's expedition, like da Gama's, made a huge profit.[7]

The third fleet, in 1502–3, was led, once again, by Vasco da Gama. Different historians offer different numbers, but it may have involved as many as twenty-five ships, all heavily armed. Upon arriving in East Africa, da Gama obtained the submission of Mombasa and extorted gold from Kilwa by threatening to bombard them. He then made for Calicut, which he also bombarded. His ships destroyed a fleet belonging to the Zamorin and to Arab merchants who sailed out to meet them and sank every Muslim ship they came across, even pilgrim ships carrying women and children. After loading up spices at Cochin and Cannanore, da Gama returned to Lisbon in February 1503, leaving behind three carracks and two caravels under Vicente Sodré, the first permanent Portuguese presence in the Indian Ocean.[8]

In 1504, Francisco de Almeida went out as the first viceroy of India at the head of a large fleet. In preparation for an attack on Muslim shipping, he captured and fortified Sofala, Kilwa, Mombasa, and Mozambique on the East African coast. By then, the Venetian and Egyptian governments realized that the Portuguese presence in the Indian Ocean had dealt a blow to their monopoly of the spice trade and had devastated their profits.[9] The Zamorin of Calcutta, the governor of Diu in Gujarat, and the Arab traders on the Malabar coast appealed to Sultan Kansuh Gawri of Egypt for help against the Portuguese. With Venetian help, the Egyptians built a fleet of galleys equipped with cannon. In 1507, under Admiral Emir Huseyn al-Kurdi, it sailed from Suez to Jiddah, the port of Mecca. The following year the Egyptians sailed out into the Indian Ocean, where they were joined by ships from Gujarat and Calicut. At the mouth of the Chaul River north of Calicut, they encountered several Portuguese ships, captured them, and killed the viceroy's son Lourenço.[10]

In early 1509, thirsting for revenge, Viceroy Almeida put together a fleet of eighteen ships manned by fifteen hundred Portuguese and four hundred Malabari sailors and sailed for Diu, where the Egyptians and Gujarati had assembled dozens of dhows, galleys, and other craft to challenge the Portuguese domination of the Indian Ocean. Though the Muslim ships outnumbered the Portuguese, none carried guns as powerful as their enemies' and their leadership was divided and hesitant. Almeida quickly destroyed the Muslim ships, capturing and killing their crews.[11]

Until 1509, the policy of the Portuguese was to send armed fleets to the Indian Ocean to destroy Muslim shipping and enemy fleets, load up with spices, and return home. In short, they conflated religious and commercial goals. This changed when Alfonso de Albuquerque arrived to succeed Almeida as viceroy. It was his actions, more than anything else, that created a lasting sea empire in the Indian Ocean.

Albuquerque had served in the East in 1503–4. He returned in 1506 and was appointed viceroy in 1509. He quickly set out to put Portuguese dominance on a permanent footing by creating a string of naval bases where a permanent Indian Ocean fleet could provision and refit and where sailors could be recruited and recover from duty at sea.[12] These bases would also serve as entrepôts for trade and as settlements for merchants and other civilian members of the Portuguese establishment in the East. From the security of these fortresses, mobile squadrons could venture out to destroy Muslim ships and choke off the flow of spices to Europe via Egypt and Venice.

Albuquerque began by attacking Goa, an island off the Malabar coast 350 miles north of Calicut with a good harbor and shipbuilding industry. After a first failure in 1510, he succeeded the next year by providing Arabian horses to the Hindu king of Vijayanagara and denying them to the Muslim sultan of Bijapur, in whose lands Goa was situated.[13] He wrote to King Manuel:

> The taking of Goa keeps India in repose and quiet. It was folly to place all your power and strength in your navy only . . . in ships as rotten as cork, only kept afloat by four pumps in each of them. If once Portugal should suffer a reverse at sea, your Indian possessions have not power to hold out a day longer than the Kings of the land choose to suffer it.[14]

Even before Goa was fully secured, Albuquerque sailed off with eighteen ships, eight hundred European soldiers, and two hundred Indian auxiliaries. His goal was Melaka, a strategic harbor that dominated the straits between Sumatra and the Malay Peninsula and was a major entrepôt in the trade between the Indian Ocean to the west and the Spice Islands and China to the east. After seizing the city in July 1511, he spared the Hindu, Chinese, and Burmese inhabitants but had the Muslim inhabitants massacred or sold into slavery; as he wrote the king: "At the rumor of our coming the [native] ships all vanished and even the birds ceased to skim over the water."[15] He immediately set to work building a fortress called A Famosa with eight-foot-thick stone walls, which was to remain in Portuguese hands for 150 years.[16]

Possession of Melaka opened the way to the east. Tomé Pires, an accountant who worked in the Portuguese factory there from 1512 to 1515, boasted: "Whoever is lord of Malacca has his hand on the throat of Venice. As far as from Malacca to China, and from China to the Moluccas, and from the Moluccas to Java, and from Java to

Malacca and Sumatra, all is in our power."[17] The Portuguese had no
difficulty finding local navigators to guide their ships in the treacherous seas beyond Melaka. Antonio de Abreu left Melaka with three
ships and Javanese pilots in 1511, heading for the Banda Islands,
where he purchased nutmeg. In 1514, a second Portuguese fleet established a factory at Ternate, the center of clove production.[18]

Gaining access to the source of the coveted spices—the cloves
and nutmeg of the Moluccas, the pepper and cardamom of Malabar,
and the cinnamon of Ceylon—was a major achievement of Albuquerque's policy. The other goal was stopping the Muslim trade between the Indian Ocean and the Mediterranean Sea. As early as
1500 King Manuel had instructed Almeida: "nothing could be more
important for our service than to have a fortress at the mouth of the
Red Sea or near it, either inside or outside as seems most convenient,
for if that is sealed then no more spices can pass through the lands
of the Sultan, and everyone in India would give up the fantasy of
being able to trade with anyone save us."[19] In February 1513, therefore, Albuquerque left Goa with twenty-four ships and seventeen
hundred European and a thousand Indian soldiers, headed for Aden,
the port that commanded Bab al-Mandab, the entrance to the Red
Sea. The attack, however, was repulsed. After failing to take Aden,
Albuquerque sailed into the Red Sea, hoping to find a suitable base.
This attempt, too, failed and, with it, a major goal of Portuguese
policy in the East remained out of reach.

Albuquerque then turned toward Hormuz, the island that commanded the entrance to the Persian Gulf and that he had controlled,
briefly, in 1508. This time, he captured the town and had a fort
built. From there, the Portuguese were able to intercept and tax
ships entering or leaving the Persian Gulf, and there they remained
for over a hundred years.[20] Albuquerque died there before he could
continue his conquests.

At this point it is worth pausing to ask how the Portuguese, with
so few ships and men, were able to defeat numerous enemies, conquer strategic places, and dominate an ocean. Leadership, courage,
and religious conviction played a part, as did greed and ferocity. But
translating such motivations into successful actions also required a
technological edge. And this edge came not from guns or ships or
training alone, but from a combination of the three.

By the time Vasco da Gama undertook his historic first voyage, Europeans were familiar with firearms, but so were the peoples of the Middle East and South and East Asia. Gunpowder originated in China in the tenth century, where it was used in firecrackers and incendiary projectiles. In the early fourteenth century, cannon—metal tubes closed at one end from which a stone or iron cannonball was hurled by the force of the exploding gunpowder—appeared in both China and Europe.[21] The earliest cannon, called bombards, were made by welding iron rods to form a cylinder and holding them together with iron hoops, like a wine barrel; such guns were loaded from the back and sealed with a breechblock. By the fifteenth century, European warships carried several such guns. Their effective range was seldom over two hundred yards, and their shot was rarely able to damage the hull of a ship. They were often placed on the castles at each end of a ship, not to shoot at other ships but to kill enemy soldiers who attempted to board during a battle.[22]

The three ships on Vasco da Gama's first voyage in 1497–98 carried between them twenty such bombards and several lighter anti-personnel guns designed for defense and display rather than attack. Cabral's fleet in 1500–1501 and da Gama's second fleet in 1502 were much more heavily armed. Their caravels carried four heavy bombards, six medium-size guns called falconets, and ten smaller swivel guns; the carracks carried eight bombards, plus several lighter guns. With them they bombarded Calicut and sank ships in the harbor. Their ships were designed as much for war as for trade.[23]

In contrast, Indian Ocean merchantmen were seldom armed. Armed ships in the Arabian Sea carried archers and swordsmen. There is no clear evidence that firearms were used before the Portuguese arrived.[24] Oceangoing dhows, built of planks held together by ropes and dowels, were too fragile to withstand the weight and recoil of cannon. Hence the merchants and coastal states were quite unprepared for the European assault.

The Portuguese did not rest content with their bombards, intimidating as these weapons were, but stayed abreast of the continuing technological progress in firearms. In the early sixteenth century, stimulated by the competitiveness of European states—what historian Philip Hoffman has called "a tournament among western European rulers that fostered military innovation"[25]—European gun makers began casting cannon of bronze that had to be loaded from the muzzle, with only a touchhole at the breech to ignite the powder. These were much costlier than built-up iron guns, but they were much stronger and could be charged with more powder to hurl an iron cannonball weighing as much as sixty pounds. Such cannon could damage the hull of a ship up to three hundred yards away. As they weighed several tons, they could not be placed in the castles or upper deck where they would make the ship top-heavy but had to be carried on the main deck. To accommodate them, portholes were cut into the sides of the ship, with hinged covers that could be raised during battles.[26]

In 1500 King Manuel gave Cabral his instructions before he sailed for the Indian Ocean: "You are not to come to close quarters with them [Muslim ships] if you can avoid it, but you are to compel them with your artillery alone to strike sail . . . so that this war may be waged with greater safety, and so that less loss may result to the people of your ships."[27] The Portuguese were able to do so because their guns had a greater range than those on Muslim ships, and because their gunners were better trained. They carried gunpowder in bags, pre-measured for quick reloading.[28] Instead of ramming and boarding, they practiced a tactic called the broadside volley, in which all the guns on one side of a ship were fired at the same target. When Albuquerque's fleet appeared off Hormuz, he issued an ultimatum. His son described what followed.

> When morning broke, and Afonso Dalboquerque plainly saw that no message from the king [of Hormuz] was forthcoming, and that this delay meant war and not peace, he ordered a broadside to be fired. The bombardiers took aim so that with the first two shots they sent two large ships which were in front of them, with all their men, to the bottom. . . . And although the Moors endeavoured to avenge themselves with their artillery, our men were so well fortified with their defences, that they did them no harm except on the upper deck, and with their arrows they wounded some people.[29]

The broadside volley depended on superior firepower, close maneuvering, and quick reloading. To maximize the effect, the Portuguese maneuvered their ships into a "line-ahead" or bow-to-stern

Figure 2.1. A sixteenth-century Spanish galleon. *Narrative and Critical History of America* (New York: Houghton, Mifflin, and Company, 1886). Courtesy of FCIT, http://etc.usf.edu/clipart.

formation, so that the guns of several ships at once could bear upon the enemy. This arrangement, first employed off the Malabar coast in 1502, became the standard fighting tactic of European navies for the next three hundred years. Portuguese dominance of the seas required seizing strategically placed coastal cities, as Albuquerque realized. To do so, they employed not only oceangoing carracks that could bombard a town but also smaller boats with which soldiers could land and attack their targets from landward.[30]

In the course of the sixteenth century, the ships that Portugal sent out and the guns they carried grew ever larger and more powerful. *Naus* and carracks gave way to galleons, floating fortresses of a thousand tons or more with many decks, complex rigging, and dozens of cannon. In any encounter with dhows or oared galleys, such ships were invulnerable. Only other European ships posed a threat to the great galleons of the *carreira da India*, as they did to the Spanish fleets that by then were bringing silver back from the New World.

What paid for these ships and the entire Estado da India were the profits of the spice trade, which were enormous; often, one ship's cargo more than paid for an entire expedition. Later, intra-Asian trade became more profitable than trade with Europe. Over time, however, this business ran into diminishing returns as Indian Ocean merchants found ways of avoiding the Portuguese and as new enemies appeared that raised the costs of defense. The first of these were the Ottomans.

The Ottoman Challenge

The Portuguese victories of the early sixteenth century are impressive only in hindsight. At the time, in Europe, the Middle East, and India, it was not Christians but Muslims who were the most aggressive and successful empire-builders. In India, the Central Asian warlord Babur (1483–1530), a descendant of Tamerlane, founded the Mughal dynasty that was to rule most of the subcontinent for two hundred years. In the Middle East, the Ottoman Turks consolidated their rule in Anatolia, conquered Syria and Egypt, defeated the Persians, and advanced into central Europe. At sea, they dominated the eastern Mediterranean.

After the disaster at Diu in 1509, Sultan Kansuh Gawri of Egypt appealed to the Ottoman sultan Bayezid II for help in building a new fleet, for Egypt lacked the necessary wood and metal. Bayezid promised to send enough materiel to build thirty galleys and three hundred cannon, along with Ottoman officers to supervise their construction. Bayezid's successor, Selim I, continued the policy of supplying Egypt. Salman Reis, a former Mediterranean corsair, was put in change of the arsenal at Suez. The fleet, composed of up to thirty galleys, several thousand men, and hundreds of firearms of all sizes, was ready to sail in the summer of 1515. Salman Reis was promoted to admiral and placed in command.[31] His first goal was to fortify Jiddah, the port of Mecca.

The Ottomans were as adept at naval warfare as at combat on land. For two centuries, however, their experience with naval matters was limited to the Mediterranean and Black seas. They had no contact with the Indian Ocean or its tributaries the Red Sea and Persian Gulf until 1517. When the Ottomans conquered Egypt that

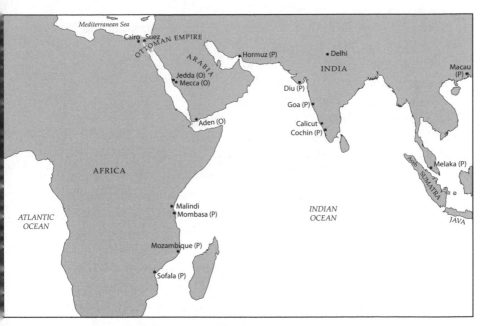

Figure 2.2. The Indian Ocean, showing Portuguese (P) and Ottoman
(O) bases and possessions circa 1550. Map by Chris Brest.

year, they inherited the Mamluks' interests in the spice trade.[32] The
Ottoman sultans appropriated the titles of "Caliph," or successor to
Muhammad, and "protector of the Holy Cities" Mecca, Medina, and
Jerusalem. As a by-product of their wars with Rhodes, Venice, and
the Balkan states, they became the standard-bearers in the permanent hostilities between Muslims and Christians. These many factors brought them into conflict with the Portuguese.

In 1517 the Portuguese sent out a new fleet under Lopo Soares de
Albergaria, with the goal of finishing what Albuquerque had begun.
He entered the Red Sea with a carrack of eight hundred tons,
twenty-two other carracks, seven galleys, and several other ships,
along with several thousand soldiers and sailors. Like Albuquerque
before him, he had great difficulty navigating the straits of Bab al-
Mandab—the "Gate of Lamentations"—owing to sandbanks,
shoals, and contrary winds. Approaching Jiddah, he found the entrance to the harbor too shallow for his larger ships. The town was
well defended by Salman Reis's huge cannon, capable of hurling

stones weighing up to a thousand pounds, for on land, the Turkish artillery was fully the equal of that of the Europeans.[33] Abandoning the attempt, the Portuguese ships sailed away, only to be caught in storms. Many crew members died of hunger, thirst, and disease. The expedition failed to take any town or fort in the Red Sea and returned to the Indian Ocean in defeat.[34]

Soon thereafter, in 1520, a new sultan succeeded Selim I: Suleyman the Lawgiver, known in the West as "the Magnificent." A brilliant strategist and war leader, Suleyman built up his fleet and army. Though his main goal was to complete the conquest of the Balkans and then to attack central Europe, he was also interested in Arabia and the Indian Ocean.

The early years of Suleyman's reign were marked by frequent skirmishes between the Portuguese and the Ottomans. In 1525, a Portuguese fleet entered the Red Sea and captured twenty-five merchant vessels. Six years later, another Portuguese fleet blockaded the Red Sea and raided the coast near Jiddah.[35] During the 1530s, tension mounted between the Portuguese and the Ottomans and their Muslim allies. In 1536, Sultan Bahadur of Gujarat, recently defeated by the Mughals, sought refuge in Diu and sent a request for help to the Ottomans. Suleyman decided to build a fleet, expel the Portuguese from the Indian Ocean, and seize Cambay in Gujarat. By July 1538, an Ottoman fleet of over sixty warships, eight thousand sailors, and sixty-five hundred soldiers sailed from Suez to Jiddah under the command of Hadim Suleyman Pasha, the Ottoman governor of Egypt. Upon arriving at Diu, his janissaries besieged the Portuguese garrison. Before the siege could succeed, Bahadur backed out of his commitment to a pan-Muslim offensive against Portugal because he was more worried about Ottoman dominance than about the Portuguese threat to his overseas trade; as he explained: "Wars at sea are merchants' affairs and no concern to the prestige of kings."[36] Upon the arrival of a Portuguese fleet of thirty-nine sailing ships and over a hundred galleys, Hadim Suleyman Pasha fled Diu and escaped back to the Red Sea.[37]

Meanwhile, Muslim pirates of the Malabar coast led by Pate Marakar had built a fleet of *fustas*, or small galleys, with four hundred guns and several thousand men. In 1538 it was destroyed by a Portuguese fleet. The next year, a Muslim fleet from Aceh in Sumatra carrying some Ottoman soldiers attacked Melaka, but was repulsed. The decade of the 1530s thus ended badly for the Muslim forces throughout the ocean. Nonetheless, the Ottomans had established control over Aden and most of Yemen, weakening the Portuguese attempts to stop the spice trade via the Red Sea and the pilgrim traffic to Mecca and Medina.[38]

In 1541, three years after the Ottoman failure to drive them out of the Indian Ocean, the Portuguese decided to counterattack. Estevão da Gama, son of Vasco, led a fleet of seventy ships and twenty-three hundred men into the Red Sea with the intention of seizing Suez, with its Ottoman fortress and arsenal. Half the fleet, however, was diverted to Massawa on the Eritrean coast, leaving only sixteen ships and 250 men to attack Suez. The expedition ended in defeat and humiliation for the Portuguese.[39]

The standoff spread to the Persian Gulf in 1546 when the Ottomans took Basra at the head of the Gulf but were blocked by the Portuguese at Hormuz, at its mouth. In 1552 the Turkish Admiral Piri Reis, coming from Suez with a fleet of twenty-four galleys and four supply ships, besieged Hormuz but was forced to lift the siege and sail on to Basra, reputedly because the sinking of a supply ship left his troops without food or ammunition.[40] The next year, the fleet, now commanded by Murad Beg, attempted to return to Suez but was intercepted off Hormuz by a Portuguese fleet under Diogo de Noronha. The resulting battle was the largest in the open sea between the Portuguese and Ottoman navies. For a time, when the wind dropped, the Ottoman galleys had the upper hand. They sank Noronha's flagship and captured another Portuguese galleon. Then the wind picked up, and the sailing ships regrouped and forced Murad to return to Basra.

A year later, in 1554, the Ottoman fleet, now commanded by Seydi Ali Reis, tried again to return to Suez but was intercepted once again by the Portuguese under Dom Fernão de Menezes and forced out into the ocean. The remaining galleys, their crews exhausted by the battle, drifted with the wind until they reached the coast of India, where they were abandoned. Though Seydi Ali Reis eventually made his way overland to Istanbul, his defeat marked the end of the Ottoman attempts to take on the Portuguese navy or capture its fortified bases.[41]

In the long series of naval encounters between the Portuguese and their Muslim enemies, we can discern a clear pattern. The Portuguese prevailed off Diu in 1509 and 1538, off Hormuz in 1551 and 1553, and off Muscat in 1554, but failed in their attempts to take Aden or penetrate the Red Sea in 1513, 1517, 1525, 1531, and 1541. Likewise, the Ottomans succeeded in dominating the Red Sea and their strategic harbors of Suez, Jiddah, and Aden, but failed to drive the Portuguese out of the Indian Ocean. This was not an unreasonable goal, given that their empire was twenty times larger, richer, and more populous than Portugal and that their armies were victorious against the Christians in the Balkans and dominated the eastern Mediterranean. What stood in their way?

One obstacle was geography. Any ship destined for the Indian Ocean had to be built in Suez or in Basra, far from sources of wood and other supplies. The enormous cost of transporting lumber all the way from Anatolia restricted ship construction.[42] Another problem was Yemen, a mountainous country that occupied a strategic location between the Middle East and the Indian Ocean. Its inhabitants were independent-minded Shi'ite Muslims who did not recognize the Ottoman sultan as caliph. In 1515 the Mamluks had planned to send a fleet to the Indian Ocean but instead diverted it to conquer Yemen. In 1526 the Ottomans sent another expedition to Yemen. Two years later, internecine warfare erupted among the Turks in Yemen and that country fell into anarchy. Likewise in 1547–48, another Ottoman expedition to Yemen captured the port of Aden, an important victory, for it thwarted the Portuguese attempts to blockade the Red Sea.[43] In 1567, an Ottoman expedition had to be postponed because of yet another Yemeni uprising.[44] To the Ottomans, as to the Mamluks before them, Yemen was as much an obstacle as a prize.

Yet the main reason the Ottomans could not sweep the Portuguese from the ocean was similar to the one that thwarted the Portuguese ambition to blockade the Red Sea: the technology of naval warfare. The traditional warships of the Mediterranean were driven by oars. The larger ones, called galleys, were manned by 144 to 200 oarsmen pulling twenty to twenty-four banks of oars. To spare the oarsmen when the wind was right, they had from one to three masts carrying lateen sails and sometimes a square sail. Smaller ships called *galliots* or *fustas* had one bank of oars on each side and one mast and sail, and were much used by corsairs and pirates. Such man-powered vessels were designed to ram or grapple an enemy vessel, allowing soldiers to board and capture it. Built for speed with long narrow hulls, they were unsafe in heavy seas and high winds. With a large crew, they could not bring along much food and water and thus could not venture too far out into the ocean. A galley with a crew of 144 oarsmen and another thirty soldiers and officers consumed ninety gallons of water a day and could only carry enough for twenty days at sea.[45]

In the fifteenth century, the states of the Middle East and South Asia were as familiar with guns as were the Europeans. Most of their artillery was used to batter down the walls of fortresses, as the Ottoman Turks famously showed in the siege of Constantinople in 1453. The Ottomans continued to use cannon with great success in their conquests of Egypt, Yemen, and Ethiopia. But they could not duplicate that success at sea, for their oared warships, to be fast and maneuverable, had to be lightly built, which made them vulnerable to cannonballs. As their sides were filled with oarsmen and their oars, what few cannon they could carry had to be placed at the bow, pointing forward so that the recoil would not overturn them.[46] Moreover, their cannon were not up to European standards; the Venetians had to melt down most of the Turkish guns that they had captured at Lepanto in 1571 "because the material was of such poor quality."[47]

What gave Europeans such an advantage in the Indian Ocean, centuries before they were able to conquer any significant land areas, was their possession of ships built to withstand long voyages on the often stormy Atlantic Ocean. Their ships, propelled by the wind, needed only a small crew and were therefore able to carry many cannon and enough food and water for weeks, even months, at sea. It is the possession of more powerful artillery, and not the numbers of ships and men, that explains the Portuguese victories at sea.

The Portuguese advantage on the ocean was balanced by their weakness in narrow seas and close to shore, where their carracks and galleons had difficulty maneuvering. This was especially true of the Red Sea, whose rocky shores, many islands, hidden reefs, and erratic winds made it dangerous for sailing ships. Here, galleys had the advantage of speed and maneuverability.

To be sure, neither the Ottomans nor the Portuguese were wedded to a specific naval technology. On a few occasions, the Ottomans employed sailing ships modeled on those of the Portuguese, although, as historian Giancarlo Casale explains, "they never took more than a supporting role in naval operations."[48] Similarly, the Portuguese included galleys, fustas, and other rowed warships in their expeditions to the Red Sea. But neither side was ever able to commit enough resources or imitate the other's technology sufficiently to tip the scales.

Hence, the two contrasting naval technologies, one designed for narrow seas with erratic winds, the other for the ocean, outweighed the motivations and skills of the protagonists, resulting in a long stalemate.

The Limits of Portuguese Power

The original goal of the Portuguese had been to monopolize the spice trade by capturing and, if possible, pillaging any Muslim ship bearing spices. After the 1530s, they changed their tactic from outright plunder to a more subtle form of parasitism, namely selling safe-conducts called *cartazes*—a form of protection money—and charging duty on goods transshipped through Melaka, Goa, and Hormuz. The chronicler João de Barros expressed the arrogant mind-set of the Portuguese when he wrote the following:

> It is true that there does exist a common right to all to navigate the seas, and in Europe we acknowledge the rights which others hold against us, but this right does not extend beyond Europe, and therefore the Portuguese as lords of the sea by the strength of their fleets are justified in compelling all Moors and gentiles to take out safe-conducts under pain of confiscation and death.[49]

Apart from their confrontations with the Egyptian and Ottoman fleets, the Portuguese encountered little opposition to their presence in the Indian Ocean. The sultan of Gujarat preferred the Portuguese to the Turks. The city-state of Calicut engaged in skirmishes with Portuguese ships for one hundred years until the two sides signed a peace treaty in 1599.[50] Though some Indian states armed their ships with cannon, they were never able to match the Europeans on the seas; according to the Indian historian Ahsan Jan Qaisar, they were "inexpert in maneuvering their ships" and "their weakness lay in their lack of skill in using guns."[51] Other than Calicut, Hindu states generally accepted the Portuguese, either for commercial reasons or as allies against Muslim states. At most, they considered the Portuguese a nuisance, as J. H. Parry explains: "Momentarily dangerous the Europeans might be; but in the eyes of a cultivated Hindu they were desperadoes, few in number, barbarous, truculent, and dirty."[52]

To most Indians, the Portuguese presence along the coasts was a sideshow compared to the invasion and conquest of their subcontinent by the Mughals. Like the Ottomans, the Mughals were a Turkish-speaking people from Central Asia who had adopted Islam. Also like the Turks, they founded an empire based on their fighting spirit and their skillful use of artillery, in short, a "gunpowder empire." When Babur invaded northern India in 1526, he brought with him cannon from Turkey served by Turkish artillerymen. By 1530, he had conquered much of northern India. Akbar (r. 1556—1605), the greatest of his successors, extended the Mughal Empire with the help of bombards weighing up to fifty tons. Like other Asian rulers, the Mughals recruited European gun founders and artillerymen; a later emperor, Aurangzeb (r. 1658–1707), even had a Christian gunners' quarter in his capital.[53]

Though they arrived in India in 1526, the Mughals did not reach the sea until forty years later. They conquered the kingdom of Vijayanagara in 1565 and Gujarat in 1572. A year later, Akbar received a Portuguese embassy; according to his chronicler Abul Fazl, "he wished that these inquiries might be the means of civilising this savage race."[54] Relations were strained because the Mughals wanted to encourage pilgrims to travel by ship from Surat to Jiddah. The

Portuguese religious authorities in Goa tried to block such travel, but the civil authorities refused to do so because such an act would endanger the trading relations on which their prosperity depended. However, they insisted that all ships belonging to Mughal princes and merchants obtain a *cartaz*, even those carrying pilgrims to Mecca. Finally in 1581, the Portuguese gave two pilgrim ships a year *cartazes* free of charge, a policy that lasted into the seventeenth century.[55]

The Mughals, meanwhile, showed little interest in the sea, considering it cheaper to pay protection money to the Portuguese than to invest in a navy of their own. In the late sixteenth century, Aurangzeb's fleet consisted of two men-of-war with a thousand men.[56] According to the Englishman John Fryar, who visited India in the seventeenth century, "If the king's fleet be but ordinary, considering so great a Monarch and these Advantages, it is because he minds it not, he contenting himself on the enjoyment of the continent and styles the Christians Lions of the sea, saying that God has allotted that unstable element for their Rule."[57] Thus there developed a reciprocal relationship between two empires, one of the land and one of the sea, based on mutual advantage.

In contrast to the Mughal Empire, the sultanate of Aceh in Sumatra was permanently hostile to the Portuguese, both for religious reasons and because their presence in Melaka, directly across the Straits, interfered with their exports of pepper to Calicut and points west. The sultan of Aceh, Alauddin Riayat Shah al-Kabar (1539–71), turned his state into a rival entrepôt in the spice trade from which Muslim traders from Gujarat could reach Java and the Spice Islands by sailing down the west coast of Sumatra. In 1539, 1547, and 1551 the Acehnese attacked Melaka, but in vain. In 1561, the sultan of Aceh asked the Ottomans to send ships, guns, and military experts. Three years later, an Ottoman envoy named Lufti arrived in Aceh, inspiring the Acehnese to turn against the Portuguese. In 1567 the Ottomans prepared to send fifteen war galleys and two sailing ships to aid Aceh, but they were diverted to deal with an uprising in Yemen. In the end, only the transports reached Aceh, bringing soldiers, cannon, ammunition, and gunners. With them, the sultan of Aceh launched an attack on Melaka, but it too failed, as did later attacks in 1570 and 1582. Despite these military reverses, the ties between Aceh and the Ottomans weakened the Portuguese hold on Indian Ocean commerce and helped revive the spice trade via the Red Sea. Soon, Aceh was shipping as much pepper to the Red Sea as Portuguese ships were carrying to Europe via the Cape of Good Hope.[58]

At their peak in the mid-sixteenth century, the Portuguese maintained over forty trading posts in the East, stretching from Sofala in southeastern Africa to Nagasaki in Japan, anchored by fortified naval bases at Mombasa, Hormuz, Goa, and Melaka. These supported a fleet of several dozen warships and over a thousand cannon.[59] Yet, for all their naval power on the ocean, the Portuguese, like the Spanish in the Atlantic, had an Achilles' heel: their merchantmen were vulnerable to corsairs.

In the mid-sixteenth century, just as the confrontation between the Ottomans and the Portuguese seemed to have reached a stalemate, a Turkish corsair named Sefer Reis discovered a way to weaken the Portuguese grip on the ocean. From the mid-1540s until 1565, his galleys attacked the Portuguese in shallow water and near coasts, then escaped by rowing against the wind. He lurked off the coast between Diu and Goa, where Portuguese merchant ships were especially numerous. The Portuguese sent out fleet after fleet in a costly but fruitless attempt to catch this guerrilla of the sea.

In 1551 a Portuguese squadron of four sailing ships and one fusta entered the Red Sea to capture him, but Sefer Reis lured the fusta into shallow waters where the sailing ships could not follow, killed the captain, Luis Figueira, and captured his crew. In 1554, with two galleys and two fustas, he captured several Portuguese merchant ships and returned to the Red Sea port of Mocha laden with treasure and prisoners. In 1558 the Portuguese sent twenty galleys into the Red Sea, but failed to catch him. Two years later, another squadron of three sailing ships and one fusta entered the Red Sea but was intercepted by four galleys under Sefer Reis, who captured two of the Portuguese ships. After that, the Portuguese did not dare enter that sea in pursuit of Muslim merchant vessels carrying spices, nor could they set up an effective blockade. Their hold on the spice trade was badly eroded by Muslim captains who had learned to avoid their thinly spread patrols. In 1564, Sefer Reis set out from Suez with a new fleet, but he died in Aden before he could attack the Portuguese.[60]

After a lapse of almost twenty years, another corsair named Mir Ali Beg took up where Sefer Reis left off. In 1581, with three armed galliots, he sacked the town of Muscat, captured three galleys, and returned to Mocha with his loot. In 1585, Hasan Pasha, the Otto-man governor of Yemen, sent him with two galliots to raid the Swa-hili coast where the Portuguese had few warships. After one of his ships had to turn back, Mir Ali Beg reached the coast with only one ship and eighty men. He received a hero's welcome at the Somali port of Mogadishu, where he was joined by twenty light coastal craft. Thus armed, he took three Portuguese ships by surprise and returned to Mocha with sixty Portuguese captives and considerable booty.

Mir Ali Beg returned in 1588–89. Once again he was warmly received by all the coastal towns he visited except Malindi, a Portu-guese ally. Upon receiving the news, the Portuguese viceroy sent a fleet out from Goa with six galleons, eleven oared warships, and nine hundred fighting men under the command of Tomé de Sousa Coutinho. When they arrived off Mombasa, the Portuguese discov-ered that the Turks and the inhabitants of the town were under attack by cannibals from the mainland called Mazimbas. Rather than be captured by the Mazimbas, Mir Ali Beg and his followers surrendered to the Portuguese. Though the threat to their position on the African coast was over, the Portuguese nonetheless moved their allies the ruling families of Malindi to Mombasa and erected a powerful fortress, Fort Jesus, that dominated the coast into the next century.[61]

Despite the predatory tactics of the Portuguese, their trade repre-sented but a fraction of the much larger trade of Asia. Throughout the sixteenth century, most Moluccan spices went to Asian, not European, customers. The same was true of all the other goods shipped between Asian countries: silver and copper from Japan, cot-ton and pepper from India, silk and porcelain from China, gold, ivory, and slaves from East Africa. Most of the commercial activities and profits of the Portuguese came from their participation in this intra-Asian trade, not from voyages to and from Europe. After mid-century, Muslim traders learned to avoid the Portuguese harbors and fleets. Thereafter, the Portuguese share of the spice trade declined while the trade to Europe via the Middle East revived.[62] By the 1560s, the old trade pattern had recovered, and more spices and other eastern products reached Europe via the Red Sea and the Persian Gulf than around Africa. Despite prohibitions from Lisbon and Goa, even high-ranking Portuguese officials colluded in the trade with the Muslims.[63]

By the mid-seventeenth century, Portugal was so weakened that it fell prey to a new Asian power, the Omani Arabs. In 1650 Sultan Ibn Saif of Oman seized the Portuguese fort at Muscat and captured some Portuguese galleons in the harbor. He ordered more warships built in Bombay and Surat. With them the Omani attacked Portuguese ships off India in the 1660s and 1670s, then turned toward the Swahili coast. By 1698, the Omani had twenty-four large warships, including a frigate with seventy-four guns and another with sixty. With them they captured Fort Jesus in Mombasa, the main Portuguese base in East Africa. Though the Portuguese recaptured it briefly in 1727–28, they were forced to withdraw to Mozambique.[64] After that, the Portuguese, at one time the most powerful maritime presence in the Indian Ocean, were reduced to Goa, Macao, Timor, and Mozambique. Their ships were even forced to buy *cartazes* from Indian corsairs.[65]

The Dutch and the English in the Indian Ocean

European naval history of the late sixteenth and seventeenth centuries is filled with descriptions of the decline of Spain and Portugal, the rise of the Netherlands and England, and the conflicts among them. From an Asian point of view, this story is not one of decline and rise but of the replacement of a small and warlike predator kingdom by two equally warlike but richer and more powerful states.

During their age of glory, the Portuguese could not have accomplished what they did without the financing of their fleets and the marketing of their spices throughout Europe by the merchants and bankers of Antwerp in the Low Countries. The Flemish and Dutch traders who came to Lisbon to purchase spices paid for them with lumber, grain, naval stores, instruments, cannon, and a plethora of other items the Portuguese lacked. Portugal, like Spain, had an insatiable demand for the cannon manufactured in the Low Countries and Germany. Some Flemish and German gun founders and gunners even found employment in the Iberian kingdoms.[66]

This symbiotic relationship worked well until Philip II, a devout Catholic, ascended the throne of the Habsburg Empire, including the Low Countries, in 1556. By that time, Calvinist and Lutheran Protestantism had made inroads into the Habsburg domains and Philip was determined to stamp them out. In 1566–67 the Netherlands—the predominantly Protestant northern section of the Low Countries—revolted. This region had prospered by encouraging all traders, whether Protestant or Catholic, and even Jewish refugees who had escaped persecution in Iberia. Unlike other European kingdoms where landowning aristocrats dominated society, in the Netherlands merchants were the ruling class.

In the course of the war that ensued, the Spanish treasury went bankrupt. When its soldiers, unpaid for months, sacked Antwerp, the merchants of that city fled to Amsterdam in the Protestant Netherlands, out of reach of the ravenous Spanish troops. Portugal became involved in this conflict when Philip inherited the throne of that kingdom in 1580. From that moment on, the Dutch became the sworn enemies of Portugal as well as Spain.

The Dutch had several advantages that propelled them into the lucrative oceanic trade of the seventeenth century. The location of the Netherlands made it the gateway between the German states and England, Scandinavia, and Iberia. Through its harbors came not only precious goods like spices and guns but also basic commodities like herring, salt, timber, and grain. Unlike the Iberian kingdoms that viewed trade as a means of enriching the government, the Dutch political climate encouraged private enterprise and social mobility.

One of the major industries of the Netherlands was shipbuilding. Its shipyards were the most mechanized and efficient in Europe. They specialized in building *fluyts*, or flyboats, cargo ships with a large hold and a small crew. Dutch ships were cheaper to build and man, and their freight costs were lower than those of any other country. By 1600, the Netherlands had the largest merchant marine in Europe.

Meanwhile, the Iberian powers that had once proudly divided up between them the world outside Europe were shown to be much weaker at sea than other Europeans had come to believe. Pirates and privateers began to attack the Spanish ships that brought back treasure from the New World. In 1577–80, Sir Francis Drake discovered just how weak the hold of Spain and Portugal on the oceans was when he circumnavigated the world, attacking Spanish towns and capturing Spanish ships along the way. In 1588 King Philip, hoping to deal a decisive blow to the newly Protestant kingdom of England, sent the Armada, the largest fleet the world had seen since the days of Zheng He, to its doom in the waters off the British Isles. Spain, suddenly weakened, took Portugal down with it.

In 1594 Philip tried to harm the Protestant powers by closing the Lisbon spice market to their traders. That was all the incentive the Dutch needed to cut out the middleman and go directly to the source. A year after the embargo, Dutch merchants sent four ships with 289 men out to the East Indies. Though only one ship and eighty-nine men returned, the venture still turned a profit. In 1598 five fleets totaling twenty-two ships sailed for the Indies, making their investors rich. Four years after that, in 1602, the merchants of Amsterdam founded the United East India Company, or VOC, one of the first of the great multinational corporations that have spread capitalism around the world.[67]

In the late sixteenth century, when the Dutch entered into competition for the lucrative Indian Ocean trade and the even richer trade with the Caribbean and Mexico, the competing naval powers built galleons, which were faster and sturdier than the carracks of the early sixteenth century. Those designed specifically for warfare became sailing fortresses, with rounded sides called "tumble-home" to hold a larger number of guns. This design also made the ships more stable by placing the weight of their cannon closer to the center of gravity and made it much more difficult for enemy ships to grapple and board them.[68]

The Dutch also made good use of an English invention: cast-iron cannon. Bronze cannon were very costly. Though tin was available in Cornwall, copper came from central Europe. Meanwhile England had abundant iron ore and the forests needed to make charcoal. The first cast-iron cannon tended to crack or explode, making them more dangerous to the gunners than to their targets. By the mid-sixteenth century, however, English iron founders had learned to make cannon that were reasonably reliable and cost one-third to one-quarter as much as bronze. In 1573 their furnaces were producing eight hundred to a thousand tons of iron a year. In the seventeenth century, the technique spread to Sweden and Russia; together they produced five thousand tons of iron a year by mid-century.[69] Liège was also a center of gun manufacturing. With their extensive trade networks, the Dutch had access to cannon from a variety of sources. In the 1620s, shipbuilders introduced wheeled gun carriages that allowed the recoil of a cannon to propel it back into the hull where it could easily be reloaded through the muzzle. A block-and-tackle system let the gunners haul it back into firing position with the muzzle sticking out of the hull.[70]

Thus, when the Dutch arrived in the Indian Ocean, it was with better ships and more cannon than the Portuguese. Their goal was to control the sources of the spices and their distribution throughout the world. To trade with the East Indies, they sailed directly from the Cape of Good Hope with the westerlies and returned with the easterly trade winds south of the equator, thereby avoiding the monsoons. This made their schedules much more flexible than those of the Portuguese. The departures of the Portuguese fleets had always been timed to take advantage of the winds in both the Atlantic and the Indian oceans: they left Lisbon in late March or early April and arrived in India toward the end of September, and left India in late December or early January in order to arrive in Lisbon in June.[71] Not beholden, like the Portuguese, to the rhythm of the monsoons, the Dutch began sending out three fleets a year, in September, late December or early January, and late April or early May. To return, their fleets left Java in late December or early January or in late February or early March. In either direction, the voyage took five to seven months, with a stop at the Cape for water.[72]

Soon after reaching the East Indies, the Dutch went to war. In 1605 they expelled the Portuguese from Ambon and the Moluccas, the source of most spices, and established a warehouse in Bantam in west Java. In constructing a maritime empire, they needed fortified bases; in this endeavor, Jan Pieterszoon Coen, governor-general from 1618 to 1629, played the same role that Albuquerque had played for Portugal a century earlier. In 1619 he obtained land near the village of Jakarta in Java and built a city he named Batavia. In 1623 the Dutch captured and massacred some English traders in Ambon, forc- ing the English to abandon the East Indies and settle in India in- stead. Under Coen the Dutch also built a fortress called Castle Zee- landia in Taiwan in order to participate in the trade between China and Japan. They blockaded Melaka in 1606, 1608, and 1615 and attacked the Portuguese ships that plied between Melaka and Goa. Despite the Portuguese efforts to strengthen the defenses of A Fa- mosa, the Dutch finally took that fortress in 1641; it remained in Dutch hands until 1795, when the British captured it.[73] In the 1650s they expelled the Portuguese from Ceylon, the source of cinnamon, and Cochin on the Malabar coast of India, a source of pepper. They also imported some farmers (*boers* in Dutch) to establish a settle- ment at the Cape of Good Hope that would provide passing ships with fresh vegetables and water.[74] They monopolized the spice trade far more effectively than the Portuguese had and profited from commodities they introduced or commercialized in the Indies, in particular sugar and coffee. They did so by maintaining a fleet of over ninety ships in eastern waters, more than the Portuguese had ever had.[75]

By the end of the seventeenth century, the Dutch controlled the seas (but very little land) in the East Indies, as well as settlements in Ceylon, South Africa, the Caribbean, and Manhattan Island. Not only did they bring pepper and spices to Europe as the Portuguese had done, they also developed the production of sugar and coffee and profited from participating in the intra-Asian trade in textiles, metals, spices, and other goods between India, Ceylon, the East In- dies, China, and Japan.

The English were slower than the Dutch at venturing into the Indian Ocean. Like the Dutch, their strength lay in the cooperation between their government and their wealthiest merchants. Neither group alone could have afforded the armed merchantmen or the warships to protect them. By the late seventeenth century, the Netherlands and England had found a way to outspend, outbuild, and outfight the Iberians.[76] London merchants formed the East India Company and received a charter from the crown in 1601, a year before the Dutch company was founded. The first English ship went out to Sumatra in 1601, returning with a cargo of pepper.

At first the English, like the Dutch, tried to trade directly with the Spice Islands. To pay for the spices, they purchased textiles in Surat, on the coast of Gujarat. In eastern waters, however, they were outnumbered and outclassed by the Dutch, who forced them out of the Indonesian archipelago. Hence they turned to the Indian Ocean, where they competed with the Portuguese Estado da India. They defeated a Portuguese squadron off Surat in 1612, and several times captured much larger Indian ships.[77] In 1618 they obtained trading privileges from the Mughals in exchange for protecting Indian pilgrim and merchant ships from the Portuguese. In 1622 they helped the Persian shah Abbas to seize Hormuz from Portugal. They acquired trading posts in Madras (now Chennai) in 1641, Bombay (now Mumbai) in 1665, and Calcutta (now Kolkata) in 1690. From these footholds on the subcontinent they would later build the world's largest and most prosperous colonial empire.[78]

China, Japan, and the Europeans

The Portuguese had imposed themselves by force in the Indian Ocean, where ports were poorly defended independent city-states. But in East Asian waters they came up against the Chinese Empire. When the first Portuguese ship to reach China arrived at the Pearl River estuary below Canton (Guangzhou) in 1514, it encountered a hostile reception, for news of the Portuguese capture of Melaka, a tributary of China, had preceded it. Five years later, a Portuguese fleet commanded by Simão d'Andrade sailed up the Pearl River to Canton and fired a salvo, an act the Chinese interpreted as an insulting defiance to their authority. Portuguese behavior got worse.

They refused to pay import duties and mistreated the officials sent onboard to collect the duty, refused to allow other ships to unload their merchandise, and kidnaped some children. The policy of terror that had served the Portuguese well in the Indian Ocean backfired in China. In 1521–22 another Portuguese fleet arrived at Tunmen, off the Pearl River estuary, but encountered a larger Chinese fleet and was chased off. After that, the Chinese government ordered local officials to discontinue trade with the Portuguese and prepared fleets of junks in case the Portuguese returned. For several decades, Portuguese smugglers traded secretly, bribing customs agents. Not until 1557 did the hostilities cool down enough for the local Chinese officials to allow the Portuguese to set up a warehouse on the small island of Macao at the mouth of the Pearl River; the imperial government in Beijing did not learn about this arrangement for another half-century.[79]

The Chinese had once been at the forefront of technological progress—witness the invention of gunpowder, paper, printing, and the compass. But, as historian Kenneth Chase has pointed out, their inventiveness coincided with periods of fragmentation and warfare, such as the Song dynasty (960–1279) and the Yuan (1271–1368).[80] In more peaceful times, however, as during the Ming dynasty (1368–1644), the Chinese government maintained an ambivalent attitude toward firearms. Chinese officials, never fond of soldiers and warfare, feared the potential of guns to cause trouble should they fall into the hands of bandits, rebellious peasants, or disloyal troops. Yet they also understood the military value of firearms and recognized the superiority of the Portuguese cannon. A high official wrote: "The Feringis [i.e., "Franks," or Europeans] are most cruel and crafty. Their arms are superior to those of other foreigners. Some years ago they came suddenly to the city of Canton and the noise of their cannon shook the earth."[81] Once they had admitted the Portuguese into Macao, they tolerated the attempts by Jesuit priests to enter China and gain converts because the Jesuits also brought with them the secrets of gun making. Hence the long series of Jesuit gun founders who proselytized Christianity and Western technology at the same time.[82]

The Dutch, eager to weaken the Portuguese wherever possible, attacked Macao in 1622 but were driven back. Like the Portuguese a century earlier, they then tried to coerce the Chinese to trade on their terms, but a series of skirmishes with Chinese war junks forced them to back off. The Dutch ships were superior to the Chinese on the high seas, but the junks could outrun the Dutch before the wind and could sail in shallow coastal waters where the Dutch could not follow. War junks were essentially merchant ships armed with pow-dered lime, lances, darts, arrows, and fire rockets. Their preferred tactic was to ram and board enemy vessels. They carried a few small cannon of poor quality and light anti-personnel pieces. A military treatise written in 1624 explained: "Fire arms can be used on large boats but the waves make aiming very hard. Chances of hitting the enemy are very slender. Even if one enemy boat should be hit, the enemy would thus not incur severe losses. The purpose of hav-ing fire arms is purely psychological, namely that of disheartening the enemy."[83]

The confrontation between the Dutch and the Chinese therefore resembled that between the Portuguese and the Turks a century ear-lier; one was dominant on the high seas, the other in coastal waters. As a Ming official explained in 1623: "The great ships and big guns upon which these foreigners [the Dutch] depend are effective at sea but not on land; moreover, their aims do not go beyond their greed for Chinese goods."[84] Yet, since the goal of the Dutch was trade, the fact that the Chinese could control their access to the ports and could cut off trade with foreigners when necessary essentially gave China the upper hand in their confrontation.[85]

After the Ming dynasty was overthrown by the Manchu in 1644, a Ming loyalist and pirate leader named Zheng Chenggong—known to Westerners as Coxinga—fled to Taiwan. In 1661 he appeared off Castle Zeelandia with hundreds of junks and an army of twenty-five thousand men. After a siege of several months, Zeelandia fell and the Dutch were expelled from Taiwan. Twenty years later, the Man-chu incorporated Taiwan into their empire.[86]

Of all the states in the world during the early modern age, Japan was the furthest from Europe by sea. Hence it was the last Asian state to be reached by the Portuguese and the one that offered them the least trade. Yet their presence in that country had a profound and lasting influence. In 1543 a storm swept a Chinese junk to the coast of Japan. Onboard were three Portuguese carrying matchlock muskets. Soon thereafter, more Portuguese began arriving, some on Chinese junks, others on Portuguese ships. Unlike the Chinese, who had known Europeans since Marco Polo, if not earlier, and who viewed them as barbarians whose technologies were useful curiosities at best, the Japanese eagerly embraced the technologies and religious ideas of the newcomers. According to the Dutch traveler J. H. van Linschoten, "The Japanese are sharpe witted and quickly learne anything they see," and the Portuguese Fernão Mendes Pinto wrote: "They are naturally addicted to war and in war take they more delight than any nation that we know."[87] The Japanese quickly learned how to manufacture infantry weapons based on the Portuguese model, especially muskets and cannon.

In 1549 the Jesuit Francis Xavier arrived and began to convert Japanese to Christianity. The tensions inherent in Japanese society at the time and the introduction of foreign ideas and techniques produced a dramatic increase in both conversions to Christianity and internecine warfare. By the early seventeenth century, the Tokugawa shoguns who ruled Japan began persecuting Japanese Christians, who numbered over a hundred thousand. During the eighty-odd years that they were tolerated in Japan, however, private Portuguese merchants often visited Japan on their own ships or on Chinese junks manned by international crews. They conducted a brisk trade, not with Europe but between Japan and China, carrying Chinese silk and porcelain and Indonesian spices to Japan and returning with silver, copper, and other products.[88]

In 1636 the Japanese *bakufu*, or military government, forbade its citizens to travel abroad or build ships. Two years later they expelled the Portuguese. Thereafter, the Japanese authorities allowed only one Dutch ship a year to land at Deshima, an island in the harbor of Nagasaki.

Conclusion

In most books on naval history, the age of sailing ships, from the sixteenth through the early nineteenth centuries, is a heroic age of battles between the Portuguese, Spanish, Dutch, English, and French fleets in the Atlantic, the Mediterranean Sea, and the Indian Ocean. The only non-Western power to appear is the Ottoman Empire, and only until its defeat at the battle of Lepanto in the Mediterranean in 1571.[89]

Seen from an Asian perspective, however, the story is quite different. Though the Portuguese dominated the Indian Ocean for a time, they never controlled it completely, and their stranglehold on the spice trade lasted barely half a century. By the seventeenth century their position was undermined by the Omani Arabs in East Africa and by the Dutch in the East Indies. After that, the Indian Ocean was contested by the Dutch, the English, and later the French. These sea empires rested on naval bases and entrepôts along the coasts of that ocean, a consequence of the fact that the port cities were either independent city-states or only loosely tied to land empires. The European hold on coastal enclaves was strong only where the local states were weak.

Where a land empire controlled the ports the situation was reversed. The Ottoman Turks were able to keep the Europeans out of the Red Sea, the Persian Gulf, and the Arabian Peninsula. In East Asia, the Chinese controlled access to the ports and kept the Portuguese and the Dutch at arm's length. In Japan, the Portuguese ran afoul of the Tokugawa *bakufu* and were summarily expelled, after which the Japanese government kept a tight rein on contacts with the West. The Europeans participated in the East Asian trading network, but never dominated or controlled it.

The balance of power between Europeans and Asians was a consequence of not only the size of the states that controlled the harbors but also the nautical environment. The Europeans had, and retained throughout this period, the mastery of the high seas, thanks to their stout and heavily armed ships. However, that superiority did not extend to coastal waters and narrow shallow seas, where oar-driven galleys and flat-bottomed junks held their own and on occasion defeated even the best-armed carrack or galleon. In short, politics, technology, and geography conspired to create a stalemate in Asian waters that lasted for three centuries. No one in those centuries could have predicted what would happen next.

Notes

1. Sanjay Subrahmanyam, *The Career and Legend of Vasco da Gama* (Cambridge: Cambridge University Press, 1997), pp. 93–121; Michael Pearson, *The Indian Ocean* (New York: Routledge, 2003), chapter 5.

2. Ronald Findlay and Kevin O'Rourke, *Power and Plenty: Trade, War, and the World Economy in the Second Millennium* (Princeton: Princeton University Press, 2007), pp. 140–51.

3. Ibid., p. 129.

4. J. H. Parry, *The Discovery of the Sea* (New York: Dial Press, 1974), pp. 166–78; Subrahmanyam, *Career*, pp. 129–37.

5. C. R. Boxer, *The Portuguese Seaborne Empire, 1415–1825* (New York: Knopf, 1969), p. 37; Subrahmanyam, *Career*, p. 160.

6. Parry, *Discovery*, p. 183.

7. Subrahmanyam, *Career*, pp. 151–82; Boies Penrose, *Travel and Discovery in the Renaissance, 1420–1620* (Cambridge, Mass.: Harvard University Press, 1960), pp. 55–58.

8. J. H. Parry, *The Age of Reconnaissance: Discovery, Exploration and Settlement, 1450–1650* (Cleveland: World Publishing, 1963), pp. 142–43; Parry, *Discovery*, p. 253; Henrique Lopes de Mendonça, *Estudios sobre navios portuguezes nos secolos XV e XVI* (Lisbon: Academia Real das Sciencias, 1892), p. 53; Subrahmanyam, *Career*, pp. 195–226.

9. Sanjay Subrahmanyam, *The Portuguese Empire in Asia, 1500–1700: A Political and Economic History* (New York: Longman, 1993), p. 66.

10. Palmira Brummett, *Ottoman Seapower and Levantine Diplomacy in the Age of Discovery* (Albany: SUNY Press, 1994), pp. 111–15; Jean Louis Baqué-Grammont and Anne Kriegel, *Mamlouks, Ottomans et Portugais en Mer Rouge: L'Affaire de Djedda en 1517, Supplément aux Annales islamologiques*, no. 12 (Cairo: Institut Français, 1988), pp. 1–2; Subrahmanyam, *Career*, pp. 255–56.

11. Saturnino Monteiro, *Batalhas e combates da Marinha Portuguesa*, vol. 1: *1139–1521* (Lisbon: Livraria Sá da Costa Editora, 1989), pp. 177–92; P. J. Marshall, "Western Arms in Maritime Asia in the Early Phases of Expansion," *Modern Asian Studies* 14, no. 1 (1980), p. 18; Peter Padfield, *Guns at Sea* (New York: St. Martin's, 1974), pp. 25–28; K. M. Panikkar, *Asia and Western Dominance* (New York: Macmillan, 1969), p. 37; Brummett, *Ottoman Seapower*, pp. 112–14.

12. Edgar Prestage, *Afonso de Albuquerque, Governor of India: His Life, Conquests, and Administration* (Watford, England: E. Prestage, 1929), pp. 27–31; Parry, *Discovery*, p. 254.

13. Parry, *Age of Reconnaissance*, pp. 143–45; Panikkar, *Asia*, pp. 39–41; Prestage, *Afonso de Albuquerque*, pp. 37–44.

14. G. R. Crone, *The Discovery of the East* (London: Hamish Hamilton, 1972), p. 54; Eila M. J. Campbell, "Discovery and the Technical Setting, 1420–1520," *Terrae Incognitae* 8 (1976), p. 14.

15. Quoted in Carlo Cipolla, *Guns, Sails and Empires: Technological Innovation and the Early Phases of European Expansion, 1400–1700* (New York: Random House, 1965), p. 137.

16. Graham Irwin, "Malacca Fort," *Journal of South-East Asian History* 3, no. 2 (Singapore, 1962), pp. 19–24; Parry, *Discovery*, pp. 254–56.

17. *The Suma Oriental of Tomé Pires: An Account of the East, from the Red Sea to Japan, Written in Malacca and India in 1512–1515*, quoted in Parry, *Discovery*, p. 256.

18. Parry, *Discovery*, pp. 256–57; Penrose, *Travel and Discovery*, pp. 62–64.

19. Quoted in Subrahmanyam, *Portuguese Empire*, p. 65.

20. Salih Özbaran, "The Ottoman Turks and the Portuguese in the Persian Gulf, 1534–1581," *Journal of Asian History* 6, no. 1 (1972), pp. 46–47; John F. Guilmartin, Jr., *Gunpowder and Galleys: Changing Technology and Mediterranean Warfare at Sea in the 16th Century*, 2nd ed. (London: Conway Maritime Press, 2003), pp. 8–9; Prestage, *Afonso de Albuquerque*, pp. 53–61.

21. Ian V. Hogg, *A History of Artillery* (Feltham, England: Hamlyn, 1974), chapter 1; Cipolla, *Guns, Sails and Empires*, pp. 21–22, 75, 104.

22. Roger C. Smith, *Vanguard of Empire: Ships of Exploration in the Age of Columbus* (New York: Oxford University Press, 1993), pp. 153–54; Geoffrey Parker, *The Military Revolution: Military Innovation and the Rise of the West, 1500–1800* (Cambridge: Cambridge University Press, 1996), pp. 84–90; Parry, *Age of Reconnaissance*, pp. 117–18.

23. Kenneth W. Chase, *Firearms: A Global History to 1700* (Cambridge: Cambridge University Press, 2003), p. 134; Smith, *Vanguard*, p. 157; Padfield, *Guns at Sea*, pp. 25–29; Parry, *Age of Reconnaissance*, pp. 115, 122, 140. The nomenclature of guns is complex and confusing, especially in translation; Hogg, *History of Artillery*, p. 28, gives two lists of names of guns, with their characteristics, one from 1574, the other from 1628.

24. Subrahmanyam, *Portuguese Empire*, pp. 109–12; Parker, *Military Revolution*, p. 105. See also Simon Digby, "The Maritime Trade of India," in Tapan Raychaudhuri and Irfan Habib, eds., *The Cambridge Economic History of India* (Cambridge: Cambridge University Press, 1981), vol. 1, p. 152; and Boxer, *Portuguese Seaborne Empire*, p. 44.

25. Philip T. Hoffman, "Why Is It That Europeans Ended Up Conquering the Rest of the Globe? Prices, the Military Revolution, and Western Europe's Comparative Advantage in Violence," http://gpih.ucdavis.edu/files/Hoffman.pdf (accessed March 9, 2008).

26. William H. McNeill, *The Pursuit of Power: Technology, Armed Force, and Society since A.D. 1000* (Chicago: University of Chicago Press, 1982), p. 100; Parry, *Age of Reconnaissance*, pp. 118–20; Padfield, *Guns at Sea*, p. 29; Smith, *Vanguard*, pp. 154–55.

27. Quoted in Chase, *Firearms*, p. 134.

28. Padfield, *Guns at Sea*, pp. 26–27; Parker, *Military Revolution*, p. 94.

29. Afonso d'Alboquerque, *Commentaries of the Great Afonso Dalboquerque*, 4 vols., trans. Walter de Gray Birch (London: Hakluyt Society, 1875–84), vol. 1, pp. 112–13.

30. Malyn Newitt, "Portuguese Amphibious Warfare in the East in the Sixteenth Century (1500–1520)," in D.J.B. Trim and Mark Charles Fissel, eds., *Amphibious Warfare, 1000–1700: Commerce, State Formation and European Expansion* (Leiden: Brill, 2006), chapter 4.

31. Baqué-Grammont and Kriegel, *Mamlouks*, pp. 2–7; Brummett, *Ottoman Seapower*, 115–18. See also Dejanirah Couto, "Les Ottomans et l'Inde Portugaise," in *Vasco da Gama e a India: Conferência Internacional, Paris, 11–13 Maio, 1998*, 3 vols. (Lisbon: Fundacão Calouste Goulbenkian, 1999), vol. 1, pp. 181–200.

32. Giancarlo Casale, "The Ottoman Age of Exploration: Spices, Maps and Conquest in the Sixteenth-Century Indian Ocean" (Ph.D. diss., Harvard University, 2004), pp. 8–9. I am indebted to Giancarlo Casale for allowing me to read and cite his dissertation.

33. Gabor Agoston, *Guns for the Sultan: Military Power and the Weapons Industry in the Ottoman Empire* (Cambridge: Cambridge University Press, 2005).

34. Andrew C. Hess, "The Evolution of the Ottoman Seaborne Empire in the Age of Oceanic Discoveries, 1453–1525," *American Historical Review* 75 (1970), p. 1910; Baqué-Grammont and Kriegel, *Mamlouks*, pp. 21–46; Guilmartin, *Gunpowder and Galleys*, pp. 9–13.

35. Casale, "Ottoman Age," pp. 85–86.

36. Ibid., p. 126.

37. Ibid., pp. 91–101; Monteiro, *Batalhas e combates da Marinha Portuguesa*, vol. 2: *1522–1538*, pp. 320–32.

38. Casale, "Ottoman Age," pp. 94–104, 334; see also Salih Özbaran, "Ottoman Naval Policy in the South," in Metin Kunt and Christine Woodhead, eds., *Suleyman the Magnificent and His Age: The Ottoman Empire in the Early Modern World* (London: Longman, 1995), pp. 59–60.

39. Casale, "Ottoman Age," pp. 110–18.

40. Özbaran, "Ottoman Naval Policy," p. 64.

41. Casale, "Ottoman Age," pp. 148–54.

42. Ibid., p. 326.

43. Ibid., pp. 53–54, 80–83, 139–40.

44. Ibid., pp. 199–200.

45. John H. Pryor, *Geography, Technology, and War: Studies in the Maritime History of the Mediterranean, 649–1571* (Cambridge: Cambridge University Press, 1988), pp. 71–77; Martin van Creveld, *Technology and War: From 2000 b.c. to the Present* (New York: Free Press, 1989), pp. 63, 127–33; Guilmartin, *Gunpowder and Galleys*, pp. 23n2, 206, 226–27; McNeill, *Pursuit of Power*, pp. 99–100; Parker, *Military Revolution*, pp. 84–85.

46. Guilmartin, *Gunpowder and Galleys*, pp. 199–207.

47. Michael E. Mallett and John R. Hale, *The Military Organization of a Renaissance State: Venice, c. 1400 to 1617* (Cambridge: Cambridge University Press, 1984), p. 400.

48. Casale, "Ottoman Age," p. 10.

49. Quoted in Ahsan Jan Qaisar, *The Indian Response to European Technology and Culture*, A.D. 1498–1707 (New York: Oxford University Press, 1982), p. 25.

50. Panikkar, *Asia*, p. 43.

51. Qaisar, *Indian Response*, pp. 44–45.

52. Parry, *Age of Reconnaissance*, p. 143.

53. Percival Spear, *A History of India*, vol. 2: *From the Sixteenth to the Twentieth Century* (London: Penguin, 1978), pp. 21–23; Qaisar, *Indian Response*, pp. 46–48; Cipolla, *Guns, Sails and Empires*, pp. 127–28.

54. Quoted in M. N. Pearson, "Portuguese India and the Mughals," in *Vasco da Gama e a India*, vol. 1, p. 233.

55. K. S. Mathew, "Akbar and the Portuguese Maritime Dominance," in Irfan Habib, ed., *Akbar and His India* (Delhi: Oxford University Press, 1997), pp. 256–65; Pearson, "Portuguese India," pp. 226–33.

56. Atul Chandra Roy, *A History of the Mughal Navy and Naval Warfare* (Calcutta: World Press, 1972), chapter 7; Qaisar, *Indian Response*, p. 45.

57. John Fryar, *A New Account of East India and Persia, Being Nine Years' Travel, 1672–81*, ed. W. Crooke, 3 vols. (London, 1909–1915), vol. 1, p. 302, quoted in Qaisar, *Indian Response*, p. 46.

58. Findlay and O'Rourke, *Power and Plenty*, pp. 152–53, 201; Subrahmanyam, *Portuguese Empire*, 133–37; Parker, *Military Revolution*, pp. 105–12; Casale, "Ottoman Age," pp. 177–78, 188, 199–204, 223, 248.

59. According to Parker (*Military Revolution*, p. 104), in 1522 the Portuguese had sixty ships and 1,073 artillery pieces.

60. Casale, "Ottoman Age," pp. 144–90.

61. Ibid., pp. 245–76.

62. A.J.R. Russell-Wood, *The Portuguese Empire, 1415–1808: A World on the Move* (Baltimore: Johns Hopkins University Press, 1998), pp. 23–32; Findlay and O'Rourke, *Power and Plenty*, p. 157.

63. Casale, "Ottoman Age," pp. 124–25, 164–65, 280.

64. Patricia Risso, *Oman and Muscat: An Early Modern History* (New York: St. Martin's, 1986), pp. 11–13, 120; Boxer, *Portuguese Seaborne Empire*, pp. 133–34.

65. Boxer, *Portuguese Seaborne Empire*, pp. 136–37.

66. Cipolla, *Guns, Sails and Empires*, p. 31.

67. Charles Tilly distinguishes between coercion-intensive empires, like the Spanish, dedicated to territorial acquisition, settlement, forced labor, and the exaction of tribute, and capital-intensive empires, like the Dutch and English, based on trading monopolies. The Portuguese evidently occupied an intermediate position. See *Coercion, Capital, and European States*, AD 990–1992 (Oxford: Blackwell, 1990), pp. 91–95.

68. Clinton R. Edwards, "The Impact of European Oceanic Discoveries on Ship Design and Construction during the Sixteenth Century," GeoJournal 26, no. 4 (1992), pp. 443–57; Van Creveld, Technology and War, p. 134.

69. Cipolla, Guns, Sails and Empires, pp. 37–43, 71–73; Van Creveld, Technology and War, p. 133.

70. Stephen Morillo, Michael Pavkovic, Paul Lococo, and Michael Palmer, War in History: Society, Technology and War from Ancient Times to the Present (New York: McGraw-Hill, 2004), chapter 20.

71. Smith, Vanguard, p. 12.

72. C. R. Boxer, The Dutch Seaborne Empire, 1600–1800 (New York: Knopf, 1965), p. 197; Parry, Age of Reconnaissance, p. 200.

73. Irwin, "Malacca Fort," pp. 27–41.

74. Boxer, Dutch Seaborne Empire, pp. 295–300.

75. On the Dutch maritime empire, see Pearson, The Indian Ocean, pp. 145–51, and Morillo et al., War in History, chapter 18.

76. Morillo et al., War in History, chapter 20.

77. Qaisar, Indian Response, p. 44.

78. Spear, History of India, vol. 2, pp. 65–68.

79. Tien-Tse Chang [Tianze Zhang], Sino-Portuguese Trade from 1512–1644 (Leiden: Brill, 1934), pp. 35–89; Derek Massarella, A World Elsewhere: Europe's Encounter with Japan in the Sixteenth and Seventeenth Centuries (New Haven: Yale University Press, 1990), pp. 22–23.

80. Chase, Firearms, pp. 32–33.

81. Quoted in Chang, Sino-Portuguese Trade, p. 51.

82. Cipolla, Guns, Sails and Empires, pp. 114–18.

83. Ibid., pp. 125–26; see also Parker, Military Revolution, p. 84.

84. John E. Wills, Jr., Pepper, Guns and Parleys: The Dutch East India Company and China, 1622–1681 (Cambridge, Mass.: Harvard University Press, 1974), p. 1.

85. Ibid., pp. 22–23. Besides the Dutch and the Chinese, a third force operated in the waters off China: pirates who sometimes sided with the Dutch and sometimes fought them; see Tonio Andrade, "The Company's Chinese Pirates: How the Dutch East India Company Tried to Lead a Coalition of Pirates to War against China, 1621–1662," Journal of World History 15, no. 4 (December 2004), pp. 415–44.

86. William Campbell, Formosa under the Dutch, Described from Contemporary Records, with Explanatory Notes and a Bibliography of the Island (1903; reprint, Taipei: Ch'eng-wen Publishing, 1967), pp. 426–55; Chase, Firearms, pp. 157–58; Boxer, Dutch Seaborne Empire, pp. 144–46; Parker, Military Revolution, pp. 112–14.

87. Cipolla, Guns, Sails and Empires, p. 127.

88. Findlay and O'Rourke, Power and Plenty, pp. 170–71; Parry, Age of Discovery, p. 191; Cipolla, Guns, Sails and Empires, pp. 112–27; Massarella, World Elsewhere, pp. 39–40.

89. See, for example, Richard Harding, *The Evolution of the Sailing Navy, 1509–1815* (New York: St. Martin's, 1995); Harding, *Seapower and Naval Warfare, 1650–1830* (Annapolis, Md.: Naval Institute Press, 1999); Robert Gardiner, ed., *The Line of Battle: The Sailing Warship, 1650–1840* (London: Conway Maritime, 1992); or Peter Padfield, *Maritime Supremacy and the Opening of the Western Mind: Naval Campaigns That Shaped the Modern World, 1588–1782* (London: J. Murray, 1999).

❀ Chapter 3 ❀

Horses, Diseases, and the Conquest of the Americas, 1492–1849

Few events in history have had such profound consequences as the opening of contacts between the Eastern and Western hemispheres and the displacement of the Native American peoples by peoples of European and African origin. To some, this is a tale of triumph; to others, it is a tale of disaster. Yet nothing in history is ever straightforward. Looked at more closely and respecting the chronology of events, the story of the encounter is one of victories and defeats on both sides.

The Spaniards who first arrived in the New World quickly conquered and occupied the larger Caribbean islands, much of Mexico and Central America, and Peru and northern Chile. Then their momentum failed them, and they were unsuccessful in what is now southern Chile, Argentina, northern Mexico, and the southwestern United States. Other Europeans were even less successful; by the eighteenth century, the Portuguese, British, Dutch, and French occupied some islands and lands near the Atlantic Ocean, but the interiors of the two American continents were still Indian territory and did not come under European control until the late nineteenth century, four centuries after Columbus. Our task is to understand not only the triumphs of the Europeans but also their failures.

Historians have traditionally dwelled on the ambitions and goals of the conquistadors, their craving for gold and silver, their desire to convert the Indians to Christianity, their eagerness to discover new lands, their search for land or for freedom from religious persecution. To be sure, such motives were very important, but they do not explain the outcome of their actions; those who failed to conquer new lands were no less motivated than those who succeeded.

Instead, to understand the consequences of their actions, we must turn to the circumstances in which they operated. These are of two sorts. One is the means at the disposal of the protagonists, in particular the animals and weapons available to both sides. The other is the environments in which their actions took place, not only land and climate but also disease ecologies and social environments, and the changes in these environments brought about by the encounter. The claim of this chapter is that animals, weapons, and diseases were far more important in determining the outcome of the encounter than the personalities, goals, and motives of the protagonists.

The First Encounter: The Caribbean

On his second voyage across the Atlantic in 1493, Christopher Columbus brought with him domesticated animals familiar to Europeans but not to Indians, in particular horses, cattle, pigs, goats, and dogs. Horses were the most valuable of the animals and the most difficult to transport. Of the twenty that embarked in Spain, only sixteen arrived safely. Every subsequent expedition brought horses. Many were lost at sea when ships were stranded for days or weeks in the equatorial doldrums and the crew had to sacrifice their horses in order to spare fresh water for humans; hence the expression "horse latitudes." By 1503, there were sixty to seventy horses on the island of Hispaniola. With these, the Spanish crown established breeding farms to supply later expeditions with mounts. From Hispaniola, horses were taken to Puerto Rico, Cuba, and Jamaica. Hernán Cortés settled in Cuba in 1513, and that is where he obtained the horses he took with him to Mexico in 1519.[1]

These horses were not the large and powerful warhorses of medieval Europe, carrying knights in full suits of armor. Instead, they were part-Arabian, obedient, quick, and maneuverable mounts bred for herding. Riders were lightly armored, with chain mail on their bodies and heavy plates only on their legs. They rode *a la jineta*, with their knees drawn up so that they could easily turn and slash with their swords. They also carried light lances up to fourteen feet long and shields made of wood and leather.

Though small by European standards, the Spanish horses were the largest and most frightening animals the Indians had ever seen. To make them even scarier, the Spaniards put bells on their necks and made them rear up and whinny. Such horses were much faster than any runner and could easily catch up with a warrior trying to escape and outrun any scout or messenger. Time and again, they allowed the Spaniards to take Indians by surprise.[2]

The Spaniards also had dogs the likes of which the New World had never seen. The Indians' dogs were small hairless creatures that did not bark and were fed corn and fattened up to eat. Spanish dogs, in contrast, were mastiffs, bulldogs, and greyhounds, ferocious animals that barked loudly. They were used as guard dogs and trained to attack Indians and capture fugitives. While the Spaniards professed horror at the cannibalism they attributed to the Indians, they did not hesitate to feed human flesh to their dogs. The Spaniards made much use of such dogs in the Antilles, especially in forests and broken terrain where horses could not go.[3]

The other advantage the Spaniards had was the quality of their weapons. Their swords, helmets, breastplates and other armor and parts of their lances, halberds, and shields were of steel, a metal unknown to the Indians. These weapons seldom broke and kept their edge well; those made in Toledo or Viscaya were considered the best in Europe. Spanish infantrymen in the Americas wore partial suits of armor covering their torsos or down to their knees. Against the weapons of the Indians—wooden swords and spears, bows and arrows, and little or no armor—the Spanish soldiers, especially their cavalrymen, had overwhelming power.

The Spaniards carried two kinds of projectile weapons: crossbows and firearms. Crossbows were awkward devices weighing up to sixteen pounds, with steel bows so strong they had to be bent with a lever or a crank and ratchet. They took a minute to load and fired a bolt weighing up to three ounces with such force that they could penetrate a suit of armor at a distance of 70 yards and kill at up to 350 yards.

Guns were less important than cold steel. In the early sixteenth century, Europeans used primitive muskets called harquebuses that weighed up to twenty pounds. Such guns fired when a match—a long cotton rope that smoldered—was touched to a small amount of gunpowder in the flash-pan on the outside of the barrel that then ignited the gunpowder in the barrel through a hole. Reloading took over a minute. The match had to be kept burning until the battle was over. Handling a heavy gun, loose gunpowder, and a burning match in the midst of battle was a dangerous business. The Spaniards also brought a few cannon, small falconets that shot two- to four-pound cannonballs, ten- to sixteen-pounders called culverins, and heavy iron bombards used in sieges.

The tactics of the Spaniards came from long experience fighting against Arabs and Berbers in the wars of the Reconquista and, more recently, against the French in the Italian Wars that began in 1494. In those wars, they broke from the ideal of man-to-man combat prevalent in the Middle Ages and began to develop the cohesive teamwork that was soon to make the Spanish *tercios* the scourge of Europe. Discipline forged a feeling of interdependence between the pikemen arranged in a square and the musketeers or crossbowmen who fired from the corners of the square, backed up by the cavalry and the artillery.

Humans and animals were not the only living beings to cross the Atlantic. More mysterious, and far more deadly, were the diseases they brought. Diseases were never (and are still not) uniformly distributed around the world. Their prevalence depended on the natural environment, including animal reservoirs like monkeys or rodents and vectors like mosquitoes or fleas. Even where the natural environment was favorable, a disease could only spread once it had been introduced. People who had never encountered a particular disease also had no immunity to it. New diseases that reached previously isolated people caused devastating "virgin-soil" epidemics. Each of the major diseases—smallpox, measles, cholera, typhoid, yellow fever, the plague, and others—was spread from its place of origin by traders, pilgrims, soldiers, and other travelers. The "Black Death" or bubonic plague that reached Europe in the 1340s came from Asia, perhaps from southern China, along with the Mongol armies that conquered much of Eurasia from Vietnam to Hungary. By the fifteenth century, diseases that had once been local had spread far and wide across Eurasia and much of Africa. A variety of regional disease pools had merged into one relatively homogeneous Eastern Hemisphere disease pool.[4]

Another, very different disease pool existed in the New World. According to environmental historian Alfred Crosby, before 1492 the American peoples were familiar with pinta, yaws, syphilis, hepatitis, encephalitis, polio, tuberculosis, pneumonia, and intestinal parasites. But they had no experience whatsoever with the diseases that ravaged the Eastern Hemisphere.[5] One reason there were so few of the Old World–type crowd diseases, despite high population densities in Mexico and Peru, is that the Native Americans had very few domesticated animals: only turkeys in Central and North America, llamas, alpacas, and guinea pigs in the Andes, Muscovy ducks in tropical South America, and dogs everywhere. As for their only domesticated herd animals, the llamas, the Andean peoples did not keep them indoors nor drink their milk, and their herds were too small to sustain diseases.[6]

Few Old World diseases were introduced from outside before 1492. The first migrants to the New World, the ancestors of the Indians, Aleuts, and Inuits, came in small numbers after years of trekking across Siberia and Alaska, environments that weeded out the sick. There was far less genetic diversity in the New World than in the Old. As a result, Native Americans had very homogeneous immune systems; bacteria and viruses that were successful in infecting one person could easily infect another without needing to mutate to adapt to different immune system antigens.[7] Later arrivals, like the Vikings who landed in eastern Canada in the eleventh and twelfth centuries, were few in number and had survived long, difficult ocean passages. There is no evidence that they carried diseases, but if they did and infected the indigenous people they encountered, the low population densities in that part of America ensured that any resulting local epidemic would burn out quickly. To Old World microorganisms, the Americas were truly a virgin soil.[8]

Historical demographers have been arguing for a very long time about the size of the population of the Americas in 1492. Estimates range from as low as 8.4 million to as high as 112.5 million. After reviewing the data and the various estimates, historical demographer Russell Thornton came up with his own estimate: "at least 72 million and probably slightly more." Most others seem comfortable with a number in this approximate range, though questions about numbers remain and are likely to remain for a long time to come.[9]

Not only was the population substantial, it was also free of most scourges that afflicted the peoples of the Old World. A Yucatan Indian recalled the days before the Spaniards arrived.

> There was then no sickness; they had then no aching bones; they had then no high fever; they had then no smallpox; they had then no burning chest; they had then no abdominal pain; they had then no consumption; they had then no headache. At that time the course of humanity was orderly. The foreigners made it otherwise when they arrived here. They brought shameful things when they came.[10]

This was an exaggeration, for the native population did suffer from the diseases listed earlier. Yet overall, says Thornton, "It is quite clear . . . that American Indians were in far, far better shape with regard to diseases than those who discovered them in 1492."[11]

Death accompanied the Europeans to the New World. The first place to be colonized was also the first to be decimated. When Columbus arrived, the island of Hispaniola was well populated; Bartholomé de Las Casas, the great critic of the Spanish colonial enterprise, thought the population was in the millions; a more recent estimate by demographer Angel Rosenblat has put it at 100,000 to 120,000.[12] Within months, the native Arawaks began dying. The causes are unclear, for the Spanish chroniclers had very hazy notions of epidemiology; perhaps it was influenza, followed by typhus and measles.[13]

Then came smallpox, the great Indian killer, a disease that was instantly recognizable. It arrived in Hispaniola in 1507 and reappeared in December 1518 or January 1519. Its arrival was delayed because most Europeans, and the African slaves they brought with them from 1503 on, had been exposed to smallpox and were therefore immune. Those who caught it just before embarking either died or recovered before reaching the New World. Only if a ship carried several non-immune persons who caught it one after the other, so that an active carrier reached the New World, or if infected scabs were somehow lodged in some cloth, could the virus survive the trip. When smallpox did, finally, cross the ocean, its effect was explosive.

Smallpox is an ancient scourge, noted as early as 1122 BCE in Chinese texts and found on the mummified skin of Egyptian pharaohs. It is a strictly human disease, transmitted directly from one person to another, usually caught by breathing the air expelled from the lungs of an infected person. Like measles, its has an incubation period of up to two weeks during which an infected, but still healthy, person can flee from the scene of sickness and infect many others. Then fever sets in, followed three or four days later by a skin eruption. The patient suffers excruciating pain, accompanied by respiratory infections, coughing, vomiting blood, and nosebleeds. If the patient does not die, he or she will be immune for life, though disfigured by pockmarks.[14]

In Europe, smallpox was endemic, a childhood disease that killed between 3 and 10 percent of the population. In the Americas, where no one was immune, it hit people of all ages, killing 30 percent of those infected. When added to the earlier epidemics, the social disruption they caused, and the mistreatment by the Spaniards who enslaved the Indians, smallpox decimated the population of Hispaniola. Las Casas estimated that "of that immensity of people that was on this island and which we have seen with our own eyes," no more than a thousand survived into the 1520s. By 1548, the chronicler Gonzalo Fernández de Oviedo thought only five hundred were left. Soon thereafter, the Arawaks were extinct as a people.[15]

The Conquest of Mexico

In the spring of 1519 Hernán Cortés landed on the Gulf Coast of Mexico and founded Villa Rica de la Vera Cruz (now Veracruz), the first Spanish settlement on the American mainland. In August of that year he marched inland with three hundred Spanish soldiers and sixteen horses, assisted by several hundred Indians from Cuba and coastal Mexico. His goal was Tenochtitlán, the capital of the Aztec Empire in the center of Mexico. The Spaniards found their way barred by thousands of Otomi warriors. Fray Bernardino de Sahagún describes what happened from the Indian perspective:

When they came to Tecoac, in the land of the Tlaxcaltecas, they found it was inhabited by Otomies. The Otomies came out to meet them in battle array; they greeted the strangers with their shields. But the strangers conquered the Otomies of Tecoac; they utterly destroyed them. They divided their ranks, fired their cannons at them, attacked them with their swords and shot them with their crossbows. Not just a few, but all of them, perished in the battle.[16]

How to explain this extraordinary outcome? The Otomies, like other Mexican Indians, did not lack courage or motives to defend their land. Rather, it was their weapons and tactics that failed them.[17]

Indian weapons were made of wood and stone; gold, silver, and copper were used for ornamentation, not for weapons. Noble warriors carried swords, or *maquahitl*, long oak sticks into which were embedded sharp pieces of obsidian; these were extremely dangerous to exposed flesh or padded cotton armor but easily broken and ineffective against steel. Commoners carried heavy clubs and shot arrows with fire-hardened or obsidian points and hurled darts with the atlatl (a curved stick) and stones with slingshots.[18]

Aztec warfare was highly ritualized. Its purpose was not to kill enemies but to engage in man-to-man combat and capture as many prisoners as possible, to be sacrificed later on the altars of the gods. Wars, preceded by elaborate declarations, began in September, after the harvest. At the forefront of their forces was an elite of highly trained and disciplined aristocratic warriors armed with *maquahitl* swords, elaborate cotton armor, and highly decorated feathered headgear. Supporting the nobles were farmers, more crudely armed and dressed. They advanced on as wide a front as possible, for only the front rank fought and they needed room to swing their swords. In order to be visible from a distance, their leaders carried banners and decorative emblems on their backs.[19]

The Spaniards, Cortés most of all, were powerfully motivated. Cortés, having defied the orders of the governor of Cuba, was aware that he was a wanted man; had he failed, he would have been brought to justice. On arrival in Mexico, he burned his ships, so

Figure 3.1. Hernán Cortés. Engraving by W. Holl, published by Charles Knight.

his soldiers knew they had to win or die. To this fear of losing, we must add the soldiers' lust for gold and the conviction that God was on their side and that it was their God-given duty to convert or enslave the Indians. But motives alone did not bring victory. Battles were decided by weapons and tactics as well as courage and determination. Bernal Díaz del Castillo, one of Cortés's soldiers, describes a battle:

> We were a full hour fighting in the fray and our shots must have done the enemy much damage for they were so numerous and in such close formation, that each shot must have hit many of them. Horsemen, musketeers, crossbowmen, swordsmen and those who used lance and shield, one and all, we fought like men to save our lives and to do our duty, for we were certainly in the greatest danger in which we had ever found ourselves.[20]

Figure 3.2. Battle of Metztitlan between the Spanish and the Otomies.
Note the weapons used by both sides and the headdress of the Indian leader.
Drawing by Daniel Headrick after Diego Muñoz Camargo, *Descripción de la
Ciudad y Provincia de Tlaxcala de las Indias y del Mar Océano para el buen gobierno y
ennoblecimiento dellas* (México: Instituto de Investigaciones Filológicas,
Universidad Nacional Autónoma de México, 1981).

In one sentence, Díaz lists horses, muskets, crossbows, swords, lances, and shields; he could have added helmets and suits of armor, as well as dogs. These, and the use the Spaniards made of them, were the keys to their victories.

When they first arrived in Mexico, the Spaniards brought with them sixteen horses, one of them a colt born onboard ship. With their swords and lances, Spanish soldiers could pierce the cotton armor and slash the wooden swords of the Aztecs, while their helmets and armor protected their heads and torsos from Aztec swords, lances, and arrows.

In early 1519, messengers from Moctezuma to the Spaniards at Veracruz brought back pictures of dogs that are said to have alarmed the emperor. According to Sahagún, the Indians reported seeing "very big dogs, with floppy ears, long hanging tongues, eyes full of fire and flames, clear yellow eyes, hollow bellies shaped like spoons, as savage as devils, always panting, always with their tongue hanging down, spotted, speckled like jaguars."[21] The terrain of Mexico was much more accessible to horses than that of Hispaniola and dogs played only a minor part in battles.

The goal of the Spaniards was not to capture Aztecs but to defeat them, if necessary by killing them. The open formation in which the Aztecs fought made them vulnerable to the Spanish cavalry. When surrounded by overwhelming numbers of Aztec warriors, the Spaniards formed a square, the infantry supporting the cavalry and defending the guns. They deliberately aimed at the Aztec leaders, easily recognized by their banners, for they knew that when a chief died, it meant defeat and the warriors would flee, making them easy targets for the pursuing cavalrymen.

The Spaniards attacked during the planting and harvest seasons, when most Indians were in the fields and the Aztec leaders had difficulty raising armies. The Spaniards felt no compunction about killing civilians, burning crops, and setting fire to towns and villages. Whatever the Aztecs' well-deserved reputation for violence, the Spanish soldiers were even more brutal, and in ways that demoralized their opponents.[22]

Nor should we forget the role played by the many Indians who fought on the Spaniards' side. Cortés's relations with the Mexicans owed much to the assistance of his interpreter Malintzin, a noble Indian woman whom the Spaniards called Doña Marina, who spoke Maya and Spanish as well as her native Nahuatl, the language of central Mexico. After the defeat of the Otomies, the people of nearby Tlaxcala, long enemies and frequent victims of the Aztecs, hastened to ally themselves with Cortés. They were eager to wreak their revenge on the Aztecs, who had long victimized them. They and other tribes furnished thousands of warriors, porters, and other auxiliaries, as well as food and shelter. So important were they that one eminent historian opined that "Mexico was not conquered from abroad but from within" and that "the Spaniards usurped the victory for which their Indian allies had fought and died."[23]

Cortés and his soldiers entered Tenochtitlán in November and were well received by the Aztec ruler Moctezuma, whom they

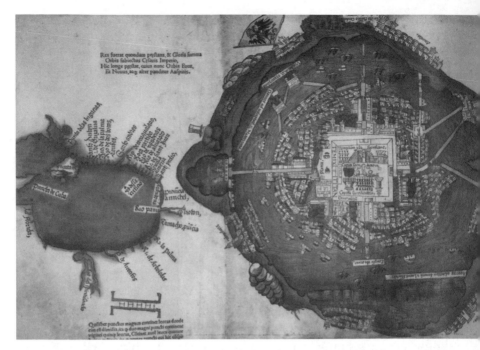

Figure 3.3. The city of Tenochtitlán before the Spanish invasion. Painting by
Dr. Atl (1930) in the National Museum of Anthropology, Mexico City.

promptly captured and held hostage. He and his followers were
taken by surprise, for the Spaniards did not resemble any warriors
they had ever encountered and did not fight at the times and in the
ritualized manner the Aztecs were accustomed to. At first, they
may have thought that the Spaniards were gods or supernatural beings. Even after they realized that the newcomers were only men,
they found it hard to believe that such a small number of men could
possibly represent a danger. Moctezuma, an indecisive man, did
not know what to make of the Spaniards. He received them as
guests and did not expect to be captured. As Aztec society was very
hierarchical, the capture of their leader left the others perplexed
and confused.

When another Spanish expedition, led by Pánfilo de Narváez,
arrived on the coast with orders to arrest him, Cortés left Pedro de
Alvarado in Tenochtitlán in command of two hundred Spaniards
and returned to the coast to confront this new rival. Using diplo-

macy, bribery, and ruthlessness, Cortés convinced most of Narváez's soldiers to switch sides and join him in conquering Mexico. This added nine hundred men and sixty horses to his forces. Along with Narváez's troops, Cortés also acquired a few falconets and bombards. Besides being heavy and clumsy, such guns and artillery pieces were useless in wet weather. Yet they were effective in Mexico, partly because of the noise they made, and partly because several Indians could be killed or wounded with one cannonball, as Bernal Díaz del Castillo noted.[24]

In Cortés's absence, meanwhile, Alvarado attacked the Aztec warriors in Tenochtitlán. When Cortés returned in June, he found himself confronted with a major uprising led by the Aztec warrior chief Cuitláhuac. Tenochtitlán was not a fortified city like those the Spaniards had seen in Spain, Morocco, or Italy. Instead, it was built on an island in the middle of Lake Texcoco, linked to the mainland by several causeways. On the night of June 30, 1520, the Spaniards, surrounded and vastly outnumbered, were forced to flee the city by one of its causeways. They fought their way out of the city under a rain of arrows and rocks from nearby rooftops and from hundreds of canoes on the lake. As Díaz del Castillo recalled:

> When I least expected it, we saw so many squadrons of warriors bear-
> ing down on us, and the lake so crowded with canoes that we could
> not defend ourselves. . . . as we went along the causeway, charging
> the Mexican squadrons, on one side of us was water and on the other
> azoteas [flat rooftops], and the lake was full of canoes so that we could
> do nothing.[25]

In their flight, the Spaniards lost half their forces, two-thirds of their horses, and all their cannon, along with almost a thousand Indian allies. Moctezuma was killed during the evacuation. It was a disaster the Spaniards called la Noche Triste, the Night of Sorrows.

But sorrowful as the night was for the fleeing soldiers, a far greater disaster befell the people of Tenochtitlán, for in their retreat, the Spaniards left behind the scourge of smallpox. The disease appeared in Veracruz in early 1520, brought by a member of Narváez's expedi-tion. By October, it had reached the Valley of Mexico and spread quickly through the population of Tenochtitlán and surrounding areas. A survivor recalled its effects:

> The illness was so dreadful that no one could walk or move. The sick were so utterly helpless that they could only lie on their beds like corpses, unable to move their limbs or even their heads. They could not lie face down or roll from one side to the other. If they did move their bodies, they screamed in pain.
>
> A great many died from this plague, and many others died of hunger. They could not get up to search for food, and everyone else was too sick to care for them, so they starved to death in their beds.[26]

The smallpox epidemic of 1520–21 was one of the single most devastating that ever struck any people. Epidemiologists calculate that it killed about half the people of central Mexico.[27] For the Spaniards, it was an act of God; as Francisco de Aguilar, a follower of Cortés, explained: "When the Christians were exhausted from war, God saw fit to send the Indians smallpox, and there was a great pestilence in the city."[28]

Not only was it a human calamity, it was also the single most important cause of the collapse of the Aztec Empire and the Spanish conquest of Mexico. For among the dead were Cuitláhuac, the leader of the Aztec forces during the Night of Sorrows who had succeeded Moctezuma as emperor, and many of his warriors. Cuitláhuac had become infected in November, just as his people were preparing their defenses, and died on December 4, after only eighty days in office. He was succeeded by Moctezuma's nephew Cuauhtémoc. Neither man had the experience or the time to consolidate the allegiance of tributary tribes. The changes of leadership in the midst of war and epidemic must have added to the confusion and fear among the people of the city.[29] The Indians allied to the Spaniards also succumbed to smallpox, of course, but when their leaders died, it was Cortés who chose the successors and provided leadership.

Once away from the city, Cortés received reinforcements from Cuba and prepared for the next campaign. He returned for the final assault on Tenochtitlán in May 1521. He brought with him 550 Spanish soldiers of whom eighty carried crossbows or harquebuses, forty horses, and nine cannon, supported by several thousand Tlaxcalans.[30] During the siege, reinforcements kept arriving from Vera-cruz, bringing more horses and cannon.

When the Spaniards returned, the Aztecs pinned their hopes on their fleet of canoes that protected their island city. Cortés's hope of penetrating the city depended therefore on a means of destroying the Aztec canoes or keeping them away from the causeways on which the cavalry and foot soldiers were to advance. To do so he called on the carpenter Martín López, assisted by Indian workers, to build thirteen brigantines. These were small flat-bottomed sailing ships that each carried twenty-six men, twelve of them armed with crossbows or harquebuses, as well as twelve men with paddles and two artillerymen. Each brig had a bronze cannon and some falconets; the somewhat bigger flagship carried two cannon. The brigs were built in sections and carried down to the lake in April 1521. A month later the battle began when the small Spanish fleet advanced onto the lake to confront the hundreds of Aztec canoes. With their four-foot freeboard and castles at both ends, the brigs towered over the canoes. When the wind rose, they were able to maneuver and shoot at, ram, or capsize the canoes.

By the second day, the brigs dominated the lake on both sides of the main causeway. In the naval battles that followed, they were aided by thousands of allied Indian canoes. To thwart them, the Aztecs planted sharpened stakes in the lake bottom, hoping to im-pale the brigs, but with little success. One brig was captured in an ambush. By the tenth day, the Spaniards had completely cut off the city from the mainland, as well as from the aqueduct that supplied the city with fresh water (the lake was brackish). Almost at will they could approach the city, enter the canals, and set fire to buildings. The brigs also served as pontoons to bridge gaps that the Aztecs had dug in the causeways, and helped supply the cavalry and infantry fighting on the causeways.[31]

Historians disagree on the relative role of the brigs and the land forces. Clinton Gardiner, historian of the brigs, has claimed that they were the main cause of the Spanish victory, while Ross Hassig, historian of the conquest, maintains that the war was ultimately won by land forces. Both are right. By the summer of 1521, the Aztecs had gotten over their fright and confusion and prepared themselves for a war to the finish, which they had reason to expect they would win, given their prowess, their numbers, and the experience they had gained against horses and firearms in previous battles. They were not prepared, however, to fight a naval war against vastly more powerful ships.

After a siege lasting three months, the Spaniards captured and killed the newly crowned emperor Cuauhtémoc. They entered Tenochtitlán on August 13, 1521. The city, once the most beautiful the Spaniards had ever seen, was in ruins, and most of its two hundred thousand inhabitants were dead. The Spaniards interpreted the epidemic among the Indians, and their own immunity, as signs of God's favor; as Vázquez de Tapia, a follower of Cortés, wrote: "there died a great quantity of men and warriors and many lords and captains against whom we would have to fight and deal with as enemies, and miraculously Our Lord killed them and removed them from before us."[32]

The epidemics that spread through the Mexican population in the 1520s and 1530s were only the beginning of their calamities. From 1520 to 1600 Mexico suffered through fourteen epidemics; Peru endured seventeen. Close on the heels of smallpox came other ills: measles, typhus, influenza, diphtheria, mumps, and others, sometimes several at once. Epidemics followed one another at four-and-a-half-year intervals, on average. The typhus epidemic of 1545–48 was even more severe than the first smallpox epidemic. Those who survived those diseases often died of pneumonia, which attacked their weakened constitutions. Between 1519 and 1600, the population of Mexico is said to have dropped from almost fourteen million to one million, a decline of 93 percent.[33]

The virgin-soil epidemics that swept through the Indian population were particularly destructive because so many people were afflicted at the same time. Medical practices that worked for ailments the Indians were familiar with, such as plunging a sick person in cold water, only aggravated the condition of those suffering from the new diseases. Smallpox and gonorrhea often caused miscarriages or left survivors sterile. Among those who remained healthy, many fled, abandoning the sick and neglecting their animals and their crops. Small children whose parents fell ill died of hunger. As the Spanish chronicler Toribio de Motolinia wrote: "Because they all fell ill at a stroke, [the Indians] could not nurse one another, nor was there anyone to make bread, and in many parts it happened that all the residents of a house died and in others almost no one was left."[34]

Having defeated the most powerful empire in North America, the Spaniards proceeded to conquer and subdue much of what is today Mexico and Central America.[35] This was more difficult than they expected. The highly organized peoples in the densely populated areas succumbed fairly rapidly, having been conquered and subdued before by the Aztecs and their predecessors, but nomadic hunter-gatherers and loosely organized societies resisted fiercely. The Chichimecas, a derogatory term the Spaniards applied to the Pames, Guamares, Zacatecos, and Guachichiles Indians who lived in the highlands between the Sierra Madre Oriental and the Sierra Madre Occidental north of Tenochtitlán, were part-time agriculturalists who depended also for their survival on gathering wild plants and hunting game. Onto this dry plateau came Spanish conquistadors looking for precious metals. In 1546, at the head of a small group of men, Captain Juan de Tolosa gave some trinkets to Zacatecos Indians and was given some silver in exchange. News traveled back to the capital and by 1550 a silver rush was on. Spanish miners, ranchers, muleteers, and missionaries and their Indian or African servants invaded the region. On the road between Mexico City (as the Spanish called their new capital built on the site of Tenochtitlán) and the new mining town of Zacatecas mule trains carried food and supplies and returned with silver.[36]

The Chichimecas were skilled warriors armed with spears and with bows that shot thin, obsidian-tipped arrows that could penetrate all but the most closely woven chain-mail armor. They ambushed travelers along the road and attacked mule trains and isolated ranches and mining camps. They took captives and maimed or tortured them to death. In retaliation, the Spaniards sent raids to kill Indians or bring back captives to sell on the slave markets of Mexico City. At first the Indians treated the Spaniards' horses, mules, and cattle like game, but soon they learned to steal or capture horses. As a result, one Spanish settler explained, "They are no longer content to attack the highways on foot, but they have taken to stealing horses and fast mares and learning to ride horseback, with the result that their warfare is very much more dangerous than formerly, because on horses, they make raids and flights with great speed."[37]

By 1570, the Chichimecas organized larger war parties. They sent spies to scout out the Spanish defenses and learned to attack at dawn. They felt bold enough to attack mining towns. As the fighting and the brutality escalated on both sides, the government hesitated between sending in a large army, as the miners and ranchers demanded, or coming to an accommodation with the Indians, as missionaries advised. Finally a new viceroy, the Marqués de Villamanrique, negotiated peace treaties with Indian leaders and the fighting died down. It had taken fifty years to impose Spanish rule on north-central Mexico, ten times longer and with far more Spanish casualties than the conquest of the Aztec Empire.

Peru and Chile

Thousands of miles to the south of Mexico was another empire, larger and richer than that of the Aztecs. The Inca Empire stretched for twenty-five hundred miles along the west coast of South America, from southern Colombia to northern Chile, and encompassed the coastal plains, the Altiplano or high plateau of the Andes, and parts of the tropical lowlands to the east of the mountains. Unlike the Aztecs who ruled through allies and tributaries, the Incas had forged a highly centralized administration centered on their capital, Cuzco, and tied together through a network of roads and messengers, with military posts planted throughout their domain. They had a large army that frequently warred against their neighbors to the north and south. They had built warehouses stocked with food, arms, clothes, and equipment. In many ways, their system of government can be compared to that of the Romans; all they lacked was writing and a means of travel faster than a human runner.[38]

Into this powerful empire came an enemy against which they had no defense: smallpox. From Mexico, it spread south to Panama. It reached the domain of the Incas in 1524 or 1525, years before the first Spaniard arrived. The first confirmable smallpox victims among the Incas were those of their emperor Huayna Capac and his captains, then on campaign in the north of his realm, who died, "their faces covered with scabs." When his heir, Ninan Cuyoche, also died, a civil war broke out between two half-brothers, Huáscar and Atahuallpa. In the midst of this succession struggle and epidemic, with their attendant misery and confusion, came the Spaniards.[39]

The Spanish expedition, led by Francisco Pizarro and his brothers Hernando and Gonzalo, was even smaller than Cortés's; only about 170 soldiers and ninety horses reached the camp of Atahuallpa at Cajamarca in northern Peru. They faced an Inca army that has been estimated at a hundred thousand. But the Spaniards were ready, for they were veterans of campaigns in Italy, the Caribbean, and Mexico. All had heard of Cortés's exploits and were determined to repeat them. The Incas, in contrast, were taken by surprise. On November 16, 1532, no sooner was the small Spanish force received by the Inca Atahuallpa than they captured him and held him ransom until his followers brought them all the gold and silver in Cuzco. From his captivity, Atahuallpa ordered the assassination of his half-brother, Huáscar. Eight months later, having only whetted their lust for precious metals, the Spaniards executed him.

Inca society was even more hierarchical than that of the Aztecs, for the Incas believed their emperor was no mere mortal but the "Child of the Sun." The death of Huayna Capac and Ninan Cuyoche and the murder of Huáscar and Atahuallpa decapitated the empire. Nonetheless, the Inca state survived longer than the Aztecs'. While the conquistadors fought among themselves, the Incas retreated to the remote region of Vilcabamba. From there, Huáscar's brother Manco organized an army of a hundred thousand or more that besieged 190 Spaniards in the old capital, Cuzco; aided by Indian allies, the Spaniards held out for almost a year, until Manco's army disintegrated. By the 1570s, Spain controlled the peoples of the former Inca empire as thoroughly as they did the peoples of Mexico and Central America.[40]

The weapons and tactics the Spaniards used in Peru were the same as those they had used in Mexico, although the mix was different. Their main weapon was horses. Horses could surprise sentries and outrun the scouts and messengers the Incas relied on to control their forces and learn of their enemies' whereabouts. In battles, horses easily frightened Inca soldiers. When booty was distributed, horsemen received much more than foot soldiers. Indeed, horses were valued at 1,500 to 3,300 gold pesos, as much as sixty swords.[41] So valuable were horses that when their shoes wore out, the Spaniards forged horseshoes out of copper or silver.

Many years later the Inca Garcilaso de la Vega, the author of the most famous work on the Inca Empire and the Spanish conquest, described the Indians' first reaction to the sight of horses.

> And so nothing convinced them to view the Spaniards as gods and submit to them in the first conquest so much as seeing them fight upon such ferocious animals—as horses seemed to them—and seeing them shoot harquebuses and kill enemies two hundred or three hundred paces away. For these two things . . . they took them for sons of the Sun and surrendered with so little resistance as they did.[42]

As in Mexico, the Spaniards wore steel armor or chain-mail shirts, and sometimes padded cloth armor adopted from the Aztecs that was much lighter than metal and did not rust. They carried steel swords and steel-tipped pikes and lances. They brought with them a few small cannon, crossbows, and harquebuses but seldom used them. In battles, horsemen, swordsmen, and pikemen fought in teams.

Against the Spanish weapons and tactics, the Incas were even less well prepared than the Aztecs had been. They had no cutting or slashing weapons. They used clubs and battle-axes with stone or bronze heads. They threw javelins with fire-hardened points and used slings to hurl rocks the size of apples. Though the peoples of the Amazon rain forest used bows and arrows, the Incas did not, for they lacked suitable wood. Their tactics, like those of the Mexicans, involved advancing in loose formations leading to hand-to-hand combat. Their weapons were dangerous to other Indians but not to the Spanish cavalry or to infantrymen with their steel helmets and armor. Manco's soldiers experimented with new weapons and tactics, such as digging pits to trip horses, throwing *bolas*, or stones, tied together to bring down horses, or hurling red-hot stones with slingshots to set fire to the thatched roofs of their enemies' houses. But in the end they had too little time to emulate the weapons of the Spaniards and assimilate their tactics.[43]

Thanks to their horses, their swords and armor, their boats, their firearms, and the germs they carried, the Spaniards defeated the two most populous, highly organized, and militarily successful states in the New World. Why, then, did they, and other Europeans who followed them, not go on to conquer the rest of the American continent, inhabited as it was by smaller states and more loosely organized tribes? To understand this paradox, let us turn to their failures and the victories of the Native American peoples.

In 1572, Luís Vaz de Camões, a Portuguese who had spent many years overseas, published the epic poem Os Lusíadas glorifying the Portuguese conquests in the Indian Ocean. At the same time, starting in 1569, a Spanish soldier by the name of Alonso de Ercilla y Zuñiga also wrote an epic poem, La Araucana. Unlike Camões, Ercilla did not celebrate the victories of the conquistadors but those of their most tenacious enemies, the Araucanian or Mapuche Indians of southern Chile. Their story, and that of other "wild" Indians, shows how ephemeral was the power that horses gave the Europeans in the Americas.

The Araucanians were hunter-gatherers and slash-and-burn farmers who grew maize, potatoes, and beans and raised dogs and llamas. They had no government, but were organized by family and clan. They practiced warfare and honored their warriors. Their weapons included bows and stone-tipped arrows, pikes and lances, slings, darts, javelins, spears, cudgels, and seven-foot-long clubs. They had no metal or cutting-edge weapons. Their shields and body armor were of leather and wood. They were among the few Indians living west of the Andes who had defeated the Incas.[44]

The Araucanians first encountered the Spaniards in 1546 when Pedro de Valdivia led some seventy Spanish soldiers and several hundred Indian auxiliaries across the Maule River, approximately 150 miles south of Santiago. In the ensuing battle, the Araucanians lost two hundred of their eight thousand warriors and retreated to the Bío-Bío River, 150 miles further south, while Valdivia returned to Santiago.

Valdivia crossed the Maule River again in 1550 and founded three towns. By then the Araucanians had had four years to organize their warriors, elect captains, and practice new tactics. Rather than engage the Spaniards in open battle, they chose to attack at night or to retreat to broken terrain where they had dug pits covered with branches to trip the Spanish horses. Where they could not avoid the horses, they arranged themselves in lines with two rows of pikemen in front covering several rows of archers. In short, they were developing new tactics to counter the superior weapons of the Spaniards. In December 1553, an Araucanian army led by the chief Caupolicán surrounded and killed Valdivia and his five hundred soldiers. After this victory, other Indians joined the Araucanians while the Spanish settlers abandoned their towns and forts and fled back north across the Maule River. When the Spaniards returned a year later under Valdivia's successor, Francisco de Villagra, they were again defeated, this time by the Mapuche chief Lautaro.[45]

This did not end the warfare. For the next ninety years, warfare was endemic in the region between the Maule and Bío-Bío rivers. Periodically, Spanish soldiers ventured south to capture slaves and Indians came north to destroy settlements and steal horses. In the process, frontier warfare was completely transformed. The most important change began when the Araucanians captured the horses brought by Valdivia, Villagra, and others. By 1566 they had acquired several hundred horses. Toward the end of the century, they also raised thousands more, for their land was ideal for pasture. They became very adept at fighting on horseback, and their cavalry was soon faster and more skilled than that of the Spanish.

To the weapons they had traditionally used they added several new ones. To bring down a Spanish cavalryman, they snared him with a kind of lasso at the end of a long pole; when he fell, they beat him with cudgels. They also learned to make use of the swords, daggers, and armor of fallen enemies by breaking them into short pieces and making spear points out of them. They sometimes obtained harquebuses, along with shot and powder. They captured six cannon from the Villagra expedition, but these were later recovered by the Spaniards. The Araucanians even experimented with fortifications made of poles similar to those the Spaniards erected.[46]

Their tactics changed with their weapons. In skirmishes with Spanish soldiers, they took advantage of their greater numbers by forming squadrons that took turns fighting until the Spaniards were too tired to lift their swords. Besides their cavalry, they also used mounted infantry that would rush to the scene of a battle, dismount to fight, then gallop away. Sometimes one horse carried two riders, one wielding a lance and the other shooting arrows. They besieged Spanish forts and surrounded them with an outer barrier of trenches lined with sharp stakes to thwart relief efforts. In short, they were as creative in their tactics as they were in their weaponry.[47]

Many of the Spanish soldiers in Chile were men exiled from Peru for crimes. They were poorly clothed and equipped and often posted in outlying forts without sufficient food or supplies. Some, driven by hunger, traded their swords, guns, and gunpowder for food. Others, mestizos of mixed European and Indian ancestry recruited to fight on the frontier, deserted to the Indians, bringing their weapons with them.[48]

The frontier war of the late sixteenth century reached a climax when Captain-General Martín García Oñez de Loyola, who had made his reputation as an Indian fighter, took the offensive. At dawn on December 23, 1598, while he was camped out at Curalaba with fifty Spanish soldiers and two hundred Indian auxiliaries, three hundred Auraucanian horsemen attacked him and annihilated his force. In the uprising that followed this defeat, the Spaniards were forced back north of the Bío-Bío River, while many, including women and children, were taken prisoner.

Though expelled from Araucania, the Spaniards continued to harass the Indians. Every summer they launched raids called *malocas* to burn Indian crops and villages and capture slaves. In defense, the Araucanians retreated into the mountain valleys, where they had larger fields and villages and where the Spanish soldiers were afraid to follow them. In retaliation, the Araucanians also launched *malocas*, raiding Spanish settlements north of the river to capture horses, metal tools, and weapons.[49]

Finally, in 1641, after ninety years of warfare, Spain signed a treaty recognizing the independence of the Araucanians. Missionaries entered Araucania, merchants traded, and whites and Indians intermarried. Yet the frontier along the Bío-Bío River was never completely peaceful. Spaniards still raided Araucanian territory for slaves and Araucanians raided Spanish settlements for horses and weapons. Low-level tensions were periodically punctuated by major uprisings and frontier wars.

Despite their prowess as warriors, the Araucanians gradually lost ground to the Spaniards over the course of the eighteenth century. The reason was demographic. During that century, the white and mestizo population of the Reino de Chile north of the Bío-Bío quadrupled. In 1791 the Araucanians were weakened by a smallpox epidemic, after which more and more whites began to settle in their territory. Yet their independence was not extinguished until the late nineteenth century. It had lasted three hundred years.[50]

Argentina and North America

Before the arrival of the Europeans and their animals, the Pampas of Argentina were poor in fauna and the inhabitants led a more primitive life than did the Araucanians. The Querandí in the north, the Puelches in the center, the Tehuelches in Patagonia, and the Pehuenches in the foothills of the Andes were nomadic hunter-gatherers who lived in animal-skin shelters and wore animal-skin clothes.[51]

In 1536 Pedro de Mendoza landed on the Río de la Plata with sixteen ships, two thousand men, and seventy-one horses and founded the settlement of Buenos Aires. At first, the Querandí received the Spaniards warmly, provided them with food, and became accustomed to their horses. By 1541, however, Spanish demands provoked an Indian uprising. Under attack and short of food, the Spaniards were forced to abandon their settlement. Some returned to Spain while others made their way upriver to Paraguay.

By the time another Spanish expedition under Juan de Garay returned to the Río de la Plata in 1580, they found that wild horses had taken over the Pampas. Some authors have claimed that these were the descendants of horses abandoned at the time of the first withdrawal, but Madaline Nichols, historian of Argentine horses, believes it more likely that they were the offspring of horses from earlier Spanish settlements in Paraguay, northern Argentina, Chile, Peru, or even Brazil.[52]

The Pampas were the ideal environment for horses. Here were vast open grasslands similar to the steppes of Central Asia where horses originated. Unlike the Great Plains of North America with their herds of bison, before the Spaniards introduced horses and cattle the Pampas had few herbivores other than deer and ostriches. Nor were there large felines to keep the herbivore population in check as in Africa. In such an environment, horses multiplied rapidly. Visitors spoke of huge herds numbering in the thousands. An English visitor in the mid-eighteenth century described the scene:

> There is likewise a great plenty of tame horses and a prodigious number of wild ones. . . . The wild horses have no owners, but wander in great troops about these vast plains. . . . Being in these plains for the space of three weeks, they were in such vast numbers, that during a fortnight, they continually surrounded me. Sometimes they passed by me in thick troops, on full speed, for two or three hours together.[53]

In the late sixteenth century, soon after the first Araucanian victories in Chile, the Indians of the Pampas also acquired horses. By the late seventeenth century, Indians as far south as Patagonia were on horseback.[54] To capture horses or kill them for food, they used the *bolas*, stones tied together with long leather cords that entangled the animals' legs, as well as lassos and lances fifteen to eighteen feet long.[55] With their horses and simple weapons, they quickly became the rulers of the Pampas.

The Pampas Indians did not just ride and eat horses, they also traded them with the Spaniards in Chile for goods they needed, such as sugar, tobacco, tea, liquor, woolen blankets, and other products of settled agriculturalists. The intermediaries in this trade were the Pehuenches of the Andes and the Araucanians of Chile. During the eighteenth century, Araucanians migrated across the mountains and joined the Pampas Indians, who soon began speaking Araucano. In the process, the Araucanians taught the Pampas Indians their organizational and fighting skills.[56]

As in Chile, their relations with the whites of Buenos Aires were mainly hostile. For two hundred years, the Indians raided the Spanish convoys and estancias to capture horses, cattle, and women (for they took no male prisoners). The Spaniards launched raids to capture horses and cattle and to punish the Indians, but to little avail. Periodically, the frontier would erupt in full-scale warfare. Against the fast-moving Indian horsemen, Spanish weapons were ineffective; in fact, Spanish gauchos preferred the Indians' lances, lassos, and *bolas*.

The boundary between European and Indian territory followed the Salado River, a hundred miles to the west and south of Buenos Aires. In the mid-eighteenth century, to protect outlying estancias, the city fathers decided to build a line of forts along the river, but they had difficulty finding men willing to be posted there. To gather salt, a commodity in short supply in their settlements, the Spaniards had to send heavily armed expeditions to the salt beds southwest of Buenos Aires. The opening of direct trade with the rest of the Spanish empire, part of the Bourbon reform program, attracted more whites to the Río de la Plata but did not break the stalemate right away. In 1796, Félix de Azara, the leader of an expedition to Patagonia, reminded Viceroy Melo de Portugal that the boundaries of the Río de la Plata were the same as they were in 1590 "because of a few annoying barbarians."[57] As in Chile, the Indians had blocked the advance of the Europeans for two hundred years.

The history of North America resembles that of the Pampas of Argentina on a larger scale. Europeans settled along the Atlantic coast and the Gulf of Mexico in the late sixteenth and early seventeenth centuries. By the late eighteenth century they had moved inland a couple of hundred miles, but the British and French colonies of North America were still very underdeveloped compared to Mexico or Peru. Meanwhile, the vast expanse of the continent remained in Indian hands until well into the nineteenth century. What needs explaining is not what the whites conquered, but what they could not, and why.

What kept Europeans out of the Great Plains that stretched from the Appalachians to the Rockies was Indians on horseback. Before they acquired horses, the few Indians who lived there grew maize, beans, and squash along the riverbanks. The open range belonged to the vast herds of bison. The Indians hunted them, but with great difficulty, to obtain hides, sinew, and bones and to supplement their largely vegetarian diet.[58]

Then came the horse. Clark Wissler, an early historian of the Plains, believed that they were the descendants of horses left behind or lost by the Spanish expeditions of De Soto and Coronado in the 1540s. Another historian, Francis Haines, later showed that this scenario was most unlikely and that the horses that bred on the Plains came from the Spanish missions in New Mexico in the early seventeenth century. The government of New Spain prohibited the sale of horses to Indians, and the military guarded their horses carefully. Yet missionaries needed horses and had Indians care for them, and as a result, some horses were lost or sold. Mexican traders were not above exchanging horses for slaves and animal hides, or trading utensils and liquor for horses.[59]

Even after the Indians obtained horses, it took awhile for them to learn to breed them and use them effectively. Not until the mid-seventeenth century did any become what was later called "horse Indians."[60] From New Mexico, horses spread throughout the Plains, from the Southwest to the Northeast. Indians in Texas had horses in the 1680s. In the early eighteenth century, horses were found throughout the southern Plains. By the late eighteenth century horses were found as far west as the Rocky Mountains and as far north as Saskatchewan. Once they obtained horses, Indians from the mountains like the Shoshone descended into the Plains and became full-time bison hunters.[61]

Horses transformed the lives of the Plains Indians as they did the Araucanians and Pampas Indians. Some, like the Mandan, the Arikara, the Pawnee, and the Wichita, remained farmers. Others— the Kiowa, Comanche, Crow, Arapaho, Cheyenne, and Sioux—became horseback-riding warriors. They gave up agriculture and took up hunting full-time, eating the meat of bison and horses and using their by-products. They could travel farther and faster and bring far more belongings with them than they had been able to before. The new way of life also attracted Indians like the Coeur d'Alene from the Rocky Mountains. Wealth was measured in the number of horses that warriors or clans possessed and their skill at acquiring more horses. The Comanches, being among the first to acquire horses and having more than other tribes, became the best horsemen of all. Intertribal warfare became endemic on the Plains. Warriors fought over bison-hunting territory, or carried out quick raids to steal horses or to avenge the loss of a horse or a fellow warrior. The Plains Indians, both sedentary and nomadic, engaged in trade, mainly to obtain horses and guns. Those who had horses and guns could then defeat their competitors and gain access to more trade goods. Trade and warfare became symbiotic.[62]

The skills used in hunting were the same ones needed in raids and intertribal warfare. Warriors carried short bows, quivers with a hundred arrows, long spears, and buffalo-hide shields. Equestrians since childhood, they learned to grasp the side of a horse with their legs while riding at full gallop shooting arrows from under the horse's neck. On raids, they approached an enemy's camp in the dead of night, dismounted, and used stealth and camouflage to enter his camp and steal his horses.[63]

For a century and a half, from the eighteenth to the mid-nineteenth century, the Plains Indians were the most skilled and dangerous cavalrymen since the Mongols. Against their weapons and tactics, the firearms the Europeans had before the 1840s were weak and ineffective. Swords and lances were almost useless against warriors who kept their distance. Loading a musket or a rifle took a minute, long enough for an Indian to shoot twenty arrows. Horse pistols had one bullet apiece. After two or three shots, a European soldier would have to dismount to reload, while the Indian warrior rode away. Historians who have studied the history of the Great Plains in this period are unanimous in claiming that the horse was the decisive factor in the long resistance of the Indians to European encroachment. As Walter Prescott Webb wrote, "At the end of the Spanish régime [in Mexico] the Plains Indians were more powerful, far richer, and in control of more territory than they were at the beginning of it."[64] In Alfred Crosby's words, "In the long view of history, the greatest effect of the horse on the Indian was to enhance his ability to resist the advance of Europeans into the interiors of North and South America."[65] And according to Bernard Mishkin,

"It is one of those accidents of history that an instrument of Spanish expansion in the New World, the horse, was an important factor in barring further expansion to the same nation."[66]

Disease and Demography

In spite of their ability to resist the advance of Europeans, the Indians of Argentina and North America were as vulnerable to diseases as their counterparts in Mexico and Peru, although the consequences of this vulnerability appeared much more slowly, for the wide dispersion of nomadic peoples over large areas kept epidemics localized and did not allow them to make so many victims at one time as in densely populated civilizations. Yet in the long run they were more irreversible. Diseases did not just affect the balance of military and political power, they transformed the ethnic composition of North and South America.

Smallpox first appeared on the Argentine Pampas in 1558–60, decimating the Indians who lived closest to the Spanish settlements, but sparing the Europeans. It may have reached the Río de la Plata from Chile, but later epidemics might have come from Brazil, where the disease was brought by the Portuguese or by their African slaves. Spaniards were not as vulnerable as the Indians, for inoculation, introduced into Spanish America in the late eighteenth century, helped keep the disease under control among whites.[67] As a result, as the white population slowly grew, the Indian population declined. The English visitor Thomas Falkner observed, "Although formerly very numerous . . . [the Indians] are now so much diminished as not to be able to muster four thousand men amongst them all."[68] Like all Native Americans, their societies were doomed in the long run. But they had held their own for over two centuries, a remarkable achievement.

As in Peru, Old World diseases preceded the arrival of the Europeans into the interior of North America by many decades. The Southeast and Midwest of the United States were once densely populated by agricultural peoples whom we call the "Moundbuilders" because of the large earthen pyramids and sinuous mounds they left behind.

By the time Hernando de Soto traveled through the region from 1539 to 1542, the Indians were already suffering from the first epidemics. Later explorers found villages, even substantial cities like Cahokia, Illinois, abandoned while the few remaining inhabitants had gone back to hunting and gathering. A Frenchman who visited Natchez, Mississippi, wrote, "Touching these savages, there is a thing that I cannot omit to remark to you, it is that it appears visibly that God wishes that they yield their places to new peoples."[69]

Much the same happened in New England, a region that was both densely populated with Indians and attractive to English settlers. The first epidemic, of plague or perhaps of typhus, hit in 1616–19 and killed about nine-tenths of the Indians living along the coast. It was followed in the 1630s and 1640s by a smallpox epidemic that swept through the St. Lawrence and Great Lakes region, killing half the peoples of the Huron and Iroquois confederations.[70] The English colonists, unlike the Spanish, wanted the land and not the labor of the Indians, but they too saw the hand of God in this catastrophe. Increase Mather wrote in 1631, "About this time the Indians began to be quarrelsome touching the Bounds of the Land which they had sold to the English, but God ended the Controversy by sending the Smallpox amongst the Indians of Saugust, who were before that exceedingly numerous."[71] Three years later, John Winthrop, the first governor of the Massachusetts Bay Colony, noted, "For the natives, they are neere all dead of small Poxe, so that the Lord hathe cleared our title to what we possess."[72]

Epidemics continued to decimate the indigenous peoples of North America throughout the eighteenth and nineteenth centuries. Smallpox killed half the Cherokees in 1738 and almost half the Catawbas in 1759.

For centuries smallpox had been a scourge around the world. Though Europeans suffered much less than Indians, they were far from immune to this dread disease. In areas of dense population, the disease was endemic and affected mainly children. Small towns and rural areas suffered from periodic epidemics. Middle Eastern, African, and Asian peoples had long developed methods of reducing the incidence of this disease through inoculation, that is, by introducing matter from a pustule of a smallpox patient into a small cut in the skin of a healthy person. Most people thus inoculated suffered only a mild case of the disease, though some died from it. This practice was imported to England in 1721, and quickly became popular in the aristocracy. By the 1770s, it was in general use among rural and small-town people; people in London and on the European continent resisted the practice much longer. In North America, inoculation was popular among whites, most of whom lived in small towns and isolated settlements and were thus vulnerable to epidemics. During the American War of Independence, General George Washington ordered the inoculation of his troops. The practice of inoculation helps explain the rise in the European population of North America in the eighteenth century.[73] In this, as in later triumphs of medicine against disease, we see a divergence between its effects on Europeans and on non-Europeans. Nowhere was this more evident than in the Americas.

Then, at the end of the eighteenth century, came another breakthrough in the age-old war between humans and infectious diseases. In the 1790s Dr. Edward Jenner, an English physician, introduced vaccination, or inoculation with cowpox. This method quickly became popular in Europe, for it was much less dangerous than inoculation of pus from a smallpox victim. It spread to the United States after 1800, thanks to the efforts of Benjamin Waterhouse and the support of Thomas Jefferson. Popular acceptance of the new method came slowly and epidemics still broke out periodically in Philadelphia, Baltimore, New York, Quebec, and other cities throughout the nineteenth century. Yet these outbreaks were localized and the patients quarantined.[74]

Among the Indians the disease proved much more devastating than among whites, and far more than it had been in the past. Thanks to horses, there was much more contact among the Indian tribes than there had been in earlier centuries, when they traveled on foot and were more isolated. Thus, what were local epidemics among European Americans became continent-wide pandemics among Native Americans. An epidemic in the 1770s and early 1780s killed 50 to 60 percent of the Cree, Arikara, Mandan, and Crow Indians; the Shoshone, Comanche, and Hidatsas were also hard-hit. White traders and soldiers on the frontier helped spread the disease by giving the Indians blankets infected with pustules. General Thomas Gage authorized reimbursements for "Sundries" used "to convey the Smallpox to the Indians." When George Vancouver explored Puget Sound in 1782–83, he found piles of bones and survivors with pockmarked faces.[75]

The first pandemic of the nineteenth century began in 1801 and spread from the Gulf of Mexico to the Northwest Coast and is said to have killed two-thirds of the Indian population of that region.[76] President Jefferson proposed to bring the benefits of vaccination to the Indians. Meriwether Lewis and William Clark, in their expedition across the continent, carried pustules of cowpox with them, to little effect. In 1832 the United States Congress appropriated twelve thousand dollars for the vaccination of Indians. Many Indians, however, were wary of whites and refused to be vaccinated, while others lived beyond the reach of European medicine.

The best-documented smallpox epidemic is the one that attacked the peoples of the Great Plains in 1837–38, for by then white fur traders, prospectors, and settlers were coming through the area in large numbers. In April 1837, the steamer *St. Peter's* left St. Louis on its annual journey up the Missouri River to resupply the Pratte and Chouteau Company's trading posts and bring back the furs and bison pelts collected since its last visit. Onboard was the smallpox virus, which ignited a pandemic that spread throughout the North American continent west of the Mississippi River, from New Mexico to northern Canada. The Assiniboine lost one-third to one-half of their people, the Arikaras half, the Blackfoot confederation half to two-thirds. The Osage, Choctaws, Comanches, Apaches, Pueblos, and Kiowas were decimated. The Mandans, once the most successful traders and farmers in the Midwest, were surrounded by Sioux warriors when the epidemic hit, so the healthy ones could not flee. Of sixteen hundred to two thousand Mandans, only one hundred survived, and their identity as a tribe vanished.[77]

An unsigned letter written from New Orleans in June 1838 describes the effects of the epidemic on the Indians.

> We have, from the trading posts on the western frontier of the Missouri, the most frightful accounts of the ravages of the small pox among the Indians. The destroying angel has visited the unfortunate

sons of the wilderness with terrors never before known, and has converted the extensive hunting grounds, as well as the peaceful settlements, of those tribes, into desolate and boundless cemeteries. The
number of the victims within a few months is estimated at 30,000,
and the pestilence is still spreading. The warlike spirit which but
lately animated the several Indian tribes, and but a few months ago
gave reason to apprehend the breaking-out of a sanguinary war, is
broken. The mighty warriors are now the prey of the greedy wolves
of the prairie, and the few survivors, in mute despair, throw themselves on the pity of the Whites, who, however, can do but little to
help them. . . . Every thought of war was dispelled, and the few that
are left are as humble as famished dogs.[78]

Nor was that the last of the disasters that visited the North American Indians during the nineteenth century. From the 1830s to the
1860s, several more epidemics struck parts of the continent. In 1849
a new disease, cholera, appeared in Indian territory, brought by
white migrants on the Oregon Trail; many tribes, already hard-hit
by smallpox, were decimated. California alone suffered four epidemics in quick succession, almost emptying that region of Indians before the great gold rush of 1849.[79]

If the population of the Americas before Columbus is a hotly debated mystery, so is the demographic impact of the encounter. Many
authors state that the population of Mexico reached its nadir of 1.6
million around 1650, a decline of 75 to 90 percent. The demographer Russell Thornton estimated that the Western Hemisphere lost
about 94 percent of its native population. The population of what
was later the contiguous forty-eight states of the United States fell
from over five million in 1492 to 250,000 at the end of the nineteenth century, a decline of 93 percent.[80] Whatever the figures, there
is no question that the epidemics that afflicted the peoples of the
Americas were the worst catastrophe that has ever befallen the
human race.

Not all of the sufferings of the Indians were the result of diseases.
They also died in wars, both against Europeans and among themselves. Many were enslaved, torn from their families, forced to mi-
grate to places where they could not survive, or forced to work under
terrible conditions. Yet none of these conditions can be counted

among the causes of the depopulation of the Americas, for the Europeans were no kinder to the peoples of the Philippines or to the slaves they imported from Africa, none of whom saw their numbers decline.

The other side of the demographic coin is that the mainland of the Americas became populated by Europeans. In some places, such as Mexico and Peru, enough Indians survived and had offspring with Europeans so that the population became largely mestizo. In the "Southern Cone"—Argentina, Chile, and Uruguay—and north of the Rio Grande, people of European origin soon outnumbered all others, transforming those lands into what Alfred Crosby called "Neo-Europes." This was an entirely different kind of imperialism from the domination that Europe imposed on Africa and South and Southeast Asia in the nineteenth century. That political and economic imperialism did not last long and did not replace one population with another. In the Americas, in contrast, imperialism was ecological—to use Crosby's term—and, from the human perspective, racial and demographic.

In the demographic history of the New World, one exception stands out: the West Indies and the tropical lowlands of northern South America and southeastern North America, where the population is substantially, in places overwhelmingly, of African descent. It was not always so. In the first century after Columbus, as the Indians died off, they were replaced by Europeans. In the early seventeenth century, the French and English who had wrested western Hispaniola, Jamaica, Barbados, and other parts of the Caribbean from Spain expected these lands to be settled by European indentured servants. They soon realized, however, that European newcomers suffered from a higher mortality rate than newly imported African slaves. Furthermore, Europeans, brought over on indentured labor contracts, became free after seven years, whereas Africans were slaves for life. By the mid-seventeenth century, it was more profitable to import African slaves than Europeans, and planters imported Africans in such numbers that it changed the composition of the population.[81] Epidemics of plague in 1647–49 and of yellow fever in 1690 contributed to the demographic change.[82] Slavery and the plantation complex that produced sugar, cotton, rice, and other tropical crops were the root cause of this transformation. Yet slavery, or something close to it, existed in the silver mines of Mexico and Peru without an influx of Africans. The explanation for this disparity lies, once again, in disease, in this case yellow fever.

The homeland of yellow fever is West Africa. It is spread by the female *Aedes aegypti* mosquito, which requires standing water to breed in and temperatures above sixty degrees Fahrenheit to feed and above seventy-five degrees to multiply. A. *aegypti* is a domestic mosquito that seldom flies more than three hundred yards from its place of birth, unless it gets a ride on a ship. Hence the disease was restricted to the humid tropics where the mosquitoes live, with occasional forays into the port cities of North America during the summer months. In areas with a sufficiently dense population of humans and forest monkeys, it is endemic. In such areas, everyone gets infected in childhood, when the symptoms are generally mild. Those who survive gain a lifetime immunity. Elsewhere, the disease dies out, then returns in epidemics. In such places, when adults are infected, they are infective for three to six days, during which an A. *aegypti* mosquito must first bite the infected person, then bite a non-immune person in the vicinity. If all goes well for the virus, it will infect other humans. For the disease to spread, it needs many mosquitoes and many non-immune humans in the same place. The case mortality is then as high as 85 percent. Outside of West Africa, yellow fever epidemics were rare but devastating, and newly arrived adult males were the most vulnerable of all. The peculiar nature of the disease played a major role in the history and demography of the tropical lowlands of the Americas.[83]

The first case of yellow fever appeared in Barbados in 1647, then quickly spread to Cuba, Yucatan, Guadeloupe, and St. Kitts. Sugar plantations were especially advantageous to the mosquitoes and the virus they carried. To spread, the disease needed large numbers of non-immune people crowded together. Besides blood, mosquitoes also consumed sucrose. And they bred best in small containers, such as the clay pots that were used to separate the crystallized sugar from the molasses for three or four months after the harvest and were left lying around the rest of the year.[84]

In every epidemic, large numbers of Europeans died and the rest fled the islands. Those who remained acquired ever more African slaves, most of whom had been immune since childhood. Occasionally, however, one would be a carrier, igniting an epidemic among the non-immunes in the Caribbean. This shaped the history of the Caribbean—both political and demographic—for the next two and a half centuries.

In the late seventeenth and eighteenth centuries, sugar replaced spices and silver as the greatest source of profits in the European colonial empires. So desirable were the areas where sugar grew well that the great powers repeatedly attacked one another for possession of the islands and tropical lowlands of the Caribbean basin. In 1655 England sent an expedition of seven thousand men that took Jamaica from Spain in a week. Within a few months of this victory, almost half the English soldiers had died and of the rest, half were sick; thereafter the English garrison lost 20 percent of its men every year. Later expeditions were not so lucky. The English attacked Guadeloupe in 1689 and Martinique in 1693, but both attempts failed because their soldiers succumbed to yellow fever. In 1694 an Anglo-Spanish force lost 61 percent of its troops during the abortive attack on French Saint-Domingue (now Haiti). In 1739 Admiral Edward Vernon took Portobelo and Chagres on the Isthmus of Panama with twenty-five thousand men and in 1741 he tried, but failed, to take Cartagena in Colombia. He then attacked Santiago de Cuba, and once again had to withdraw, having lost three-quarters of his men. In 1762 Admiral George Pocock and fourteen thousand men took Havana after a nine-week siege, only to find 41 percent of his men had died and another 37 percent were ill, leaving only a fifth of his men fit to bear arms; soon thereafter Britain returned Havana to Spain. Thus did yellow fever protect Spain's strongpoints in the Caribbean from its enemies.

In one place, however, yellow fever turned against the colonial rulers. By the end of the eighteenth century, Saint-Domingue, the richest of all the sugar islands, had a population that was overwhelmingly of African origin. In the early 1790s, the idea of freedom, introduced by the French revolutionaries, set off an uprising against the white settlers and plantation owners, soon led by the former slave

Toussaint Louverture. First the British, then the French tried to suppress the slave revolt. In 1794 British troops occupied the major ports but lost fifty thousand men, mostly from yellow fever. In 1802 Napoleon sent an army of twenty-five thousand to suppress the revolt and reimpose slavery, but most died of yellow fever and the rest were defeated by the Haitian revolutionaries. Few returned.[85]

Conclusion

What conclusions can we draw from the encounters between Europeans and Native Americans? In particular, to what extent do technological and environmental factors help us understand the contrast between the fate of the Aztecs and Incas, on the one hand, and that of the Indians of Chile, Argentina, and the Great Plains, on the other?

The Aztecs and Incas were military peoples who ruled over many other subject peoples. Their societies were very hierarchical, with an all-powerful ruler, an aristocracy, and a population of farmers. They built great cities, temples, and roads. In many ways, their civilizations resembled those of ancient Egypt, Babylonia, Rome, and Han China. Until the Spaniards arrived, they were also isolated from one another and from technologically more advanced peoples. Hence, when they encountered strange people from across the ocean, they were taken by surprise. Worse yet, they faced other and far more destructive enemies than Europeans: horrific diseases against which they had neither immunities nor remedies. Their societies, once decapitated by the capture or death of their rulers, were thrown into confusion.

The Indians of northern Mexico, the Great Plains, and southern South America, in contrast, had no emperors and no governments. The autonomy of their warrior clans allowed them to recover from setbacks. They were of course vulnerable to the same diseases as other Indians, and their numbers were depleted by smallpox, but not as rapidly. Their low population density and scattered habitations gave them an epidemiological advantage over the dense populations of central Mexico and the Peruvian Altiplano, for most Old World ills were crowd diseases that had a less severe and more drawn-out impact on hunter-gatherers than on agricultural and urban peoples.[86] The Araucanians and the Indians of the Pampas and the Great Plains also benefited from the fact that their territory was invaded in stages separated by long intervals. Unlike the Aztecs and Incas, they had both the time and the flexibility to adapt their warfare to new circumstances. They acquired new weapons and developed new tactics. In particular, they stole and raised horses, the most powerful weapon of the Spaniards. By the late sixteenth century, they had mastered the Spanish art of war and were able to resist the European advance until the late nineteenth century.

The technology and the diseases that Europeans brought to the Americas gave them the power to conquer Mexico and Peru. The fact that they did not—indeed could not—conquer the last two-thirds of the Americas until well into the nineteenth century demonstrates how ephemeral was the power that their technology had given them over the other peoples of the Americas.

Notes

1. John J. Johnson, "The Introduction of the Horse into the Western Hemisphere," *Hispanic American Historical Review* 23 (November 1942), pp. 587–610.

2. Pablo Martín Gómez, *Hombres y armas en la conquista de México* (Madrid: Almena, 2001), pp. 67–70; Johnson, "Introduction of the Horse," p. 599.

3. John Grier Varner and Jeanette Johnson Varner, *Dogs of the Conquest* (Norman: University of Oklahoma Press, 1983), pp. 4–8, 61–66; Alberto Mario Salas, *Las armas de la conquista* (Buenos Aires: Emecé, 1950), pp. 159–66.

4. William H. McNeill, *Plagues and Peoples* (Garden City, N.Y.: Doubleday, 1976), chapter 3: "Confluence of Disease Pools, 500 B.C.–A.D. 1200." In exchange, the Americas gave the Old World syphilis; see Kristin H. Harper et al., "On the Origin of the Trepanematoses: A Phylogenetic Approach," *PLoS: Neglected Tropical Diseases* (January 2008) in http://www.plosntds.org (accessed January 28, 2008).

5. Alfred Crosby, *Ecological Imperialism: The Biological Expansion of Europe, 900–1900* (Cambridge: Cambridge University Press, 1986), pp. 197–98; Suzanne Austin Alchon, *A Pest in the Land: New World Epidemics in a Global Perspective* (Albuquerque: University of New Mexico Press, 2003), p. 15.

6. Jared Diamond, *Guns, Germs, and Steel: The Fates of Human Societies* (New York: Norton, 1997), pp. 212–13; McNeill, *Plagues and Peoples*, p. 201.

7. Elizabeth Fenn, *Pox Americana: The Great Smallpox Epidemic of 1775–1782* (New York: Hill and Wang, 2001), pp. 25–28.

8. The same is true of the inhabitants of islands that had been isolated from the Old World disease pool for a long time, such as the Polynesians or the Guanches of the Canary Islands; see Crosby, *Ecological Imperialism*, pp. 92–94 and chapter 10.

9. Russell Thornton, *American Indian Holocaust and Survival* (Norman: University of Oklahoma Press, 1987), pp. 22–25; John W. Verano and Douglas H. Uberlaker, eds., *Disease and Demography in the Americas* (Washington, D.C.: Smithsonian Institution Press, 1992), pp. 171–74. On the dangers of speculative demography, see David Henige, *Numbers from Nowhere: The American Indian Contact Population Debate* (Norman: University of Oklahoma Press, 1998), and Woodrow Borah, "The Historical Demography of Aboriginal and Colonial America: An Attempt at Perspective," chapter 1 in William M. Denevan, ed., *The Native Population of the Americas in 1492*, 2nd ed. (Madison: University of Wisconsin Press, 1992).

10. *The Book of Chilam Balam of Chumayel*, ed. and trans. Ralph L. Roy (Washington, D.C.: Carnegie Institute of Washington, 1933), p. 83, quoted in Noble David Cook and W. George Lovell, "Unraveling the Web of Disease," in Noble David Cook and W. George Lovell, eds., *"Secret Judgment of God": Old World Disease in Colonial Spanish America* (Norman: University of Oklahoma Press, 1992), p. 213.

11. Thornton, *American Indian Holocaust*, pp. 22–25, 40.

12. Angel Rosenblat, "The Population of Hispaniola at the Time of Columbus," chapter 2 in Denevan, *Native Population*. Noble David Cook estimates put the precontact population of Hispaniola at circa half a million; see *Born to Die: Disease and New World Conquest, 1492–1650* (Cambridge: Cambridge University Press, 1998), pp. 23–24.

13. Kenneth F. Kiple and Brian T. Higgins, "Yellow Fever and the Africanization of the Caribbean," in Verano and Uberlaker, *Disease and Demography*, p. 237; Cook and Lovell, *"Secret Judgment,"* pp. 213–42.

14. On smallpox, see Donald R. Hopkins, *The Greatest Killer: Smallpox in History* (Chicago: University of Chicago Press, 2002), pp. 204–5; Sheldon J. Watts, *Epidemics and History: Disease, Power and Imperialism* (New Haven: Yale University Press, 1997), chapter 3; Alfred W. Crosby, *The Columbian Exchange: Biological and Cultural Consequences of 1492* (Westport, Conn.: Greenwood Press, 1972), pp. 45–47; and McNeill, *Plagues and Peoples*, 206–7.

15. Robert McCaa, "Spanish and Nahuatl Views on Smallpox and Demographic Catastrophe in Mexico," in Robert I. Rotberg, ed., *Health and Disease in Human History* (Cambridge, Mass.: MIT Press, 2000), p. 175; Crosby, *Columbian Exchange*, pp. 44–49; Cook, *Born to Die*, pp. 23–24.

16. Fray Bernardino de Sahagún, *Codex Florentino*, quoted in Miguel Léon-Portilla, ed., *The Broken Spears: The Aztec Account of the Conquest of Mexico* (Boston: Beacon Press, 1992), pp. 38–39.

17. On these aspects, see Ross Hassig, *Aztec Warfare: Imperial Expansion and Political Control* (Norman: University of Oklahoma Press, 1988), p. 242.

18. On Aztec weapons and armor, see ibid., pp. 75–86, and Martín Gómez, *Hombres y armas*, pp. 22–24.

19. Hassig, *Aztec Warfare*, pp. 237–38; Martín Gómez, *Hombres y armas*, pp. 18–25.

20. Bernal Díaz del Castillo, *The True History of the Conquest of New Spain*, trans. and ed. Alfred Percival Maudslay, 5 vols. (London: Hakluyt Society, 1908–16), vol. 2, p. 127.

21. Quoted in Salas, *Las armas de la conquista*, p. 159.

22. Alan Knight, *Mexico: From the Beginning to the Spanish Conquest* (Cambridge: Cambridge University Press, 2002), pp. 229–31; Martín Gómez, *Hombres y armas*, pp. 59–64, 114–16.

23. Ross Hassig, *Mexico and the Spanish Conquest* (New York: Longman, 1994), p. 149.

24. William W. Greener, *The Gun and Its Development*, 9th ed. (New York: Bonanza, 1910), pp. 54–61; Hassig, *Mexico*, p. 38; Hassig, *Aztec Warfare*, pp. 237–38; Martín Gómez, *Hombres y armas*, pp. 99–101. On sixteenth-century artillery, see Joseph Jobé, *Guns: An Illustrated History of Artillery* (Greenwich, Conn.: New York Graphic Society, 1971), pp. 29–32.

25. Díaz del Castillo, *True History*, vol. 2, pp. 244–46.

26. Sahagún, *Codex Florentino*, quoted in Léon-Portilla, *Broken Spears*, p. 93.

27. Hanns Prem, "Disease Outbreaks in Central Mexico during the Sixteenth Century," in Cook and Lovell, *"Secret Judgment,"* pp. 24–26; McCaa, "Spanish and Nahuatl Views," p. 169.

28. Crosby, *Columbian Exchange*, p. 48.

29. Hassig, *Mexico*, pp. 101–2.

30. On the siege of Tenochtitlán, see Martín Gómez, *Hombres y armas*, pp. 146–51, and Hassig, *Mexico*, pp. 121–49.

31. The major work on the Spanish ships is Clinton H. Gardiner, *Naval Power in the Conquest of Mexico* (Austin: University of Texas Press, 1956), especially pp. 129–79. See also Martín Gómez, *Hombres y armas*, pp. 121–52, and Hassig, *Mexico*, 121–49.

32. McCaa, "Spanish and Nahuatl Views," p. 193.

33. Noble David Cook, "Impact of Disease in the Sixteenth-Century Andean World," in Verano and Uberlaker, *Disease and Demography*, p. 210; Alfred Crosby, *Germs, Seeds and Animals: Studies in Ecological History* (Armonk, N.Y.: M. E. Sharpe, 1994), pp. 100–101; Crosby, *Columbian Exchange*, pp. 38–43; Prem, "Disease Outbreaks," pp. 20–48; Thornton, *American Indian Holocaust*, pp. 44–45; McCaa, "Spanish and Nahuatl Views," pp. 184–97; McNeill, *Plagues and Peoples*, pp. 209–10.

34. McCaa, "Spanish and Nahuatl Views," pp. 190–91.

35. The original account of the conquest is by Bernal Díaz del Castillo. The most famous (though dated) English-language account of the conquest is William H. Prescott, *History of the Conquest of Mexico* (New York: Hooper, Clark and

Company, 1843, with many later editions). The best recent work is Hassig, *Mexico*, but see also Martín Gómez, *Hombres y armas*, and León-Portilla, *Broken Spears*.

36. Philip W. Powell, *Soldiers, Indians, and Silver: The Northward Advance of New Spain* (Berkeley: University of California Press, 1952), pp. 10–14; John F. Richards, *The Unending Frontier: An Environmental History of the Early Modern World* (Berkeley: University of California Press, 2003), p. 354.

37. Quoted in Powell, *Soldiers, Indians, and Silver*, p. 50.

38. John F. Guilmartin, Jr., "The Cutting Edge: An Analysis of the Spanish Invasion and Overthrow of the Inca Empire, 1532–1539," in Kenneth Andrien and Rolena Adorno, eds., *Transatlantic Encounters: Europeans and Andeans in the Sixteenth Century* (Berkeley: University of California Press, 1991), pp. 41–48; John Hemming, *The Conquest of the Incas* (New York: Harcourt Brace Jovanovich, 1973), pp. 5–27.

39. On the smallpox epidemic in the Andes, see Cook, "Impact of Disease," pp. 207–8; Crosby, *Columbian Exchange*, pp. 52–55; Hopkins, *The Greatest Killer*, pp. 208–10; and Diamond, *Guns*, pp. 77–88.

40. Hemming, *Conquest of the Incas* passim. On the events of Cajamarca, see Diamond, *Guns*, pp. 67–81.

41. Hemming, *Conquest of the Incas*, p. 112; Salas (*Las armas de la conquista*, p. 138) gives the cost of a horse as 1,000 to 4,000 gold pesos, compared to eight pesos for a sword, three for a dagger, and one for a lance.

42. Inca Garcilaso de la Vega, *Comentarios reales de los Incas*, ed. Carlos Araníbar (Lima and Madrid: Fondo de Cultura Económica, 1991), vol. 1, p. 158.

43. Guilmartin, "The Cutting Edge," pp. 41–61; Hemming, *Conquest of the Incas*, 107–16, 192–95.

44. Ricardo E. Latcham, *La capacidad guerrera de los Araucanos: Sus armas y métodos militares* (Santiago de Chile: Imprenta Universitaria, 1915), pp. 4–25; A. Jara, *Guerre et société au Chili: Essai de sociologie coloniale: La transformation de la guerre d'Araucanie et l'esclavage des Indiens, du début de la conquête espagnole aux débuts de l'esclavage légal* (Paris: Institut des études de l'Amérique latine, 1961), pp. 51–61; John M. Cooper, "The Araucanians," in Julian H. Stewart, ed. *Handbook of South American Indians* (Washington, D.C.: GPO, 1946–59), vol. 2, pp. 687–760; Sergio Villalobos R., *Vida fronteriza en la Araucanía: El mito de la Guerra de Arauco* (Barcelona and Santiago de Chile: Editorial Andrés Bello, 1995), p. 27; Brian Loveman, *Chile: The Legacy of Hispanic Capitalism* (New York: Oxford University Press, 1988), pp. 53–59.

45. Jaime Eyzaguirre, *Historia de Chile* (Santiago: Zig-Zag, 1982), pp. 69–73; "Araucanos," in *Enciclopedia Universal Ilustrada Europeo-Americana*, vol. 5, p. 1233; "Chile" in ibid., vol. 17, p. 345; Latcham, *La capacidad guerrera*, pp. 12–33.

46. Louis de Armond, "Frontier Warfare in Colonial Chile," *Pacific Historical Review* 33 (1954), pp. 126–28; Latcham, *La capacidad guerrera*, pp. 27–35, 48–50; Jara, *Guerre et société*, p. 62.

47. Armond, "Frontier Warfare," pp. 125–28; Patricia Cerda-Hegerl, *Fronteras del Sur: La región del Bío Bío y la Araucanía chilena, 1604–1833* (Temuco, Chile: Ediciones Universidad de la Frontera, 1996), pp. 13–14; Jara, *Guerre et société*, pp. 63–68; Latcham, *La capacidad guerrera*, pp. 36–38.

48. Armond, "Frontier Warfare," pp. 129–31.

49. Sergio Villalobos R., *La vida fronteriza en Chile* (Madrid: Editorial MAPFRE, 1992), pp. 13–14; Jorge Pinto Rodríguez, *Araucanía y pampas: Un mundo fronterizo en América del Sur* (Temuco, Chile: Ediciones Universidad de la Frontera, 1996), pp. 15–21; Cerda-Hegerl, *Fronteras del Sur*, pp. 13–29; Armond, "Frontier Warfare," pp. 130–32; Hemming, *Conquest of the Incas*, p. 461; Latcham, *La capacidad guerrera*, pp. 50–51.

50. Fernando Casanueva, "Smallpox and War in Southern Chile in the Late Eighteenth Century," in Cook and Lovell, *"Secret Judgment,"* pp. 183–212; Villalobos, *Vida fronteriza*, pp. 8–15, 35–36; Latcham, *La capacidad guerrera*, pp. 60–68; Cerda-Hegerl, *Fronteras del Sur*, pp. 15–17; Cooper, "The Araucanians," vol. 2, p. 696.

51. Rómulo Muñiz, *Los indios pampas* (Buenos Aires: Editorial Bragado, 1966), pp. 20–21; Alfred J. Tapson, "Indian Warfare on the Pampas during the Colonial Period," *Hispanic American Historical Review* 42 (February 1962), pp. 2–3.

52. Madaline W. Nichols, "The Spanish Horse of the Pampas," *American Anthropologist* 41, no. 1 (1939), pp. 119–29; Dionisio Schoo Lastra, *El indio del desierto, 1535–1879* (Buenos Aires: Editorial Goncourt, 1977), pp. 23–26; Carlos Villafuerte, *Indios y gauchos en las pampas del sur* (Buenos Aires: Corregidor, 1989), pp. 16–18; Prudencio de la C. Mendoza, *Historia de la ganadería argentina* (Buenos Aires: Ministerio de Agricultura, 1928), pp. 11–14.

53. Tapson, "Indian Warfare on the Pampas," p. 5n21.

54. Robert M. Denhardt, *The Horse of the Americas*, rev. ed. (Norman: University of Oklahoma Press, 1975), pp. 171–75; Villafuerte, *Indios y gauchos*, p. 19; Muñiz, *Los indios pampas*, p. 36; Nichols, "Spanish Horse of the Pampas," pp. 127–29.

55. Felix de Azara, *The Natural History of the Quadrupeds of Paraguay and the River La Plata*, trans. W. Perceval Hunter (Edinburgh: A. & C. Black, 1838), pp. 13–14; Villafuerte, *Indios y gauchos*, pp. 20–26; Muñiz, *Los indios pampas*, pp. 36–38; Mendoza, *Historia*, p. 13; Tapson, "Indian Warfare on the Pampas," pp. 5–6.

56. Salvador Canals Frau, *Las poblaciones indígenas de la Argentina: Su orígen, su pasado, su presente* (Buenos Aires: Editorial Sudamericana, 1953), pp. 534–38; Frau, "Expansion of the Araucanians in Argentina," in *Handbook of South American Indians*, vol. 2, pp. 761–66; Cooper, "The Araucanians," vol. 2, p. 688; Tapson, "Indian Warfare on the Pampas," p. 6; Villafuerte, *Indios y gauchos*, p. 26; Nichols, "Spanish Horse of the Pampas," p. 129.

57. Tapson, "Indian Warfare on the Pampas," pp. 1–27.

58. Peter Farb, *Man's Rise to Civilization as Shown by the Indians of North America from Primeval Time to the Coming of the Industrial State*, 2nd ed. (New York: Bantam, 1978), p. 113.

59. Clark Wissler, "The Influence of the Horse in the Development of Plains Culture," *American Anthropologist* 16, no. 1 (1914), pp. 1–25; Francis Haines, "Where Did the Plains Indians Get Their Horses?" *American Anthropologist* 40, no. 1 (1938), pp. 112–17.

60. Bernard Mishkin, "Rank and Warfare among the Plains Indians," *Monographs of the American Ethnological Society*, no. 3 (Lincoln: University of Nebraska Press, 1992), pp. 5–6; Bradley Smith, *The Horse in the West* (New York: World, 1969), p. 16.

61. Frank Raymond Secoy, *Changing Military Patterns on the Great Plains (17th Century through Early 19th Century)* (Locust Valley, N.Y.: Augustin, 1953), pp. 20–38; Frank Gilbert Roe, *The Indian and the Horse* (Norman: University of Oklahoma Press, 1955), pp. 72–122; Theodore Binnema, *Common and Contested Ground: A Human and Environmental History of the Northwest Plains* (Norman: University of Oklahoma Press, 2001), pp. 86–106. See also Farb, *Man's Rise to Civilization*, p. 115; Denhardt, *The Horse*, pp. 92–111; and Smith, *The Horse in the West*, p. 14.

62. Colin G. Calloway, "The Inter-tribal Balance of Power on the Great Plains, 1760–1850," *Journal of American Studies* 16 (April 1982), pp. 25–48.

63. Walter Prescott Webb, *The Great Plains* (New York: Grosset and Dunlap, 1931), pp. 58–67; Mishkin, "Rank and Warfare," pp. 10–12, 57–60; Roe, *Indian and the Horse*, pp. 219–32.

64. Webb, *The Great Plains*, p. 138.

65. Crosby, *Columbian Exchange*, p. 104.

66. Mishkin, "Rank and Warfare," p. 5.

67. Dauril Alden and Joseph C. Miller, "Out of Africa: The Slave Trade and the Transmission of Smallpox to Brazil, 1560–1831," in Rotberg, *Health and Disease*, pp. 203–30; Hopkins, *The Greatest Killer*, pp. 215–19.

68. Quoted in Tapson, "Indian Warfare on the Pampas," p. 4.

69. Quoted in Crosby, *Ecological Imperialism*, pp. 209–15.

70. Ibid., p. 202; Crosby, *Columbian Exchange*, pp. 40–41.

71. Quoted in Thornton, *American Indian Holocaust*, p. 75.

72. Quoted in Crosby, *Ecological Imperialism*, p. 208.

73. McNeill, *Plagues and Peoples*, pp. 249–51.

74. Hopkins, *The Greatest Killer*, pp. 262–69.

75. Fenn, *Pox Americana*, pp. 88–89, 210–23; Crosby, *Germs, Seeds and Animals*, p. 98; Crosby, *Ecological Imperialism*, p. 203; Thornton, *American Indian Holocaust*, pp. 91–94; Calloway, "Inter-tribal Balance of Power," pp. 41–43.

76. Esther W. Stearn and Allen E. Stearn, *The Effect of Smallpox on the Destiny of the Amerindian* (Boston: Bruce Humphreys, 1945), p. 74.

77. R. G. Robertson, *Rotting Face: Smallpox and the American Indian* (Caldwell, Idaho: Caxton Press, 2001), pp. 239–311; see also Thornton, *American Indian Holocaust*, pp. 94–96. Hopkins (*The Greatest Killer*, p. 271) says only twenty-seven Mandans survived.

78. Stearn and Stearn, *Effect of Smallpox*, pp. 89–90.

79. Hopkins, *The Greatest Killer*, pp. 273–74; Calloway, "Inter-tribal Balance of Power," p. 46.

80. Thornton, *American Indian Holocaust*, p. 42.

81. Philip D. Curtin, *The Rise and Fall of the Plantation Complex* (Cambridge: Cambridge University Press, 1990), pp. 79–81.

82. David Watts, *The West Indies: Patterns of Development, Culture and Environmental Change since 1492* (Cambridge: Cambridge University Press, 1987), pp. 215, 225, 353.

83. John R. McNeill, "Ecology, Epidemics and Empires: Environmental Change and the Geopolitics of Tropical America, 1600–1825," *Environment and History* 5, no. 2 (1999), pp. 175–84; McNeill, "Yellow Jack and Geopolitics: Environments, Epidemics, and the Struggles for Empire in the American Tropics, 1640–1830," in Alf Hornborg, J. R. McNeill, and Joan Martínez-Alier, eds., *Rethinking Environmental History: World-System History and Global Environmental Change* (Lanham, Md.: Altamira Press, 2007), pp. 199–217; Kiple and Higgins, "Yellow Fever," p. 239. Oddly, yellow fever was unknown in tropical Asia and rarely seen in Europe; see Philip Curtin, *Death by Migration: Europe's Encounter with the Tropical World in the Nineteenth Century* (Cambridge: Cambridge University Press, 1989), pp. 17–18, 130.

84. J. McNeill, "Ecology," pp. 175–79; Kiple and Higgins, "Yellow Fever," pp. 239–45.

85. J. McNeill, "Ecology," pp. 180–81.

86. Hopkins, *The Greatest Killer*, pp. 213–14.

Chapter 4

The Limits of the Old Imperialism:
Africa and Asia to 1859

In the Americas, the conquistadors and later Europeans benefited not only from their temporarily superior technology but also from their greater resistance to the diseases that decimated the native populations. European eagerness to conquer was not limited to the New World, however. Monarchs, merchants, and missionaries were also attracted to Africa and India. Yet the history of the encounters between their inhabitants and the European interlopers contrasts sharply with that of the Americas. In India, European empire-builders were successful, but against increasing odds. In Afghanistan and sub-Saharan Africa, they failed. And in Algeria and the Caucasus, their success came at a huge price. The five cases described in this chapter illustrate the limits of the Old Imperialism of the early modern period, before the Industrial Revolution gave Western imperialism a new impetus.

Sub-Saharan Africa to 1830

Sailing along the west coast of Africa in the 1430s, the Portuguese reached Cape Verde in the 1440s, the Cape of Good Hope in 1488, and the east coast of the continent in 1497. Yet it was not until the mid-nineteenth century, some four centuries later, that Europeans succeeded in penetrating the interior of the continent. This extraordinary delay cannot be explained by a lack of motivation, for the Portuguese and other Europeans were just as eager to acquire African gold and silver, their leaders just as interested in conquering new lands and peoples, and their missionaries just as desirous to convert the heathen as the Spaniards were in the Americas. What stopped them was the African disease barrier.

Biologically speaking, in the New World, Europeans were an invasive species that occupied the lands left vacant by the demographic catastrophe that felled the Indians. In Africa, in contrast, they encountered an ecological barrier that was as impenetrable as the Americas were open and inviting, and remained so until the mid-nineteenth century. The history of Africa and the Americas, containing between them half the land area on the planet, was determined in large part by invisible microorganisms.

Africa is not only the homeland of the human race but also that of many diseases that have evolved in symbiosis with human hosts. Africans, in constant contact with other parts of the Eastern Hemisphere, suffered from many of the same diseases, such as smallpox and measles, as the peoples of Eurasia and the Mediterranean; only tuberculosis and pneumonia were less prevalent than in Eurasia. Yellow fever, though widespread, was an endemic and relatively benign childhood disease. In addition, Africans suffered from much higher infection rates of yaws, Guinea worm, trypanosomiasis (sleeping sickness), onchocerciasis (river blindness), schistosomiasis (liver flukes), and, most of all, malaria.[1] Three diseases—malaria, yellow fever, and trypanosomiasis—had a particularly strong impact on European imperialism before the nineteenth century.

Malaria comes in four varieties, depending on the *Plasmodium* or protozoan that causes it. *Plasmodium vivax*, common throughout Eurasia (and later the Americas), causes a debilitating but rarely fatal form of malaria. *Plasmodium malariae*, prevalent mainly around the Mediterranean, provokes a more serious and persistent fever, as does *Plasmodium ovale* in isolated parts of East Africa. *Plasmodium falciparum*, found throughout tropical Africa, is by far the deadliest of the four. This protozoan is transmitted by the female *Anopheles* mosquito, which lives only in tropical regions, both wet and dry. Eight to twenty-five days after an infected mosquito has punctured a person's skin, the fast-multiplying plasmodia cause liver, kidney, or respiratory distress. Among those infected for the first time—both new arrivals and newborn Africans—the fatality rate ranges from 25 to 75 percent. Unlike yellow fever, recovery from malaria does not provide immunity but only a resistance that diminishes over time. Persons constantly reinfected can remain healthy, if weakened; those who escape reinfection gradually lose their resistance and become vulnerable again, though not as much as the first time. Africans who carry the sickle-cell trait—a genetic response to malaria—are less vulnerable than people who do not.[2]

West Africa is also the homeland of yellow fever. This disease is restricted to the humid tropics where the *Aedes aegypti* mosquito lives. Most inhabitants of West Africa gained immunity in childhood, but newcomers to the continent, such as visiting Europeans, suffered from periodic but catastrophic epidemics. Yellow fever played an important role in the Caribbean and the tropical lowlands of the Americas in the seventeenth and eighteenth centuries. There, yellow fever epidemics were rare but devastating, and adult males newly arrived from Europe were the most vulnerable of all. As we saw in chapter 3, this disease was responsible for the Africanization of tropical America.[3]

Like yellow fever, trypanosomiasis is a disease of both humans and animals. The trypanosome, a protozoan, is transmitted by the tsetse flies *Glossina palpalis* near water courses and *G. tachinoides* in drier areas. Once in the body, the trypanosomes multiply in the bloodstream and lymph system. After several months they invade the cerebrospinal fluid and gradually destroy the nervous system, resulting in death some two years later. This disease helps explain the low human population densities in the environments that supported tsetse flies, the rain forests and moister savannas.

Human trypanosomiasis affected Africans and newcomers equally, and there is no evidence of epidemics from this early period. Another variety of trypanosomiasis called nagana, however, played a part in preventing a European invasion before the nineteenth century by killing cattle and horses. To this day, it is difficult to raise these animals in large parts of Africa. That is why the herds of native wild ungulates that are immune to this disease—zebras, kudus, wildebeests, and others—have not met the fate of the American bison and been replaced by cattle. It also explains the lack of pack and farm animals and their manure and the resulting low protein diet and poverty of the people who live in tsetse-infected areas.[4]

In the 1440s, the Portuguese built a fort at Arguim on the coast of what is now Mauretania. It was the first of many establishments built along the African coasts by the Portuguese and later by Dutch, French, and English traders. But before the nineteenth century, only the Portuguese tried to penetrate the interior as the Spaniards had done in the Americas.

The first place in Africa where the Portuguese attempted to conquer a colony was Angola. In 1485 Diogo Cão and several of his men paddled up the Congo River to the Yalala rapids but had to turn back when several men died *da doença* (of disease), probably malaria.[5] In the 1490s, the Portuguese established good relations with the kingdom of Kongo, even converting King Affonso I and his court to Christianity. Meanwhile, private traders were purchasing slaves all along the coast.

King Sebastião, who came of age in 1568, dreamed of finding in the interior of Africa a bonanza like the ones his rivals the Spaniards had discovered in the Americas. In 1571 he decided to send Paulo Dias de Novaes, the grandson of Bartholomeu Dias, to conquer territory south of the Congo River. Dias arrived in Angola in 1575 with a hundred settler families, four hundred soldiers, and twenty horses and founded the town of Luanda. He was poorly received by the private Portuguese slave traders who did business on the coast and by the Ngola, or king, of the Mbundu. War broke out in 1579. The following year Dias advanced up the Kwanza River and built a small fort at Makunde, sixty miles inland. Three years later another fort was built at Massangano, eighty miles up the river from Luanda. After Dias left, all further advance was halted. The remaining settlers encountered fierce resistance from the Mbundu and Imbangala warriors. By 1590 the first attempt to colonize the interior of Angola had failed.

In 1592 Portugal, now under the Spanish king Philip II, tried again, sending Francisco de Almeida with seven hundred men. After much opposition from slave traders, Jesuit priests, and Mbundu warriors, the Portuguese soldiers marched into the interior, found no silver or anything else of value, and retreated to the coast. From then until the nineteenth century, the Portuguese colony of Angola consisted of two coastal settlements, three small forts, a tenuous hold on a hundred miles of river, and little else. Instead of colonizing the land, the Portuguese stayed on the coast and engaged in sporadic fighting with their African neighbors or sent their sons by African women into the interior to purchase slaves to be shipped to Brazil.[6] In the end, their hold on Angola was more like that of the Spaniards in southern Chile than in Mexico or Peru.

Part of the reason for their lack of success was the resistance of the Africans. In some ways, the Angolan style of warfare resembled that of the Aztecs. Battles began with a volley of arrows, followed by hand-to-hand combat. Soldiers fought in open order, spread out so that they could use their swords. Unlike the Aztecs, however, Africans had iron swords and ax and spearheads. The Portuguese were equipped with steel swords and body armor like the Spaniards. Thus, Portuguese infantrymen and their African auxiliaries had no particular advantage over the Angolan warriors they encountered. Their cavalry, consisting of a dozen or so horses imported from Brazil, had no impact, probably because the horses soon died of nagana. Without pack or draft animals, armies had to carry their own supplies, greatly restricting their mobility.

Firearms were even less useful in Angola than in the Americas. Harquebuses and muskets were notoriously inaccurate and ineffective against dispersed formations of warriors. Artillery pieces were heavy and difficult to transport without draft animals; besides, they were of little use against people who had no fortified towns. By the seventeenth century Angolans could purchase muskets, but they were reluctant to use them, for their goal was to capture their enemies for sale to slave traders, not to injure or kill them.[7]

What weakened the Portuguese were the diseases they faced. Those who survived preferred to stay on the coast and send their agents into the interior. The advantages that the Spanish possessed in Mexico and Peru—horses, firearms, steel, and the disease ecology—were absent or of little benefit to the Portuguese in Angola. Instead of carving out a territorial empire based on peasant labor and precious metals, the Portuguese, like the other Europeans who followed them, used Africa as a source of slaves for their American colonies.

The weakness of the Portuguese in the African environment was even more striking in their other attempt at colonizing Africa: Mozambique. Their involvement in that land dates back to 1506, when a Portuguese fleet under Pero d'Anhaia built a small fort in Sofala, on the southern coast. A year later, they occupied Mozambique Island, further north. These naval bases, along with others at Malindi and Mombasa, were established to supply the ships of the *carreira da India* and to attack the Muslim merchants who traded along that coast.

At Sofala, the Portuguese officials obtained gold, but not enough to pay for the spices they wished to purchase in India and in the Spice Islands, where local merchants had no interest in European goods. Meanwhile, Portuguese fortune hunters and deserters drifted inland from Sofala to escape the government's forts and seek gold in the realm of Mwene Mutapa. In the course of the sixteenth century, the trickle of gold and the tales brought back from the interior by these *sertanejos* about gold and silver mines fired the imagination of the Portuguese crown.[8]

In 1568 Francisco Barreto, a nobleman close to the court, persuaded King Sebastião to let him lead an expedition to Mwene Mutapa to avenge the death of a Jesuit missionary and find the fabled gold mines. Barreto's expedition reached Mozambique in May 1570. After dawdling for a year and a half, he left Mozambique Island in November 1571 with a thousand soldiers, as well as horses, camels, and oxen pulling carts; it was the largest military force ever sent overseas by Portugal. By the time they reached Sena, 130 miles up the Zambezi River, men and horses were dying. Rather than attribute their deaths to an act of God, the surviving Portuguese accused the local Muslim merchants of poisoning their comrades and horses, and put them to death.[9]

Despite the losses, Barreto pushed on, fighting a battle against several thousand Mongas warriors. With their muskets and cannon, the Portuguese routed the Africans but gained only fifty cows. So many Portuguese and their animals were dying that they had to turn back without setting eyes on Mwene Mutapa or its mines. Barreto departed for the coast, leaving Vasco Homem in command of the remaining four hundred soldiers. When he returned in May 1573, he found that 150 more members of the expedition, including most of the officers, had died, while none remained in good health. Two weeks later, Barreto himself died. Homem and his remaining 180 men returned to the coast, most of them sick.

These setbacks did nothing to quell the government's craving for precious metals. A year after returning from Sena, Vasco Homem left Sofala, this time cutting across land to avoid the dangerous Zambezi valley. After a few near-mutinies and skirmishes with African warriors, he returned to Sofala with little gold or silver to show for his efforts. A third expedition up the river was equally futile, with almost none of the three hundred members returning alive.[10]

Nor were these the last attempts by Portugal to find an Eldorado in Africa. In 1609 Diogo Simões Madeira led an expedition into the interior, but was stopped by the Chombe whose army had eight thousand soldiers and 150 muskets obtained from Portuguese traders. In 1631 another uprising killed hundreds of Portuguese, leaving the outposts of Tete and Sena on the Zambezi reduced to thirteen and twenty Portuguese, respectively. In 1680 Portugal sent another seventy-eight men, women, and children as settlers to Mozambique, but they, too, succumbed to fever. Finally, a war with Changamire Dombo from 1684 to 1693 expelled the Portuguese, including the *sertanejos*, from most of the interior.[11]

As in Angola, the Portuguese failed to conquer the interior of Mozambique or even to keep settlers in or near the two ports of Sofala and Mozambique Island. Part of the reason was the resistance of the Africans. Like the Araucanians, they had iron weapons and adapted to the new kind of warfare introduced by Europeans, even on occasion obtaining and using muskets. Also like the Araucanians but unlike the Incas and Aztecs, their society consisted of fragmentary chieftaincies that could not be defeated by seizing one supreme leader. But the main reason was environmental; the diseases the Europeans encountered, malaria and nagana, weakened them far more than it weakened Africans. As the chronicler João de Barros wrote in 1557,

> But it seems that for our sins, or for some inscrutable judgement of God, in all the entrances of this great Ethiopia [Africa] that we navi-

gate along, He has placed a striking angel with a flaming sword of deadly fevers, who prevents us from penetrating into the interior to the spring of this garden, whence proceed these rivers of gold that flow to the sea in so many parts of our conquest.[12]

The African disease environment thwarted not only the Portuguese but all other Europeans who attempted to penetrate Africa. From the late seventeenth to the early nineteenth century, several European countries maintained trading posts along the West African coast, mainly to purchase slaves. The traders on the coast were almost as vulnerable as those who penetrated inland, and more so than slaves. Among the seventy-three European employees of the Royal African Company who arrived on the Gold Coast (now Ghana) in 1695–96, seven (or 10 percent) died in the first four months and thirty-one (42 percent) in the first year. In 1719–20, of sixty-nine who landed, twenty-nine (42 percent) died in the first four months and forty-four (64 percent) in the first year. Statistics at other trading posts on the West African coast show similar death rates. Overall, of every ten Europeans sent to Africa by the company, six died in their first year, two more between their second and seventh year, and only one returned to Britain. Yet the company never lacked for applicants, for it offered better wages than an unskilled man could earn in Britain and kept quiet about the dangers.[13] As the directors wrote in 1721,

> We are sorry to find the Mortallity & sickness has been so great amongst you, wch we presume might be occasioned by the rains coming so soon after your arrival, but as we hope those who have survived are by this time pretty well seasoned, we hope they as well as those since sent over will enjoy their healths better & be more capable of rendring the Compa. the service they were appointed for.[14]

Those who went inland were even more likely to die. In 1777–79, on his expedition to Delagoa Bay, William Bolts lost 132 out of 152 men. In 1805, Mungo Park's expedition to the upper Niger lost three-quarters of his men, including himself.[15] Captain James Tuckey ascended the Congo River in 1816, but died along with half his team. On Hugh Clapperton's expedition to the Niger in 1825–27, four-fifths of the men died.[16]

The death toll among Europeans in Africa continued well into the nineteenth century. Between 1819 and 1836, the death rate among British troops in Sierra Leone was 483 per thousand per year, in other words almost half the men died every year. On the Gold Coast, the annual death rate in 1823–26 reached 668.3 per thousand, that is, two-thirds died every year. Mortality in the Gambia, Senegal, and other parts of the coast was not much lower. In comparison, the death rate among British troops in Europe and North America was 15 to 20 per thousand; in Bengal it was 71.41; in the West Indies 85 to 130; and in the Netherlands East Indies, 170.[17] As historian Philip Curtin has pointed out, "The West African 'fever' environment was probably the most dangerous in the world to outsiders."[18] That is the reason Africa remained for Europeans a "dark continent," mysterious and impenetrable for four hundred years.

India to 1746

The European experience in India differed radically from that of both Africa and the Americas. The subcontinent was part of the Eurasian disease environment, which meant that Indians and Europeans were about equally susceptible to the same diseases. Though cholera was endemic in India and not in Europe, everyone in India was vulnerable. Until the eighteenth century, India was also at roughly the same level as Europe technologically. Indians had horses, iron and steel weapons, and firearms; only in ships did Europeans have an advantage, as we saw in chapter 2. Nonetheless, in the course of the eighteenth and early nineteenth centuries, India came, piece by piece, under British domination and proved to be as profitable to Britain as Mexico and Peru had been to Spain in earlier times. No simple ecological or mechanical explanation can help us understand this case of imperial expansion. Instead, to understand the history of India in the early modern era we must turn to explanations that stress the political and the sociocultural aspects of technology.

The Central Asian warlord Babur had invaded and conquered northern India in the early sixteenth century. For the next two centuries, most of the Indian subcontinent was dominated by the Mughals, a dynasty of Persian language and Muslim faith. Along with the Ottomans in the Middle East, the Safavids in Persia, and the Russians, theirs was one of the "gunpowder empires" of Eurasia in the early modern period.

Like most Indian rulers, the Mughals had little interest in the sea and tolerated the presence of European ships and towns along their coasts because of the trade that they brought. Portuguese attempts to expand their precarious footholds into the interior ended in failure. The English built a small fort in Madras in 1640, as did the French a few years later at Pondicherry, not to protect their traders from Indians but from the Dutch and Portuguese. The English settlement on the island of Bombay, founded in 1662, was precarious from the start. An attempt by Sir Josiah Child in 1688–90 to seize a piece of the mainland ended in a humiliating failure. Otherwise, until well into the eighteenth century the British East India Company, like other European merchant companies, pursued a policy of trade and friendly relations with the Mughals and avoided territorial ambitions or military commitments. The result was a standoff, with the Mughals in possession of most of the land and the Europeans dominant at sea.[19]

Things changed dramatically in the eighteenth century. The Mughal emperor Aurangzeb, who reigned from 1658 to 1707, alienated his Hindu subjects by trying to impose Islamic law and customs on them. Early in his reign, Maratha raiders established a separate state in the western Deccan. Aurangzeb's successors proved unable to rule effectively. As the Mughal Empire began to disintegrate, provincial governors and warlords rose up in defiance. One chief minister, Asaf Jah, moved from Delhi to the Deccan, where he became the nizam (or ruler) of Hyderabad. In 1738–39 Marathas seized the western provinces, while a Persian army invaded and occupied Delhi. Nine years later, Afghans invaded the north. By 1750 the Mughal Empire was reduced to a fraction of its former extent, barely holding onto Bengal and Hindustan in the Ganges valley, while the small successor states fell to fighting among themselves. Into the power vacuum left by the collapse of the Mughal Empire stepped the French and the British.[20]

Before we turn to the European interventions, let us consider the art of war in India under the Mughals and their successors. The Mughals' preferred arm was the cavalry, but they were also proficient in the use of heavy cannon to breach the walls of towns and fortresses. Since the sixteenth century, Indian states had imported guns from Europe and had employed Turks or Europeans as gun founders and gunners. Despite this diffusion of technology, Indian artillery was never very effective. Cannon tended to be large and heavy, requiring an elephant or twenty yoke of oxen to transport. Their poorly made mounts made them difficult to aim and their gunpowder was of low quality and prone to deteriorate. They were very cumbersome to reload and could not fire more than four times an hour, with breaks for lunch. In short, they were designed for sieges, not battles.[21]

Nor were infantry firearms much better. Until well into the eighteenth century, most Indian soldiers had handmade matchlocks that were difficult to load and fire and quickly wore out.[22] Most infantrymen were peasants recruited for seasonal campaigns. They provided their own weapons, clothes, and equipment and received no training. They were seldom paid regularly and were poorly motivated to fight. Their loyalty, such as it was, was to their *jagirs* (or landowners) rather than to the prince on whose behalf they fought.[23]

The strength of Indian armies lay in their heavy cavalry. Indian spears and swords used by cavalrymen were the products of a long tradition of careful and laborious forging and were considered better than British swords.[24] Horse breeders in Central Asia and on the Deccan plateau provided mounts for huge numbers of aristocratic cavalrymen who enriched themselves from raiding and plundering during military campaigns. Thus Indian armies consisted of aggregates of individual warriors beholden to their leaders, rather than members of bureaucratic fighting organizations.

Indian tactics differed greatly from those in European wars because warfare served a very different purpose in India. Established powers like the later Mughals turned military campaigns into traveling processions in which the ruler, his court, and thousands of hangers-on moved ponderously across the countryside, using their heavy cannon to intimidate lesser princes and dispensing large bribes to win over their enemies' loyalty.[25] Military historian Channa Wickremesekera describes Indian tactics as follows:

> The armies of Indian rulers consisted of numbers of individuals skilled in arms, but not subjected to anything approaching a uniform system of discipline and control. The root cause lay in a segmented structure of political control which favoured military units which, in terms of command and control, approximated to war bands rather than to a disciplined army. . . .
>
> Once battle was joined it turned into a series of fights around the leading chiefs on either side. Indeed, the fall of leaders quite often decided the fate of battles since leader-less troops, deprived of their principal reason for being there, promptly dispersed to their villages. . . . the close relationship between political power and military command meant that military confrontation itself often took the form of negotiation, bribery acting as the chief weapon. A hefty sum could usually induce a rival commander to switch his allegiance during the heat of battle or the commander of a garrison to deliver the fort.[26]

European intervention in Indian affairs began as a spin-off of the internal power struggles in India and of the War of Austrian Succession (1740–48) in which Britain and France found themselves on opposite sides. In 1746 a contingent led by Joseph Dupleix, governor of the French town of Pondicherry, seized the British outpost of Madras. Mahfuz Khan, son of the governor of Carnatic, where the two towns were located, attempted to take Madras with a force of ten thousand cavalrymen but was soundly defeated by three hundred Frenchmen and seven hundred Indian auxiliaries led by a Swiss officer named Paradis.[27] Buoyed by that success, Dupleix got involved in the succession struggles for nawab of Hyderabad and of Carnatic. In 1751 the British joined the fray when Robert Clive, a clerk for the East India Company, led a small contingent of two hundred British troops and three hundred Indian auxiliaries and attacked Arcot, the capital of Carnatic. After a siege of fifty days, Arcot fell and, with it, the Carnatic army's artillery. Three years later the French government recalled Dupleix, and the British East India

Company found itself in possession of an important part of southeastern India.[28] What had turned the tide in Britain's favor was the
military genius of Robert Clive, the support of the Royal Navy, and
the greater resources at their disposal. But it was more than it. It was
also, and most important, the result of their bringing to India a new
art of war known as the Military Revolution.

The Military Revolution

The sixteenth and seventeenth centuries were an unusually warlike
period for Europe. The spread of and improvements in firearms were
accompanied by changes in tactics, logistics, and other aspects of
warfare. Infantrymen firing crossbows and harquebuses were not sufficient, in themselves, to end the dominance of heavy cavalry that
had lasted until the fifteenth century, for they were vulnerable during the time they reloaded their unwieldy weapons. During the Italian Wars of the early sixteenth century, Swiss mercenary captains
began surrounding their musketeers with pikemen to shield them
while they reloaded. The tactic was taken up by the Spanish *tercios*,
squares of up to three thousand men that could dominate a battlefield but were difficult to maneuver and supply.

In the 1590s, Prince Maurice of Nassau, commander of the Netherlands army, introduced the volley performed by square units of 550
men. In this tactic, the front row of musketeers, protected by pikemen, fired in unison, then retreated to the back of the unit to reload,
while the next rank advanced to fire in turn. Maintaining a steady
rate of fire required ten musket lines to quickly perform a complicated drill while maneuvering under enemy fire. To achieve this end,
Maurice drew up elaborate drill manuals describing each movement
and imposed intensive training and discipline on his troops so that
they would perform the necessary gestures instinctively in the heat
of battle. To be ready to lead such units in battle, he made his officers
attend specialized military schools.

In the seventeenth century, European armies abandoned the
matchlocks that had been fired by holding a smoldering rope up to
the touchhole at the back of the barrel. Matchlocks were difficult

and dangerous to load, for the soldier had to handle the gun, loose gunpowder, and the burning match, all at the same time. In their place, they introduced flintlocks, in which a spark from a flint ignited the powder when the soldier pulled the trigger. Flintlocks were easier and safer to use and could be fired much more rapidly. When bayonets were attached to the barrels, they could be fired and then used as pikes when the troops closed in on the enemy.

In the 1630s the Swedish king Gustavus Adolphus added one more element to this revolution: field artillery.[29] Not only did his army include heavy siege guns, he also supplied it with lighter, more maneuverable cannon manned by gunners trained to load and fire their weapons up to three times a minute. With four such guns per battalion, these pieces were used in close support of the infantry. With his armies of musket-wielding infantry supported by quick-firing artillery and light cavalry, Gustavus Adolphus swept through Germany, making Sweden, briefly, one of the great powers of Europe.

These innovations had repercussions throughout society. They required constant training, in peacetime as well as during wars, hence standing armies. From heroic warriors on horseback, officers were turned into educated tacticians; artillery officers, in particular, had to be educated in mathematics and ballistics, opening the door to the sons of middle-class families. This provoked resentment among aristocratic cavalry officers whose forefathers had dominated European battlefields since the Middle Ages. But such resistance was outweighed by a long tradition of disciplined infantry stretching from the Greek phalanxes and Roman legions to the Spanish *tercios*, and even more so by the victories of the new armies equipped with flintlocks and field guns. In the frequent wars of the sixteenth and seventeenth centuries, armies grew in size, some of them tenfold. To defend against powerful siege guns, cities and kingdoms had to replace their medieval fortresses with new *trace italienne* fortifications with sloping walls and a complicated geometry that could prevent enemies from scaling the walls.

To support such costly forces and installations required far more money than states had ever needed before. Only a complex and highly organized bureaucratic apparatus could encourage the growth of the state's economy and extract a steady stream of tax revenues from it. Spain, despite its shipments of gold and silver from the Americas, went bankrupt waging wars against the rest of Europe. By the eighteenth century, only France, Britain, and Prussia could maintain the armies needed to fight the new wars, and only France and Britain could carry their wars overseas.[30]

The coincidence of the Military Revolution in Europe and political chaos in India made possible something that would have been unthinkable before the eighteenth century: the European conquest of the subcontinent.

For over two centuries the Europeans in the Indian Ocean were restricted to their coastal enclaves, for they had too few soldiers to confront the armies of the Indian princes. To undertake military campaigns in the interior, the French and British needed more soldiers than they could muster from among the small numbers of resident Europeans. The solution was to recruit Indians as sepoys, or native soldiers, and train them in the European-style drill. The first to do so were the French at Mahé in the 1720s. The French victory at Madras in 1746 proved the effectiveness of the new tactics.

The British in India, faced with the French threat, first tried to recruit Indo-Portuguese settlers and even African slaves to augment their small number. Then the East India Company gave Clive and another agent, Major Stringer Lawrence, permission to follow the French example and recruit and train Indian sepoys. In doing so the company possessed an advantage over the French, for it was more profitable than its French Compagnie des Indes and could pay its troops better and more consistently, thereby ensuring their greater loyalty. By the mid-1750s, it had ten thousand sepoys in the Carnatic.[31]

Gradually, the training of sepoys began to resemble that of European troops, with tight discipline and standardized drill. Officers had to accommodate the special needs of Indian troops, such as providing food that did not offend their religions and avoiding putting Hindus on ships. Yet the British never fully trusted their Hindu and Muslim soldiers, preferring "martial races" like Turks, Arabs, or Nepali Gurkhas and, above all, Europeans.[32]

Indian soldiers in European employ were issued better weapons than the matchlock muskets commonly used in India. Starting in the 1740s, first the French and then the British began importing flintlocks and training their sepoys to use them. When used with paper cartridges containing pre-measured amounts of powder, these muskets could be fired twice as fast as matchlocks. The introduction of socket bayonets did away with the need for pikemen to protect the musketeers during reloading. British guns in particular—the Brown Bess mass-produced in Birmingham—were both cheaper and more reliable than guns made in India.[33] With proper training, European troops and sepoys formed hollow squares that could keep up a steady rolling fire followed by advances with bayonets.[34]

It was during the Carnatic Wars of the 1740s and 1750s that Indians first encountered field artillery. European cannon were lighter and easier to load than the siege guns common in India. They were also fitted onto sturdy carriages with strong wheels that horses could pull at a soldier's pace over bumpy terrain. They carried ammunition boxes and were equipped with screws to adjust their elevation. They could also be armed with grapeshot, which was very effective against cavalry.[35]

The consequences of the Military Revolution in India were quickly felt, as small numbers of disciplined European troops with field guns defeated much larger armies of less disciplined and less motivated Indian soldiers. As Robert Clive wrote to Prime Minister William Pitt in 1759,

> So small a body as two thousand Europeans will secure us against any apprehensions from either the one [this country's government] or the other [the people]; and, in the case of their daring to be troublesome, enable the Company to take the sovereignty upon themselves. . . .
>
> A small force from home will be sufficient, as we always make sure of any number we please of black troops, who, being much better paid and treated by us than by the country powers, will very readily enter into our service.[36]

Plassey and After

Clive was referring to his victory at Plassey in 1757. A year after the twenty-year-old Siraj-ud-Daula succeeded his grandfather as nawab of Bengal, he attacked Fort William, the British settlement of Calcutta, with thirty thousand infantry, twenty thousand cavalry, four hundred elephants, and eight cannon. In response, Clive arrived from Madras with a small force of Europeans and sepoys. Before confronting Siraj-ud-Daula on the battlefield, he enticed the nawab's chief general, Mir Jafar, to defect by promising to make him nawab in Siraj-ud-Daula's place. Hindu bankers, tired of the exactions and bad government of the Mughals, joined the conspiracy.

Plassey was not a battle but a rout. Siraj-ud-Daula had an army of fifty thousand against Clive's eight hundred European soldiers and 2,200 sepoys. When the battle began, Mir Jafar refused to advance. The rest of Siraj-ud-Daula's army panicked in the face of Clive's artillery and fled. In this, among the most lopsided battles in history, seven European soldiers and sixteen sepoys died, and thirteen Europeans and thirty-six sepoys were wounded, while the nawab's army lost five hundred men.[37]

The results were out of proportion to the effort. Mir Jafar became nawab of Bengal, but as a puppet of the British. In 1764 the British fought another battle at Buxar in which they defeated the nawab of Oudh and the forces of the Mughal emperor Shah Alam II. As a reward, the East India Company obtained the *diwani* or right to collect taxes in Bengal and Bihar, the two richest provinces of India. Clive and his friends bled the provinces for their own enrichment, but they also used the state's funds to build up a permanent army with British officers and Indian soldiers, the army that enabled the company to conquer the rest of India.[38]

Plassey was exceptional. After that, British victories were never so easy or caused so few casualties. In the wars of the eighteenth century—four with Mysore, three with the Marathas—the British had the advantage not only of tactics and weapons but also of the greater loyalty of their troops, bought with the wealth of Bengal. Their opponents had little popular support, for their rulers were either outsiders themselves—Muslims ruling Hindus—or high-caste Brahmins and Kshatriyas ruling masses of peasants and artisans. The British also had considerable support from Hindu merchants and bankers. Historians Ronald Findlay and Kevin O'Rourke explain: "Oriented toward trade and the market, this 'new middle class' . . . could form a convenient, if uneasy and contentious, partnership to further its commercial interests and activities with the East India

Company, organizing the procurement and supply of its exports from India, collecting its revenues and even providing it with loans whenever necessary."[39]

The first opponent the British encountered after Buxar was Haidar Ali, the able and ambitious ruler of Mysore. To train his troops in the European style, he hired French soldiers disbanded after the fall of Pondicherry. He attacked the British in 1767 but was defeated. He rebuilt his army until he had twenty thousand sepoys. In 1780, with French help, he defeated a British force, but when the French withdrew their support he was defeated again. When he died in 1782, his son Tipu Sultan fought on for two years, then signed a peace treaty with the British governor Warren Hastings. In 1790 Tipu Sultan attacked the British at Travancore, but lost and was forced to cede half his territory. He tried one more time to organize an army with the help of French soldiers stranded in India by the French Revolution, but was killed at the battle of Seringapatam in 1799. After that last Mysore War, the British annexed much of southern India and effectively controlled the rest.[40]

The repeated defeats of Haidar Ali and Tipu Sultan reveal the difficulties of transferring military technology from one culture to another. After Plassey, the rulers of India scrambled to adopt the new European military system. They bought European-style muskets and improved their artillery. They had long employed Europeans, Turks, and other foreigners to cast and handle their cannon; they now recruited European adventurers or deserters from the trading companies to train some of their troops in the new art of war. However, they did not do so quickly enough. The foreign mercenaries they hired were unreliable and often incompetent. Nor did the states develop the modern bureaucratic structures needed to support, administer, and command large standing armies. None of the states of India had achieved the close integration among rulers, warriors, and merchants that characterized the most successful European states of the seventeenth and early eighteenth centuries.[41] In short, they had premodern armies with a modern veneer.[42]

The Marathas did not begin as a territorial state but as a mobile cavalry that lived from raiding. In the late seventeenth century, their leader, Sivaji, bought guns and ammunition from Europeans on the west coast and created a highly effective fighting force that thwarted the attempts of the Mughal emperor Aurangzeb to bring them to heel. In the eighteenth century, as the Mughal Empire crumbled, the Marathas gained control of much of western India from Maharashtra to Punjab. They also adopted many of the European ways of war. Their soldiers dressed and paraded like the sepoys of the East India Company, and many of them carried flintlock muskets. Yet they seem to have succumbed to some of the same customs that had weakened the later Mughals, such as going to war with huge baggage trains, camp followers, women, and other luxuries that slowed their movements.[43] Their artillery was a mixed lot of different calibers firing rough iron balls that wore out the barrels. William Henry Tone, an English visitor in the 1790s, commented that their guns were "tolerably well cast, but the carriages are in general very clumsily and badly constructed. A march of a few days shakes the carriages to pieces."[44]

The British confronted the Marathas twice. In the first Anglo-Maratha War (1775–82), Warren Hastings, the governor of Bengal, marched across the peninsula and, using diplomacy and bribes, signed a peace treaty with the Marathas that left Bombay in British hands but otherwise changed little in their relations. After that experience, the Maratha leader, Mahadji Sindia, decided to modernize the Maratha army. For this, he hired a Frenchman, Benoit de Boigne, to reform the state's finances. At the end of the century, the Maratha army had over twenty thousand infantry, over sixty thousand cavalry troops, and a mobile field artillery as good as that of the British. By then, de Boigne had been replaced by another Frenchman, Perron, with other Frenchmen serving in various capacities.

Meanwhile, the East India Company was also improving its army. Alone among the powers in India, it possessed an efficient bureaucracy able to extract enough revenue from the population to support a standing army of 120,000 sepoys led by English officers and Indian sergeants. It also adopted certain useful Indian tactics, such as an irregular light cavalry for reconnaissance and skirmishing and camels and elephants for special tasks.[45] When war broke out in 1802, Lord Mornington, the governor of Bengal, attacked the Marathas in the Deccan, while General Gerard Lake advanced toward Delhi. At that point, the British offered amnesty to all mercenaries in Maratha service, causing a mass exodus of Europeans, leaving the army leaderless.[46] British bribes also bought the defection of many Maratha units. As for the others, as one historian explained, "Some just waited to see the outcome of events in the hope of throwing in their lot with the winners. This was, after all, another aspect of the traditional South Asian military labour market. Soldiering for a living was dangerous at the best of times and only survivors could take their pay back home."[47] The British victories at Assaye (September 22, 1803) and Laswari (November 1, 1803) destroyed the Maratha power and left most of western India in British hands. But despite the financial weakness of the Marathas, which lost them the loyalty of their troops in the competition with the wealthier East India Company, their defeat was purchased at a high cost in money and required an army of 27,500 men. The gap between the British and their rivals for control of the subcontinent was narrowing.[48]

Reaching the Limit: Afghanistan and the Punjab

By the second decade of the nineteenth century, the East India Company had gained control of the Indian subcontinent up to the Indus and Sutlej rivers. Beyond lay potentially troublesome neighbors, but also tempting targets. Sind, the lower reaches of the Indus, and Punjab, the upper reaches, were rich agricultural regions. Beyond them lay the desert of Baluchistan and the mountains of Afghanistan, historically a source of danger to the rest of India.

In the late 1830s and afterward, Great Britain and Russia were engaged in the "Great Game," a rivalry for control of Asia, or so they thought. When Persia sent an army to besiege Herat in western Afghanistan, the redoubtable Viscount Palmerston and his cabinet and the governor-general of India, Lord Auckland, believed it to be the opening move of a Russian invasion of India. To counter the perceived Russian threat, they decided on a regime change to replace Dost Mohammed, the emir of Kabul and titular ruler of the Afghans, with the more amenable Shah Shuja. To do so, Auckland decided to invade Afghanistan.

As military officers knew, fighting the mountain peoples of Afghanistan presented a more difficult challenge than campaigning in the fertile plains of India. One of them, Lieutenant Colonel Claude Wade, wrote to Lord Auckland:

> There is nothing more to be dreaded or guarded against, I think, in our endeavour to re-establish the Affghan monarchy than the overwhelming confidence with which Europeans are too often accustomed to regard the excellence of their own institutions and the anxiety that they display to introduce them in new and untried soils. . . . The people of these countries are far from ripe for the introduction of our highly refined system of Government or of Society, and we are liable to meet with more opposition in the attempt to disturb what we find existing than from the exercise of our physical force.[49]

Auckland tried to enlist the support of Ranjit Singh, the ruler of the Punjab, but the wily Sikh refused, thereby closing off the Khyber Pass—the traditional route between India and Central Asia—to the British invasion. Instead, the invaders had to enter Afghanistan from the south, via Baluchistan and the Bolan Pass. The Army of the Indus, as it was named, consisted of 9,500 men of the Bengal Army, 5,600 men of the Bombay Army, and Shah Shuja's 6,000 Afghan soldiers. Accompanying the troops were three to five times that many camp followers, along with eight thousand horses, thirty thousand camels, and a huge baggage train carrying every conceivable necessity of life, including foxhounds. Evidently, the British were acquiring the same habits that had slowed down the Mughal and Maratha armies, namely the transformation of an army into a mass migration.

The advancing army soon ate up its provisions and found itself short of food in the desert of Baluchistan. Even before reaching the Bolan Pass, the troops were on half rations. The camels and oxen, well suited to the plains of northern India, died in the mountains between Baluchistan and Afghanistan. Baluchi mountaineers stole animals and picked off stragglers. Baggage carts had to be abandoned, field guns manhandled over precipices, and ammunition blown up for lack of draft animals. By the time it reached the fertile valley around Kandahar five months later, the once-magnificent

Army of the Indus was reduced to "a wretched rabble of worn-out and dispirited men."[50]

The Army of the Indus spent two months in Kandahar to recover and replenish their horses and their supplies, then advanced on Kabul. After some desultory fighting, the British entered Kabul in August 1839 and immediately acted as if they were in an Indian city. They established a cantonment or military housing area, organized polo matches and other festivities, and brought their wives and concubines. While the foreign occupation began to look permanent, the puppet regime of Shah Shuja alienated the inhabitants. By 1840, more and more Afghans began to side with Dost Mohammed's son Akbar Khan. In October 1841, the British political agent Mcnaughton decided to cut in half the subsidy being paid to the Ghilzye tribe to allow supplies through the Khyber Pass. In retaliation, the Ghilzye attacked the supply caravans, cutting off the British in Kabul from India. A month later, the uprising spread to Kabul. In early January 1842, the situation having become untenable, the British decided to evacuate. Over seven hundred British troops and civilians, over three thousand Indian sepoys, and twelve thousand camp followers fled in panic. Many died of exposure in the snowdrifts and sub-zero weather. Others were killed by snipers or massacred when they fell into Afghan hands. One hundred and five British men, women, and children were taken prisoner by Akbar Khan's soldiers. Only one man, Dr. William Brydon, reached the safety of the British garrison at Jalalabad, halfway to the Khyber Pass.[51]

How to explain such a disaster for Britain and such a victory for the Afghans? Part of the reason, no doubt, is the incompetence of the British military and political officers. They had been trained in the rich flatlands of India or Europe where supplies were abundant and large numbers of men and animals could maneuver with ease. In India, the combination of European officers and Indian soldiers had always proved remarkably effective. In Afghanistan, in contrast, neither Europeans nor Indians had any experience with desert and mountain warfare. Like Napoleon's armies in Spain and Russia, they were not prepared for difficult environments.

The fact that Afghans were experienced warriors contributed to the British disaster. They seldom could, or even tried, to fight in the

Figure 4.1. "Remnants of an Army" by Elizabeth Butler, The Tate Gallery, London. The painting depicts Dr. William Brydon, the lone survivor of the British expedition to Afghanistan in 1839–42, arriving at Jalalabad in January 1842.

open. Instead, they used classic guerrilla tactics, such as hiding behind rocks to ambush their enemies or perching on mountainsides overlooking narrow passes and shooting down with their long-barreled smoothbore muskets called *jazails*. They knew the land and could melt away into the countryside when overpowered. The British advantages of drill and discipline that were so effective in open terrain and had served them so well in India were less useful in the broken mountainous environment of Afghanistan. Their smooth-bore flintlock muskets were not significantly better than those of the Afghans.

The First Anglo-Afghan War did not end with the debacle of January 1842. Governor-General Auckland's successor, Lord Ellenborough, vacillated between the need to redeem British honor and the cost of a new campaign. Though they had lost the Kabul garrison, the British still had troops in Jalalabad and Kandahar and were assembling troops in Peshawar, on the Indian side of the Khyber Pass. Finally in April Lord Ellenborough ordered a new "Army of Retribution" commanded by General Pollock to invade Afghanistan. Borrowing a tactic from the Afghans, Pollock had his soldiers, many of them Gurkhas from Nepal, climb the slopes overlooking the pass and shoot down at the Ghilzye and Afridi tribesmen, who let the army reach Jalalabad without opposition. In August 1842, Pollock in Jalalabad and General Nott in Kandahar set out for Kabul. They rescued 121 prisoners of the Afghans, the last survivors of the Kabul garrison. They entered the city in September, looted and burned the bazaar, then returned to India before the winter set in.[52]

The Army of Retribution may have saved the honor of Britain, in the limited way in which that term was understood among the Victorians. But the Afghan adventure had undermined Britain's reputation for invincibility among the Indians. The first to challenge the British after the debacle were the Sikhs, a people with a long-standing military tradition who inhabited the Punjab. Their ruler, Ranjit Singh, had reorganized his army in the early nineteenth century. He hired French and Italian veterans of the Napoleonic Wars to introduce European training and tactics. Thirty-five thousand of his seventy-five thousand men were regular soldiers led by European officers. Their equipment was also modernized. He introduced flintlock muskets equivalent to the British Brown Bess. In addition to heavy cannon, he equipped his forces with mortars and howitzers. His foundries cast over five hundred field guns on the French pattern, nine-pounders pulled by horses that were as mobile as those of the British. In his artillery, French was the language of command. In imitation of the French, his army was reorganized into brigades, each with three or four infantry battalions, an artillery battery, and up to six thousand cavalrymen. The result was an army that matched that of the East India Company.[53] Despite the strength of his army, Ranjit Singh knew better than to tangle with the British. Until his death in 1839, he remained a loyal ally of the East India Company and a buffer against the Afghans.

After he died, his army became unruly and eager for booty. In 1845, just three years after the Afghan debacle, the Sikhs invaded British India. Though their army was not as disciplined as it had been under Ranjit Singh, it was nonetheless a formidable opponent. It took two wars, in 1845–46 and in 1848–49, for the army of the East India Company to overcome them. Even then, the British did not defeat the Sikhs in February 1849 through better discipline or tactics, but because they had a larger army and more cannon.[54] Eight years later, the Indian Rebellion (or Sepoy Mutiny, as the British preferred to call it) came close to chasing the British out of northern India. The days of easy victories à la Plassey were long gone.

Reaching the Limit: Algeria, 1830–1850

Unlike the British foray into Afghanistan, the French actually conquered and retained Algeria. Yet their experience illustrates the limits of the old imperialism even more vividly than the British debacle in Afghanistan, for it involved the largest military buildup since the days of Napoleon.

When French troops landed near Algiers in June 1830, they faced a weak adversary. Algeria was technically a province of the Ottoman Empire ruled by a dey, or governor, appointed by the sultan. His army consisted of five thousand Turkish janissaries, another five thousand half-Turkish soldiers, and fifty thousand Algerians of dubious loyalty.

The French army that disembarked near Algiers consisted of thirty-seven thousand men, nine-tenths of them infantrymen armed with smoothbore flintlock muskets, the same guns that had equipped the French Revolutionary and Napoleonic armies. They were very well trained and disciplined, many of them veterans of the Napoleonic Wars. Standing in three ranks, they could fire off three shots every two minutes. The cavalry carried sabers and lances. The artillery corps brought dozens of cannon, howitzers, and mortars and five hundred Congreve rockets. In the battle that took place outside the walls of Algiers, the French easily prevailed, thanks to their firepower.[55]

When the Turks departed, the country was left in chaos. Three years later Abd al-Qadir, a young man still in his twenties, returned from a pilgrimage to Mecca determined to create a modern state along the lines of Muhammad Ali's Egypt that he had visited and admired. He formed an army of volunteers made up of 8,000 infantrymen, 2,000 cavalrymen, and 240 artillerymen with twenty cannon. Irregular troops were also called up when needed, but they returned to their farms during the plowing, planting, and harvest seasons. They used classic guerrilla tactics: raiding French-held villages and outposts, ambushing marching columns, attacking their rear guards and stragglers. Excellent horsemen, they avoided pitched battles.[56]

At first, Abd al-Qadir's troops were armed with locally made *mokhala* or *jazail* flintlock muskets, with poor-quality powder and bullets and a maximum range of two hundred meters. In March 1834 he signed a secret agreement with General Desmichels, the French governor of Oran, who gave him four hundred muskets and allowed him to import weapons, powder, and sulfur in exchange for a promise of peace. A year later, at the head of an army of ten to twelve thousand men, most of them cavalry, he defeated a French army of 2,300, a stunning victory the French called the "disaster of Macta." In May 1837 General Thomas-Robert Bugeaud, the French commander in Oran, signed the Treaty of Tafna, promising to sell Abd al-Qadir three thousand guns and fifty tons of gunpowder, and conceding his authority over almost two-thirds of Algeria. Abd al-Qadir also bought eight thousand guns from Britain via Morocco, as both nations disliked the French presence in Algeria. By 1840 he had set up arsenals with the help of Spanish and French workers making eight guns a day, as well as bullets and gunpowder using local saltpeter and imported sulfur.[57] At the apogee of his power, Abd al-Qadir was a formidable force who ruled western Algeria and confined the French to a few coastal towns and their surroundings.

That year, Bugeaud was appointed both governor-general and commander in chief of the French army in Algeria. A veteran of the Napoleonic campaigns, Bugeaud recognized that warfare in Algeria resembled the guerrilla war that had weakened the French armies in Spain. As he said, "We must forget those orchestrated and dramatic battles that civilized peoples fight against one another and realize that unconventional tactics are the soul of this war."[58] Though he had signed the Treaty of Tafna four years earlier, this time he was determined to crush Abd al-Qadir and the Algerian resistance; as he put it: "A treaty is no longer valid when it ceases being useful."[59]

Figure 4.2. The Algerian leader Abd al-Qadir. Portrait by Ange Tissier (1842)
in Musée national du Château de Versailles.

He recognized the skill and strength of his enemies: "All are warriors; there is not one who does not know perfectly how to ride a
horse; all have a horse and a gun; all of them fight, from the eighty-
year-old man to the child of fifteen, and this population of four million souls has no fewer than five to six thousand warriors, all of them
very skilled as individuals."[60]

To achieve his objective, Bugeaud had to reform the French army in Algeria, a task he had begun, but not completed, during his earlier posting in 1836–37. He found the soldiers demoralized, many of them suffering from malaria. He built hospitals and improved the comfort and living conditions of the troops. He also aimed to make his army more mobile; previously, when the army moved, it had moved slowly, burdened by many cannon and a heavy baggage train.[61] He insisted that his troops carry only their weapons and ammunition, with the rest of their equipment to be brought by mules. He also recruited native auxiliaries called *zouaves* or *spahis* (French for sepoys). He trained his troops to operate in flexible formations adapted to the terrain and circumstances and to march forty to fifty kilometers a day for up to five days.[62]

Bugeaud's tactics were equally revolutionary. On the principle that "I wage war, not philanthropy," his campaigns were marked by pillage, torture, rape, murder, and other atrocities. His troops attacked civilians and destroyed the towns, killed the livestock, burned the crops, and cut the fruit trees of tribes that did not submit. They searched for secret underground granaries and prevented peasants from sowing and harvesting their crops.[63]

Such methods, aimed at defeating not only the leadership but the entire population of Algeria, required enormous numbers of troops to occupy every town and region of the country. In 1836 the French army in Algeria numbered some 30,000. When Bugeaud started his campaign, it rose to 65,000. Bugeaud demanded, and got, ever more soldiers. By 1844–45 he had 80,000, and by the end of the war, the French army in Algeria numbered over 100,000. It took one-third of the entire French army to overcome the Algerian resistance.[64]

Against such a large enemy army, Abd al-Qadir could no longer stage formal battles but had to resort to guerrilla warfare. Bugeaud's tactics and the severe winter of 1841–42 weakened his forces. The following year, Britain, having resolved its diplomatic difficulties with France, banned the sale of arms to Abd al-Qadir through Morocco. In 1845–46 he launched his last, futile offensive from Morocco. A year later, the sultan of Morocco forced him back to Algeria, where he surrendered to the French.[65]

Though resistance continued sporadically until the 1880s, in the end, France did conquer Algeria. But the size of the army that this conquest required was so large that it is doubtful that France would have mounted such an effort in any other part of the world.

Russia and the Caucasus

From the sixteenth through the nineteenth centuries, Russia expanded enormously, with periodic reversals. Some of its conquests were at the expense of its European neighbors—Swedes, Finns, Lithuanians, Poles, and Ukrainians—but others were achieved against the Ottoman Empire, Persia, and Tatar Khanates, successors to the Golden Horde of the Mongols. Surprisingly, the most difficult of all its conquests was not against these powerful states but against the mountain tribes of the Caucasus.

The first sustained push toward the east began during the reign of Tsar Ivan the Terrible, from 1533 to 1584. Using newly acquired cannon, Ivan attacked the Tatars of Kazan in 1552 and Astrakhan, on the Caspian Sea, in 1556. Twenty-five years later, in 1581, a renegade Cossack named Yermak Timofeevich crossed the Ural Mountains with 840 men and a few cannon and defeated the Tatars of Sibir. This opened the gates to an invasion of Siberia, a vast and almost empty forest into which Russian pioneers and adventurers, like the French coureurs des bois in Canada, rapidly advanced in search of precious furs. So rapid was their advance that they reached Okhotsk, on the sea of that name, in 1637, and the Pacific Ocean, five thousand miles from Moscow, shortly thereafter.

Meanwhile, the Caucasus, very much closer to European Russia, remained out of reach. This rugged mountain range was inhabited by peoples speaking many different languages and practicing different religions. Russia got involved in the region in the course of a war against Persia (1804–13). As a result of this war, Russia acquired the small Christian kingdom of Georgia and parts of Armenia, as well as the right to keep a navy on the Caspian Sea. After a war with the Ottoman Empire in 1828–29, Russia also acquired the east coast of the Black Sea. It now surrounded the Caucasus on all sides.

Defeating the mountain tribes of Chechnya, Dagestan, and Circassia proved to be far more difficult than the Russians expected. In 1785, when Russian troops invaded Chechnya, the religious leader Sheykh Mansur declared a holy war until the Russians retreated. Between 1804 and 1812, Russia won wars against Persia and the Ottomans and annexed most of Georgia, but the mountains between Georgia and Russia remained unconquered.[66] In the 1820s, the Ghazi Muhammad united the Muslims of the Caucasus in another jihad against Russia. His successor, the Imam Shamil, was as skillful and tenacious a warrior as Abd al-Qadir and his followers proved to be in maintaining their independence as the Berbers and Arabs of Algeria.[67] In 1838 Russia sent an army of 155,000 men under General von Grabbe against him. Though the Russians surrounded him, he managed to escape and rally his supporters, and it cost the Russians three thousand casualties. By 1841 he had built an arsenal and foundries and was receiving money and munitions from Britain, then involved in the "Great Game" against Russia. To dislodge him, the Russian government sent a force of ten thousand soldiers to the Caucasus in 1842, but it was forced to retreat after losing one-fifth of its men, while another force of twenty thousand lost four thousand men before escaping. Three years later, yet another Russian army of eighteen thousand led by the viceroy of the Caucasus, Prince Vorontsov, retreated after suffering four thousand casualties.[68] The Russian tactic of sending long columns of poorly trained soldiers into the mountains proved as ineffective as the British foray into Afghanistan.

From 1853 to 1856, the Crimean War distracted Russia from its campaigns in the Caucasus. When the war ended, it returned to the offensive with a 250,000-man army under the command of General Alexander Bariatinski, the largest colonial expedition in history. This time, the Russians did not rush into the mountains to be ambushed by Caucasian guerrilla fighters. Instead, they built roads, cleared forests, destroyed villages, and constructed a network of forts. By 1858, the inhabitants, decimated and exhausted by the Russian juggernaut, began abandoning Shamil. In April 1859, surrounded on Mount Gunib with four hundred followers, Shamil surrendered. A majority of the Muslim population fled to the Ottoman Empire rather than live under infidel rule. Sporadic uprisings continued until 1864, when Tsar Alexander II declared the war over and the Caucasus fully incorporated into the Russian Empire.[69]

Some authors have attributed the Russian victory to their weapons. The historian of the Caucasian campaign, John Baddeley, notes, "Russians were for the first time armed with rifles, a fact not to be forgotten in estimating their successes from this time forward."[70] However, there is reason to doubt that rifles were common enough to make a difference. The Russian army was, by European standards, very badly equipped. The procurement of weapons was a low priority and the army depended on state arsenals, which were poorly run; in addition, officers found their illiterate peasant soldiers hard to train on more complicated weapons. During the Crimean War, while French and British soldiers carried modern rifles, almost all Russian soldiers used smoothbore muskets, the same kinds of guns used in the war against Napoleon. The Russian government tried to purchase new guns from the American Samuel Colt and from gun makers in Liège but were not able to import them in time. Not until 1866 did the Russian army begin switching to modern breech-loading rifles.[71] Instead of superior weapons, what gave the Russians an advantage over the Caucasian mountaineers was the sheer number of troops involved and the scorched-earth tactics they employed.

Conclusion

The experience of Europeans in Africa and Asia, like their experience in the Americas, illustrates the possibilities and the limitations of imperialism in the early modern period. In India, unlike the Americas, the advantages that the Europeans had were less the result of superior hardware than of their organization, financing, tactics, and skills. Even there, the advantage only became significant because of the political chaos of eighteenth-century India. Sub-Saharan Africa, Afghanistan, Algeria, and the Caucasus, in contrast to India, show the limitations of European power. In the one case diseases, in the others the mountainous terrain and the tactics of the inhabitants were obstacles to the ambitions of the European imperialists.

Comparing the Americas with Africa and Asia reveals another interesting pattern. The Europeans were successful in conquering highly structured and organized societies like those of the Aztecs, the Incas, and the Mughals and their successors. They had difficulty, however, in their attempts to conquer more loosely organized and widely dispersed peoples, whether it be the nomadic societies of North and South America or the peoples of Angola, Mozambique, Afghanistan, Algeria, and the Caucasus. Such societies, being less highly structured, were less vulnerable to setbacks and reverses. More of their men were used to hunting and fighting, they knew the terrain, and they had more time to acquire weapons and adopt hit-and-run tactics. In the face of guerrilla warfare, European imperial ventures reached diminishing returns.

How, then, can we explain the extraordinary explosion of territorial conquests in the second half of the nineteenth century? Explanations based on the motivations of the protagonists cannot satisfy, for the Europeans were as determined to win before 1859 as any imperialist after 1860, if not more so. Nor can one argue that later objects of imperial expansion were less eager to defend their lands than the Afghans, Caucasians, and Algerians had been. Rather, to understand the New Imperialism of the late nineteenth century, we must turn to the new means placed at the disposal of Western imperialists by the Industrial Revolution.

Notes

1. Philip D. Curtin, *The Rise and Fall of the Plantation Complex: Essays in Atlantic History* (Cambridge: Cambridge University Press, 1990), p. 38.

2. Philip D. Curtin, *Disease and Empire: The Health of European Troops in the Conquest of Africa* (Cambridge: Cambridge University Press, 1998), pp. 5–9; Curtin, *Rise and Fall of the Plantation Complex*, pp. 38–39, 80–81; Michael Colbourne, *Malaria in Africa* (London: Oxford University Press, 1966), p. 13.

3. Kenneth F. Kiple and Brian T. Higgins, "Yellow Fever and the Africanization of the Caribbean," in John W. Verano and Douglas H. Uberlaker, eds., *Disease and Demography in the Americas* (Washington, D.C.: Smithsonian Institution Press, 1982), p. 239.

4. Rita Headrick, *Colonialism, Health and Illness in French Equatorial Africa, 1885–1935* (Atlanta: African Studies Association Press, 1994), pp. 42, 67–68.

5. René-Jules Cornet, *Médecine et exploration: Premiers contacts de quelques explorateurs de l'Afrique centrale avec les maladies tropicales* (Brussels: Académie royale des sciences d'outre-mer, 1970), p. 7.

6. David Birmingham, *Trade and Conflict in Angola: The Mbundu and Their Neighbors under the Influence of the Portuguese, 1483–1790* (Oxford: Clarendon Press, 1966), pp. 12–28; James Duffy, *Portugal in Africa* (Cambridge, Mass.: Harvard University Press, 1962), pp. 49–50; C. R. Boxer, *Four Centuries of Portuguese Expansion, 1415–1825* (Berkeley: University of California Press, 1969), pp. 29–31.

7. John Thornton, "The Art of War in Angola, 1575–1680," *Comparative Studies in Society and History* 30, no. 2 (1988), pp. 360–78.

8. Thomas H. Henriksen, *Mozambique: A History* (London: Collings, 1978), pp. 26–36.

9. Eric Axelson, *Portuguese in South-East Africa, 1488–1600* (Johannesburg: C. Struik, 1973), pp. 152–58; Terry H. Elkiss, *The Quest for an African Eldorado: Sofala, Southern Zambezi, and the Portuguese, 1500–1865* (Waltham, Mass.: Crossroads Press, 1981), pp. 39–40; Malyn D. D. Newitt, *A History of Mozambique* (Bloomington: Indiana University Press, 1995), pp. 56–57; Henriksen, *Mozambique*, p. 38.

10. Richard Gray, "Portuguese Musketeers on the Zambezi," *Journal of African History* 12 (1971), pp. 531–33; Axelson, *1488–1600*, pp. 158–61; Elkiss, *The Quest*, pp. 40–41; Newitt, *History*, pp. 57–58.

11. Malyn D. D. Newitt, *Portuguese Settlement on the Zambezi: Exploration, Land Tenure, and Colonial Rule in East Africa* (New York: Africana Publishing, 1973), pp. 36–38; see also Eric Axelson, *Portuguese in South-East Africa, 1600–1700* (Johannesburg: Witwatersrand University Press, 1960), pp. 157–64; Henriksen, *Mozambique*, p. 43; and Gray, "Portuguese Musketeers," pp. 532–33.

12. João de Barros, *Decada Primeira*, livro 3, cap. xii (Lisbon, 1552), quoted in Boxer, *Four Centuries*, p. 27.

13. K. G. Davies, "The Living and the Dead: White Mortality in West Africa, 1684–1732," in Stanley L. Engerman and Eugene D. Genovese, eds., *Race and Slavery in the Western Hemisphere: Quantitative Studies* (Princeton: Princeton University Press, 1975), pp. 83–98.

14. Ibid., p. 96.

15. Dennis G. Carlson, *African Fever: A Study of British Science, Technology, and Politics in West Africa, 1787–1864* (Canton, Mass.: Science History Publications, 1984), pp. 5–9.

16. Philip D. Curtin, *The Image of Africa: British Ideas and Actions, 1780–1850* (Madison: University of Wisconsin Press, 1964), pp. 165, 181, 483-87; Michael Gelfand, *Rivers of Death in Africa* (London: Oxford University Press, 1964), pp. 18–20, and Gelfand, *Livingstone the Doctor, His Life and Travels: A Medical History*

(Oxford: Blackwell, 1957), pp. 3, 12. See also Carlson, *African Fever*, pp. 11–14, and Cornet, *Médecine et exploration*, chapter 2.

17. Philip D. Curtin, *Death by Migration: Europe's Encounter with the Tropical World in the Nineteenth Century* (Cambridge: Cambridge University Press, 1989), pp. 7–8; see also Curtin, *Disease and Empire*, pp. 3–4.

18. Philip D. Curtin, "Epidemiology and the Slave Trade," *Political Science Quarterly* 82, no. 2 (June 1968), pp. 210–11.

19. Bruce Lenman, "The Transition to European Military Ascendancy in India, 1600–1800," in John A. Lynn, ed., *Tools of War: Instruments, Ideas, and Institutions of Warfare, 1445–1871* (Urbana: University of Illinois Press, 1990), pp. 105–12; Lenman, *Britain's Colonial Wars, 1688–1783* (New York: Longman, 2001), pp. 83–91.

20. On the disintegration of the Mughal state, see Ronald Findlay and Kevin H. O'Rourke, *Power and Plenty: Trade, War, and the World Economy in the Second Millennium* (Princeton: Princeton University Press, 2007), pp. 262–64.

21. Channa Wickremesekera, *Best Black Troops in the World: British Perceptions and the Making of the Sepoy, 1746–1805* (New Delhi: Manohar, 2002), pp. 44–45, 78–79; Jos Gommans, "Warhorse and Gunpowder in India, c. 1000–1850," in Jeremy Black, ed., *War in the Early Modern World, 1450–1815* (London: UCL Press, 199), pp. 105–28; Bruce P. Lenman, "Weapons of War in Eighteenth-Century India," *Journal of the Society for Army Historical Research* 36 (1968), pp. 35–42; Lenman, "Transition," pp. 119–20.

22. Ahsan Jan Qaisar, *The Indian Response to European Technology and Culture, A.D. 1498–1707* (New York: Oxford University Press, 1982), pp. 46–57; Charles R. Boxer, "Asian Potentates and European Artillery in the 16th–18th Centuries: A Footnote to Gibson-Hill," in Charles R. Boxer, ed., *Portuguese Conquest and Commerce in Southern Asia, 1500–1750* (London: Variorum, 1985), pp. 158–60; Lenman, "Weapons," p. 34; Carlo Cipolla, *Guns, Sails, and Empires: Technological Innovation and the Early Phases of European Expansion, 1400–1700* (New York: Random House, 1965), pp. 105–11; Surendra Nath Sen, *The Military System of the Marathas*, 2nd ed. (Calcutta: K. P. Bagchi, 1958), pp. 106–7.

23. Kaushik Roy, "Military Synthesis in South Asia," *Journal of Military History* 69 (July 2005), pp. 657–60.

24. Lenman, "Transition," p. 39.

25. Gayl D. Ness and William Stahl, "Western Imperialist Armies in Asia," *Comparative Studies in Society and History* 19, no. 1 (January 1977), pp. 9–13; Gommans, "Warhorse and Gunpowder in India," pp. 106–17.

26. Wickremesekera, *Best Black Troops*, pp. 34, 45.

27. T. A. Heathcote, *The Military in British India: The Development of British Land Forces in South Asia, 1600–1947* (Manchester: Manchester University Press, 1995), pp. 25–27; Lenman, *Britain's Colonial Wars*, pp. 92–100; Lenman, "Transition," p. 114.

28. On the siege of Arcot, see H. S. Bhatia, *Military History of British India, 1607–1947* (New Delhi: Deep and Deep, 1977), p. 156; and Geoffrey Moorhouse, *India Britannica* (London: Granada, 1983), pp. 35–36.

29. On the development of European artillery, see Cipolla, *Guns*, pp. 73–74.

30. On the Military Revolution in Europe, see William H. McNeill, *The Pursuit of Power: Technology, Armed Force, and Society since A.D. 1000* (Chicago: University of Chicago Press, 1982), pp. 68, 91–94; Geoffrey Parker, *The Military Revolution: Military Innovation and the Rise of the West, 1500–1800* (Cambridge: Cambridge University Press, 1996), pp. 18–24; David B. Ralston, *Importing the European Army: The Introduction of European Military Techniques and Institutions into the Extra-European World, 1600–1914* (Chicago: University of Chicago Press, 1990), pp. 3–9; and Wickremesekera, *Best Black Troops*, pp. 51–66.

31. Wickremesekera, *Best Black Troops*, pp. 91–94, 109, 117, 131; Heathcote, *Military in British India*, p. 29; Lenman, "Transition," p. 114; Lenman, "Weapons," pp. 33–34; Lenman, *Britain's Colonial Wars*, pp. 92–99.

32. Wickremesekera, *Best Black Troops*, pp. 158–61, 169–74.

33. Gommans, "Warhorse and Gunpowder in India," pp. 118–19. On the steady improvements and declining costs of European firearms, see Philip T. Hoffman, "Why Is It That Europeans Ended Up Conquering the Rest of the Globe? Price, the Military Revolution, and Western Europe's Comparative Advantage in Violence," http://gpih.ucdavis.edu/files/Hoffman/pdf (accessed March 9, 2008).

34. Kenneth Chase, *Firearms: A Global History to 1700* (Cambridge: Cambridge University Press, 2003), pp. 25–26, 200–201; Roger A. Beaumont, *Sword of the Raj: The British Army in India, 1747–1947* (Indianapolis: Bobbs-Merrill, 1977), p. 67; William W. Greener, *The Gun and Its Development*, 9th ed. (New York: Bonanza, 1910), p. 66; Wickremesekera, *Best Black Troops*, pp. 118–25, 164; Lenman, "Weapons," pp. 37–39.

35. Lenman, "Weapons," pp. 35–37.

36. Quoted in Beaumont, *Sword of the Raj*, p. 4.

37. H. H. Dodwell, "Clive in Bengal, 1756–60," in H. H. Dodwell, ed., *The Cambridge History of India*, vol. 5: *British India, 1497–1858* (Delhi: S. Chand, n.d.), pp. 149–50; George B. Malleson, *The Decisive Battles of India: From 1746 to 1849, Inclusive* (London: Reeves and Turner, 1914), pp. 35–71; James P. Lawford, *Britain's Army in India: From Its Origins to the Conquest of Bengal* (Boston: Allen and Unwin, 1978), pp. 201–16.

38. Percival Spear, *A History of India*, vol. 2: *From the Sixteenth Century to the Twentieth Century* (London: Penguin, 1978), pp. 84–86; Wickremesekera, *Best Black Troops*, pp. 135–39; Lenman, "Transition," pp. 119–23.

39. Findlay and O'Rourke, *Power and Plenty*, p. 271.

40. Wickremesekera, *Best Black Troops*, pp. 68–78, 135–46; Spear, *A History of India*, vol. 2, pp. 90–102; Heathcote, *Military in British India*, pp. 42–49; Roy, "Military Synthesis," pp. 668–69.

41. It is well to remember, however, that most European states, for example, Poland and the smaller states of Germany and Italy, had not done so either. There is a rich literature (albeit outside the scope of this book) on European state formation and its relation to economic growth, military modernization, and empire-building. See, for example, Charles Tilly, *Coercion, Capital, and European States, A.D. 990–1992* (Cambridge: Basil Blackwell, 1992); Paul Kennedy, *The Rise and Fall of the Great Powers: Economic Change and Military Conflict from 1500 to 2000* (New York: Random House, 1987); McNeill, *The Pursuit of Power*; and Findlay and O'Rourke, *Power and Plenty.*

42. E. R. Crawford, "The Sikh Wars," in Brian Bond, ed., *Victorian Military Campaigns* (London: Hutchinson, 1967), pp. 35–36; Wickremesekera, *Best Black Troops*, pp. 67–73; Lenman, "Transition," pp. 114–16; Roy, "Military Synthesis," pp. 60–65.

43. Randolph G. S. Cooper, *The Anglo-Maratha Campaigns and the Contest for India: The Struggle for Control of the South Asian Military Economy* (Cambridge: Cambridge University Press, 2003), pp. 20–40; Sen, *Military System of the Marathas*, pp. 96–109.

44. William Henry Tone, *A Letter to an Officer on the Madras Establishment: Being an Attempt to Illustrate Some Particular Institutions of the Maratta People* (London: J. Debrett, 1799), quoted in Sen, *Military System of the Marathas*, p. 103.

45. Roy, "Military Synthesis," pp. 682–88.

46. Cooper, *Anglo-Maratha Campaigns*, pp. 45–56; Wickremesekera, *Best Black Troops*, pp. 69–75, 147–57.

47. Cooper, *Anglo-Maratha Campaigns*, pp. 244–46.

48. Ness and Stahl, "Western Imperialist Armies in Asia," pp. 15–17; Roy, "Military Synthesis," pp. 669–76.

49. Quoted in James A. Norris, *The First Afghan War, 1838–1842* (Cambridge: Cambridge University Press, 1967), pp. 255–56.

50. John H. Waller, *Beyond the Khyber Pass: The Road to British Disaster in the First Afghan War* (New York: Random House, 1990), p. 142. On the march to Kandahar, see pp. 133–45, and Norris, *First Afghan War*, pp. 242–67.

51. Waller, *Beyond the Khyber Pass*, pp. 200–260.

52. Ibid., pp. 257–79; Norris, *First Afghan War*, pp. 398–416.

53. Fauja Singh Bajwa, *Military System of the Sikhs, during the Period 1799–1849* (Delhi: Motilal Banarsidass, 1964), pp. 235–38; Hugh C. B. Cook, *The Sikh Wars: The British Army in the Punjab, 1845–1849* (Delhi: Thomson Press, 1975), pp. 17–36; Steven T. Ross, *From Flintlock to Rifle: Infantry Tactics, 1740–1866* (Rutherford, N.J.: Fairleigh Dickinson University Press, 1979), p. 170; Bhatia, *Military History of British India*, pp. 169–72; Waller, *Beyond the Khyber Pass*, pp. 124–25; Roy, "Military Synthesis," pp. 677–79.

54. On the Sikh Wars, see Donald F. Featherstone, *Victorian Colonial Warfare, India: From the Conquest of Sind to the Indian Mutiny* (London: Cassell, 1992), pp. 39–98; Ross, *From Flintlock to Rifle*, p. 171; Ness and Stahl, "Western Imperialist

Armies in Asia," pp. 18–19; Roy, "Military Synthesis," pp. 683–84; and Bajwa, *Military System of the Sikhs*, pp. 151ff.

55. George Benton Laurie, *The French Conquest of Algeria* (London: W. H. Allen, 1880), pp. 27–36.

56. Abdelkader Boutaleb, *L'émir Abd-el-Kader et la formation de la nation algérienne: De l'émir Abd-el-Kader à la guerre de libération* (Algiers: Éditions Dahlab, 1990), pp. 92–93; Jacques Frémeaux, *La France et l'Algérie en guerre: 1830–1870, 1954–1962* (Paris: Economica, 2002), pp. 98–100, 186–87.

57. Charles-André Julien, *Histoire de l'Algérie contemporaine*, vol. 1: *La conquête et les débuts de la colonisation (1827–1871)* (Paris: Presses Universitaires de France, 1964), pp. 53, 79, 182; Raphael Danziger, *Abd al-Qadir and the Algerians: Resistance to the French and Internal Consolidation* (New York: Holmes and Meier, 1977), pp. 25, 117, 226–27, 246; Boutaleb, *L'émir Abd-el-Kader*, pp. 92–94; Frémeaux, *La France et l'Algérie en guerre*, pp. 98–100; Laurie, *French Conquest of Algeria*, pp. 90–101.

58. Douglas Porch, "Bugeaud, Galliéni, Lyautey: The Development of French Colonial Warfare," in Peter Paret, ed., *Makers of Modern Strategy from Machiavelli to the Nuclear Age* (Princeton: Princeton University Press, 1986), p. 378.

59. Boutaleb, *L'émir Abd-el-Kader*, pp. 143–44.

60. Frémeaux, *La France et l'Algérie en guerre*, p. 97.

61. Antony Thrall Sullivan, *Thomas-Robert Bugeaud, France and Algeria, 1784–1849: Politics, Power, and the Good Society* (Hamden, Conn.: Archon Books, 1983), pp. 83–88; Frémeaux, *La France et l'Algérie en guerre*, pp. 103–5; Laurie, *French Conquest of Algeria*, p. 21.

62. Sullivan, *Thomas-Robert Bugeaud*, pp. 85–90; Frémeaux, *La France et l'Algérie en guerre*, pp. 103–7, 196–97; Porch, "Bugeaud, Galliéni, Lyautey," p. 378; Boutaleb, *L'émir Abd-el-Kader*, p. 147.

63. Sullivan, *Thomas-Robert Bugeaud*, pp. 87–88; Boutaleb, *L'émir Abd-el-Kader*, pp. 147–48; Frémeaux, *La France et l'Algérie en guerre*, pp. 196–97, 210–12.

64. Frémeaux, *La France et l'Algérie en guerre*, pp. 101–2, 158–59; Boutaleb, *L'émir Abd-el-Kader*, p. 146; Laurie, *French Conquest of Algeria*, pp. 208–9; Julien, *Histoire de l'Algérie contemporaine*, vol. 1, p. 178; Danziger, *Abd al-Qadir*, p. 235.

65. Boutaleb, *L'émir Abd-el-Kader*, pp. 168–72, 195; Frémeaux, *La France et l'Algérie en guerre*, pp. 180–84; Danziger, *Abd al-Qadir*, pp. 230–34; Sullivan, *Thomas-Robert Bugeaud*, pp. 94–95.

66. Muriel Atkin, "Russian Expansion in the Caucasus to 1813," in Michael Rywkin, ed., *Russian Colonial Expansion to 1917* (New York and London: Mansell, 1988), pp. 162–86.

67. Shamil is still a hero among the Chechens and many anti-Western Muslims around the world; see, for example, Muhammad Hamid, *Imam Shamil: The First Muslim Guerrilla Leader* (Lahore, Pakistan: Islamic Publications, 1979).

68. Eric Hoesli, *A la conquête du Caucase: Epopée géopolitique et guerres d'influence* (Paris: Syrtes, 2006), pp. 21–98; Firuz Kazemzadeh, "Russian Penetration of the Caucasus," in Taras Hunczak, ed., *Russian Imperialism from Ivan the Great to the*

Revolution (New Brunswick, N.J.: Rutgers University Press, 1974), pp. 256–59; Andrei Lobanov-Rostovsky, *Russia and Asia* (Ann Arbor, Mich.: G. Wahr, 1965), p. 118; Hugh Seton-Watson, *The Russian Empire, 1801–1917* (Oxford: Clarendon, 1967), p. 293; Philip Longworth, *Russia's Empires: Their Rise and Fall, from Prehistory to Putin* (London: John Murray, 2005), p. 207.

69. Kazemzadeh, "Russian Penetration of the Caucasus," pp. 261–62; Seton-Watson, *The Russian Empire*, pp. 416–17; John F. Baddeley, *The Russian Conquest of the Caucasus* (London: Longmans Green, 1908; reprint, Mansfield Centre, Conn.: Martino, 2006), pp. 458–82; Hoesli, *A la conquête du Caucase*, pp. 99–106.

70. Baddeley, *The Russian Conquest of the Caucasus*, p. 460n1. See also Nicholas V. Riasanovsky, *A History of Russia*, 2nd ed. (New York: Oxford University Press, 1969), p. 431.

71. Joseph Bradley, *Arms for the Tsar: American Technology and the Small-Arms Industry in Nineteenth-Century Russia* (DeKalb: Northern Illinois University Press, 1990), pp. 46–50, 83–103.

❈ Chapter 5 ❈

Steamboat Imperialism, 1807–1898

By the mid-nineteenth century, Western imperialism seemed to have reached its limits. Three centuries after Cortés, over half of the Americas were still Indian territory. In Asia, the British advance was stopped by the Afghans. Sub-Saharan Africa, the Middle East, and East Asia were off-limits to Europeans. The French conquest of Algeria required as many years as Napoleon's conquest of Europe, as did Russia's conquest of the Caucasus.

Then, starting in the 1830s, the old barriers began to crumble. Motives that had been dormant found a new energy. More important, advances in three areas of technology—steamboats, medicine, and weapons—gave Western nations new powers over nature. These, in turn, provided empire-builders with powers over non-Western peoples that they had not had before, allowing them to act upon their motives and realize their ambitions. Let us look first at the impact of steam navigation.

Our understanding of imperialism is distorted by mapmakers' habit of coloring land areas by political ownership, leaving the seas blue. Yet expansion and conquest can occur on water as well as on land. Seas and oceans have been as hotly contested, and sometimes as dominated, as land masses; this was true in the sixteenth century when Portugal carved itself a sea empire in the Indian Ocean, and in the eighteenth and nineteenth when Britannia ruled the waves. It is still true today, though maps do not show it.

In the age of sail, however, European naval domination could only reach so far. The sailing ships that ruled the seas were ill suited to sailing up rivers and in shallow waters and vulnerable to the galleys and junks of non-Western states and to the forts and guns along their coasts and riverbanks. Hence the liquid frontier in the waters

off Arabia and East Asia and the obstacles to navigation that had
hindered the advance of Europeans into Africa and parts of the
Americas.

Then came the Industrial Revolution. Of the many innovations
it brought forth, the first to tip the balance in favor of the West was
the application of steam power to navigation. An early proponent
of steamboat imperialism explained:

> By his [James Watt's] invention every river is laid open to us, time
> and distance are shortened. If his spirit is allowed to witness the suc-
> cess of his invention here on earth, I can conceive no application of
> it that would receive his approbation more than seeing the mighty
> streams of the Mississippi and the Amazon, the Niger and the Nile,
> the Indus and the Ganges, stemmed by hundreds of steam-vessels,
> carrying the glad tidings of "peace and good will toward men" into
> the dark places of the earth which are now filled with cruelty.[1]

Asians did not see it that way. The Persian prophet Siyyid Ali
Muhammad (1819–1850), one of the founders of the Baha'i religion,
traveled by ship from Persia to Mecca, a long and difficult trip. Because of this experience,

> He supplicated the Almighty to grant that the means of ocean travel
> might soon be speedily improved, that its hardships might be re-
> duced, and its perils be entirely eliminated. Within a short space of
> time, since that prayer was offered, the evidences of a remarkable
> improvement in all forms of maritime transport have greatly
> multiplied, and the Persian Gulf, which in those days hardly pos-
> sessed a single steam-driven vessel, now boasts a fleet of ocean liners
> that can, within the range of a few days and in the utmost comfort,
> carry the people of Fárs on their annual pilgrimage to the Hijáz.
>
> The people of the West, among whom the first evidences of this
> great Industrial Revolution have appeared, are, alas, as yet wholly
> unaware of the Source whence this mighty stream, this great motive
> power, proceeds—a force that has revolutionised every aspect of their
> material life. . . . In their concern for the details of the working and
> adjustments of this newly conceived machinery, they have gradually
> lost sight of the Source and object of this tremendous power which

the Almighty has committed to their charge. They seem to have sorely misused this power and misunderstood its function. Designed to confer upon the peoples of the West the blessings of peace and of happiness, it has been utilised by them to promote the interests of destruction and war.[2]

Whether their purpose was to bring "glad tidings of 'peace and good will toward men'" or "destruction and war," steamboats increased both humans' power over nature and the power of those who possessed them over those who did not.

Steamboats in North America

Historians agree that Robert Fulton's *North River*—popularly known as the *Clermont*—was the first commercially successful steamboat. Like all successful inventions, however, it was preceded by numerous attempts, a few of them technically successful but all hampered by one flaw or another.[3] The *Pyroscaphe*, launched by the French marquis Claude de Jouffroy d'Abbans on the River Saône in 1783, was the first boat powered by a steam engine—for fifteen minutes, until the hull cracked under the weight of the engine. Four years later the American John Fitch launched the steamboat *Experiment* on the Delaware River, but its engine was too weak to push it upstream. In Virginia, James Rumsey built a steamboat propelled by a jet of water, but it failed financially. In Scotland, William Symington's tugboat *Charlotte Dundas* successfully steamed on the Forth and Clyde Canal 1802, but was quickly retired for fear that the wash from its paddle wheels might damage the canal banks.[4]

Then came Robert Fulton's *North River*, completed in 1807, on which he steamed up the Hudson River from New York City to Albany in thirty-two hours and back in thirty. It was an instant sensation, for the trip by stagecoach involved several days of great discomfort. American steamboats filled a demand for transportation in a country with many rivers, very few good roads, and more restless and adventurous people than were to be found in longer-settled parts of the world. Fulton and his financial backer, Robert "Chancellor"

Figure 5.1. Robert Fulton's steamboat *North River* (or *Clermont*) in 1807.

Livingston, immediately sought to turn their good fortune into a commercial empire. Livingstone held a monopoly on steam naviga-tion on the Hudson River, which he and Fulton tried to extend to New York Harbor. They planned to build more steamboats, not only for the Hudson but also for Long Island Sound, the Chesapeake Bay, and the Delaware and other rivers.[5] They also had their eye on the Ohio and Mississippi rivers. Livingston, who had negotiated the pur-chase of the Louisiana Territory from France, kept an interest in the region, and his brother Nicholas moved to New Orleans to practice law. Soon after steamboats had become familiar sights in eastern waters, Livingston and Fulton sent a business associate, Nicholas Roosevelt, to survey the Ohio and Mississippi rivers. In October 1811 Fulton's steamer *Orleans* left Pittsburgh, arriving in New Or-leans, two thousand miles downstream, in January of the following year; later, it was put into service on the lower Mississippi between New Orleans and Natchez.[6] It was copied in 1813 by Daniel French's *Comet*, followed by Fulton's *Vesuvius* and French's *Enterprise* in 1814.

In 1816 another rival, Henry Shreve, built the *Washington*, a steamer of a radically new design. Until then, steamboats built by Fulton and French had their engines in the hull and drew three or

Figure 5.2. Model of the stern-wheel steamer *Red Bluff*, built in 1894 in the
San Francisco Maritime Museum. Note the extremely shallow draft.

more feet of water. The hull of Shreve's steamer, in contrast, was as
wide and flat as that of a barge and drew two feet of water or less.
Since the engine would not fit into the hull, it was placed on deck
and a second deck was built above it for passengers and freight. Instead of side wheels, it carried a single large wheel at the stern that
could be raised to avoid obstacles. To provide the power needed to
overcome the fast current of the Ohio River yet not take up too much
room, Shreve installed a high-pressure engine with steam pressure of
up to 150 pounds per square inch, a daring, if dangerous, design. This
boat could navigate the Ohio and Mississippi year-round, floating
over sandbanks that would have grounded a deeper hull. The *Washington* steamed up the Mississippi and Ohio rivers from New Orleans
to Louisville in twenty-five days, a voyage that would have taken
several months in a keelboat or canoe. Shreve's design set the pattern
for the classic Mississippi river boats from that day on.[7]

Thereafter, the number of steamboats grew dramatically. By 1820
sixty-nine were operating on the Ohio and Mississippi rivers; by
1830, 187; by 1840, 557; and by 1850, 740.[8] Steamboat historian
Louis Hunter estimated that their engines produced three-fifths
of all the steam power in the United States and were a major con-
tributor to the nation's industrialization.[9] Until coal was introduced,
they were fueled with wood cut along the riverbanks; at their peak,
the Midwestern steamers consumed some seventy square miles of
forest a day.[10]

The arrival of steamboats became a daily event in the towns along the river. Without steamers, major cities such as Cincinnati, Louisville, Memphis, St. Louis, and Baton Rouge would have remained small settlements at best. As for New Orleans, by the 1820s it was the second most important port in the nation, with a thousand steamers arriving every year.

As the Ohio-Mississippi region was invaded by white Americans and their black slaves, steamboats on the Ohio and Mississippi rivers helped bring about an economic, political, and demographic revolution. The United States was transformed from an Atlantic into a continental nation. By 1830 the ten states of the region had 3.5 million inhabitants, a quarter of the total U.S. population; by 1850, the total had risen to 8.3 million, or 36 percent of the total.[11] As one enthusiastic Cincinnatian wrote in 1840, "Of all the elements of the prosperity of the West, of all the causes of its rapid increase in population, its growth in wealth, resources and improvement, its immense commerce and gigantic engines, the most efficient has been the navigation by steam."[12]

To be sure, this was not simply the result of steamboats. It would have happened anyway, for the East Coast population was growing, immigrants were flooding in from Europe, and the demand for cotton, wheat, corn, pork, and other agricultural products was making such migrations lucrative and attractive. But the dramatic increase in the speed of travel and the drop in the cost of freight that steamboats brought changed the Midwest faster than anyone could have foreseen.

When the migrants invaded the region, the land they found was inhabited by Indians. The latter were unable to put up much resistance, being for the most part farmers whose survival depended on their crops. Overwhelmed by the numbers of immigrants and decimated by diseases, they were forcibly removed from their ancestral lands and driven off to "Indian Territory" in Oklahoma and beyond.

The United States coveted the lands beyond the Mississippi Valley. The Mississippi-Missouri watershed had been purchased from France in 1803 and first explored by Meriwether Lewis and William Clark between 1804 and 1806. This land was both teeming with wildlife—bison mainly—and accessible by boat, though just barely.

The inhabitants were not docile farmers but hunters whose skills as warriors surpassed those of European Americans. Once the pressures of settlement and the continental ambitions of the United States reached the Plains, both settlers and soldiers sought the help of the most advanced technology they could find. In the area watered by the Missouri River and its tributaries, this included steamboats. Here was a classic case of steamboat imperialism.

Unlike the Mississippi and Ohio rivers, which arise in well-watered lands and seldom dry up (though they are subject to ferocious floods), the Missouri drains the High Plains and the eastern Rockies, where rain and snowfall are seasonal and erratic. The river and its tributaries sometimes flooded violently, covering their flood plains for miles, and sometimes shrank to shallow sluggish streams wandering amid shifting sandbanks. Even more than their eastern counterparts, these Midwestern rivers carried snags, trees torn from their banks that could rip the hull of a boat and sink it in minutes. To the riverboat crews and their passengers, these were dangerous waters.

The first steamer to navigate the Missouri was the *Western Engineer*, built in 1818 at a government arsenal on the Allegheny under the direction of Major Stephen H. Long. In 1819 it steamed down the Ohio River and up the Mississippi to St. Louis. Five other steamers leased from private owners joined it, but proved to be too weak to fight the current or suffered mechanical breakdowns. The *Western Engineer* alone steamed 650 miles up the Missouri to Council Bluffs, where Colonel Henry Atkinson's troops built a fort named after him.[13]

This voyage was part of the Yellowstone Expedition that was sent to discover the source of the Missouri River and, in the words of Secretary of War John C. Calhoun, "convince Indians and British alike of our ability to assert and maintain control over so remote a region."[14] The British, so soon after the War of 1812, were still seen as dangerous rivals for control of the fur trade of the upper Midwest. One exuberant booster got so carried away at the thought of opening the Missouri to steamboat traffic that he foresaw "a safe and easy communication to China, which would give such a spur to commercial enterprise that ten years shall not pass before we shall have the rich productions of that country transported from Canton to the

Columbia, up that river to the mountains, over the mountains and down the Missouri and Mississippi, all the way (mountains and all) by the potent power of steam."[15]

In 1825, while Atkinson used keelboats to go upriver to the confluence of the Missouri and Yellowstone rivers, and beyond into Montana, a small steamer, the *Virginia*, explored the upper Mississippi and Minnesota rivers.[16] The Indians realized that these were not just explorations but the advance guard of an invasion. The election of Andrew Jackson in 1828 led directly to the Indian Removal Act of 1830, putting an official stamp on the ethnic cleansing that had been going on for many years. When the Sauk and Fox Indians, led by Black Hawk, tried to return to their homelands in northwestern Illinois in 1832, they were met not only by U.S. Army infantry and cavalry units but also by the steamboats *Enterprise*, *Chieftain*, and *Warrior*, the latter armed with a cannon able to shoot a six-pound cannonball.[17] Thereafter, steamboats regularly plied the rivers of the western Plains, carrying prospectors and hunters—and their diseases—upriver and returning with buffalo skins.[18]

Meanwhile, what European Americans called the Indian Wars continued in the valley of the Missouri. By the mid-century, hundreds of migrants were making their way west every year along the Oregon and Mormon trails, lured by tales of land and gold to be had in the Far West and in California and Oregon. Along the way, they competed with the Indians for bison and grazing lands. In 1862, as troops and weapons were redeployed from the northern Plains to the battlefields of the Civil War, the Plains Indians saw an opportunity to recover their stolen lands. A massacre of white settlers in Minnesota brought down retaliation by the U.S. Army. In the spring of 1863, as General Sibley and two thousand men moved west from the Minnesota Valley toward the Missouri, General Alfred Sully and another two thousand soldiers attempted to ascend the Missouri. Low water, however, detained their steamboats, depriving them of needed supplies. The next year Sully returned with eight steamers and established Fort Pierre in what is now South Dakota and Fort Rice in North Dakota. Tribes such as the Winnebago who had not participated in the massacres and battles nonetheless fell victim to ethnic cleansing, or, as an early historian of these events put it, "the hand of vengeance fell upon them as upon the others." Those who survived were transported by steamboat into the Dakotas.[19]

When prospectors found gold in the Black Hills of the Dakota Territory and in what later became Montana and Idaho, the trickle of migrants became a torrent. Most of the prospectors and their heavy equipment relied on steamboats to carry them up the Missouri to Fort Buford at the mouth of the Yellowstone River, the highest point that steamboats could reach. Steamers left St. Louis in early April, just as the snows were melting and the river rising, and reached Fort Buford two months later. Beyond that, miners used mules and ox-drawn wagons to reach the mines. Until 1866, only half a dozen steamboats had reached Fort Buford annually. That year, the total jumped to thirty-one, and the next year, thirty-nine steamers arrived, carrying ten thousand passengers.[20]

The Sioux, meanwhile, escaped and continued to fight a guerrilla war against the river forts and the migrants crossing the Plains.[21] In 1876 they rose up once again under Chief Sitting Bull and defeated General Custer at the Battle of Little Bighorn. In 1877 General Miles went to war against the Nez Perce Indians led by Chief Joseph, using steamboats to patrol the Missouri River to prevent the Indians from escaping to Canada.[22]

With the exception of the *Warrior*, steamboats were not designed for warfare. Instead, they supplied transportation for the soldiers, tradesmen, settlers, and miners who invaded the northern Plains. They carried passengers, animals, weapons, equipment, and food upriver; on the return they carried furs, bison pelts, gold, and passengers. The greatest danger they ran was striking a snag, getting stuck on a sandbank, or suffering a mechanical problem, all of which happened with appalling frequency. Indians shot at steamboats from the riverbanks, but with little effect; as historian Joseph Hanson has explained, "there were still hovering along the banks of the Missouri a great many hostiles whose attacks upon passing steamers and even upon the military posts of the region were frequent and annoying." Annoying, perhaps, but not dangerous: "the pilot-house of the *Louella*, like that of every upper-river boat was sheathed with boiler iron against which the bullets of the savages might patter harmlessly."[23]

Without steamboats, it would have been much more difficult and dangerous to supply the many forts and towns built along the Missouri River in the decades after 1818. As General Sully remarked,

> It is impossible to estimate the great value in the military operations of this important line of communications. Forts and cantonments were strung all along the river from Fort Randall to Fort Benton, and all of them, as well as the troops in the field, depended for their support upon the riverboats. The conquest of the Missouri Valley would have been a very different matter had the government been deprived of this important aid in its operations.[24]

Steamers in South Asia

The potential of steamboats, so apparent to Americans, was no secret to Europeans. To be sure, Europeans were slower to adopt this new technology than were their transatlantic cousins because Europe had more roads and no nearby "wilderness" begging to be invaded. In Europe, therefore, steamboats supplemented existing means of transportation. The first commercially successful steamer in the Eastern Hemisphere was Henry Bell's *Comet*, built in 1812 to carry passengers between Glasgow and Greenock on the River Clyde in Scotland. Thereafter, steamboats and steamships multiplied, especially in the waters around the British Isles. By one estimate, there were 628 steamers in Britain in 1837, still fewer than in the United States.[25]

Where the introduction of steamboats marked a turning point in history was on the frontiers of empire. And the first place to feel its effects was the capital of Britain's overseas possessions, Calcutta. The first steamboat east of Europe was a pleasure boat built for the nawab of Oudh in 1819. It was followed by the *Pluto*, built in 1822 as a harbor dredge and converted two years later to a paddle-steamer, and the *Diana*, a 132-ton side-wheeler with two sixteen-horsepower engines launched in Calcutta in 1823 and used mainly as a tugboat pulling oceangoing East Indiamen on the River Hooghly.[26] In 1825 they were joined by the *Enterprize*, a sailing ship with an auxiliary steam engine that arrived from England.

With these vessels, the East India Company inaugurated a new kind of war: river warfare. In 1824 the governor-general of India, Lord William Amherst, decided to punish King Bagyidaw of Burma for invading Assam. To do so, he had to mount an amphibious attack, for Burma is surrounded by mountains on three sides and could not easily be reached except up the valley of the Irrawaddy River. At first, the campaign went very poorly for the British, whose brigs and schooners had difficulty navigating on the meandering Irrawaddy. The Burmese fought with fast river-galleys called praus propelled by up to a hundred double-banked oars. This was the classic confrontation between blue-water and brown-water navies that had thwarted the European advance for three hundred years.

Then the East India Company brought in its steamers *Enterprize*, *Pluto*, and *Diana* and later two more, the *Irrawaddy* and the *Ganges*.[27] The *Enterprize* was employed as a transport to ferry troops and supplies from Calcutta to Burma. The converted steam-dredge *Pluto* was armed with two six-pounder cannon and other weapons and took part in the attack on the Arakan coast. The star of the campaign was the little *Diana*, armed with swivel guns and Congreve rockets. The Burmese sent praus and fireboats into the path of British ships, but the *Diana* could outrace the praus and avoid the fireboats. As for the Burmese muskets, spears, and swords, they were no match for British naval artillery. In February 1826, Burma was forced to sign the treaty of Yandabu, by which it lost Assam and the Arakan and Tenasserim coasts. In the process, the East India Company and its officials in India learned the military value of steamboats.[28]

The Ganges River had long been the lifeline of Hindustan, in constant use by country boats. Like on the Missouri, navigation was impeded by shallow water and shifting sandbars and by wide variations between rainy autumns and the dry spring and early summer seasons. As its control over Hindustan grew tighter and its tax revenues more lucrative, the Indian government felt it essential to improve communications. In 1828, therefore, a new governor-general, Lord William Bentinck, sent Captain Thomas Princep to survey the river for use by steamers.

Beginning in 1834, the *Lord William Bentinck*, a 120-foot-long iron stern-wheeler pulling an accommodation boat for passengers, steamed between Calcutta and Allahabad, some six hundred miles away. The voyage upriver took twenty days in the rainy season and twenty-four in the dry season, with a return trip of eight and fifteen days, respectively. In the next few years it was joined by several other steamers. All of them were made of iron in Britain, shipped out in parts to India, and assembled in Calcutta.

The service was mainly used by the East India Company to transport its officials and magistrates and its important documents and to ship the taxes it collected in upper Hindustan down to Calcutta. The company showed no inclination to provide service for ordinary freight or passengers. A ticket between Calcutta and Allahabad cost as much as crossing the Atlantic; only bishops, planters, and Indian princes could afford to pay their own way. As for freight rates, they excluded everything but the most valuable goods, such as indigo, silk, opium, and lac. The contrast with the surge in steamboat service on the Mississippi—between colonialism and colonization— could not be more glaring.[29]

On the Indus River, steamboats were rarer than on the Ganges, for there was no British-run city at its mouth as in the case of the Ganges, and the East India Company did not acquire Sind (the lower Indus Valley) until 1843 and Punjab (the upper valley) until 1849. Furthermore, the river was even more treacherous and erratic than the Ganges. The first to attempt to send a steamer up the Indus was Agha Mohamad Rahim, a Persian merchant in Bombay. In 1835 he sent out a small steamer, the *Indus*, but its engine was too weak for the current and it had to be pulled upstream; it returned to Bombay a year later. By 1840 four steamers operated on the river, but their performance was disappointing, for they drew too much water and their engines were too weak for the current.[30]

Routes to India

The British community of Calcutta was not content to have a few steamers to use locally. Their most pressing demand was for better communication with their homeland. This desire was not driven solely by nostalgia but also by a booming trade; in particular, British cotton exports to India grew from 817,000 yards in 1814 to almost 24 million in 1824.[31] In 1822 the British naval officer and steam enthusiast James Henry Johnston came to Calcutta to seek funding for a steamship that could connect Britain and India. In response, several eminent Calcuttans formed a "Society for the Encourage-ment of Steam Navigation between Great Britain and India," offer-ing a prize to the first ship that could steam twice between Britain and Bengal in no more than seventy days per voyage. Governor-General Lord Amherst pledged 20,000 rupees, the nawab of Oudh 2,000, and Calcutta businessmen another 47,903, for a total of 69,903 rupees, or over five thousand pounds. Armed with these pledges, Johnston returned to London, where he raised enough money to build the *Enterprize*, a 464-ton vessel with two sixty-horse-power engines. It was the first steamer to venture out into the ocean. Unfortunately, steam technology was not yet up to such a task, and the ship took 113 days to reach Calcutta, having run out of coal halfway there.[32]

The failure of the *Enterprize* to fulfill the demands of the Calcutta steam committee did not dampen the eagerness of the British com-munity in India to improve their communications with their home-land. They were encouraged by the new governor-general, Lord Ben-tinck, who came up the Hooghly River to Calcutta onboard the *Enterprize*. During the seven years of his governor-generalship (1828–35) he supported steam navigation for moral as well as com-mercial reasons. In his eyes, steam power was "the great engine of working [India's] moral improvement. . . . in proportion as the com-munication between the two countries shall be facilitated and short-ened, so will civilized Europe be approximated, as it were, to these benighted regions; as in no other way can improvement in any large stream be expected to flow in."[33]

Communications between Europe and India could follow three possible routes: around Africa; via Syria, Mesopotamia, and the Per-sian Gulf; or via Egypt and the Red Sea. The ocean route avoided the volatile Middle East but took a sailing ship six to nine months. During the Napoleonic Wars, the Royal Navy had cleared all enemy ships from this route, but it still suffered from storms and shipwrecks. The other two, called Overland Routes because they required a land passage, were suited only for passengers and mail. While much shorter, they were vulnerable to the vicissitudes of Ottoman and Egyptian politics and the frequent chaos in the lands they traversed. Of the two, the Red Sea, with its notoriously fickle winds and dangerous rocks, was too dangerous for sailing ships, though potentially open to steamers. The East India Company and travelers who wished to avoid the long and tedious ocean route preferred the Persian Gulf, which was safer to navigate and where the Bombay Marine protected merchant ships from pirates. The voyage from Bombay to Basra, at the head of the Gulf, took thirty to seventy-five days, depending on the season; from Basra travelers continued on camels or horses for another three to six weeks to Aleppo or Istanbul, then on by ship, horse, or stagecoach to Britain. It was an arduous but picturesque trip lasting five or six months.[34]

The choice of routes was not just a matter of geography and taste; it was also matter of politics. The East India Company in London, the government of India, and the British communities of Calcutta and Madras preferred the familiar Cape route. Since Britain was a thousand miles closer to Bombay than to Calcutta by the ocean route, and even closer via the Middle East, the merchants of Bombay realized they had much more to gain from steam navigation than did their counterparts and government officials in Calcutta. Yet they knew that no steamer could carry enough fuel for the long voyage around Africa. They therefore preferred the Overland Routes.

In 1823 and again in 1825–26, Mountstuart Elphinstone, the governor of Bombay, proposed a steamboat service via the Red Sea, but was ignored by the Court of Directors of the East India Company. His successor, Sir John Malcolm, ordered the Bombay Marine (as the Indian Navy was called until 1830) to survey the Red Sea and establish coal depots between Bombay and Suez, in preparation for a possible steam service. He ordered two engines from England and had a steamer built in Bombay of Indian teak. He named it *Hugh Lindsay*, in honor of the chairman of the East India Company who had forbidden its construction.[35]

The *Hugh Lindsay* left Bombay on May 20, 1830, and reached Suez after a thirty-three-day voyage, of which twelve were occupied loading coal at Aden. The letters it carried reached England in a record fifty-nine days. Technically, it was a success, for it opened up the possibility of transporting passengers, mail, and precious freight between Britain and India in two months or less. However, the round-trip voyage between Bombay and Suez cost seventeen hundred pounds, mostly for coal that had to be shipped out from Britain. The price was so high that the company forbade further trips.[36] Despite repeated prohibitions, the Bombay presidency sent the *Hugh Lindsay* on four more trips to Suez in the following three years, some as short as twenty-two days, all to carry a handful of passengers and one or two dozen letters.

The Bombay and Calcutta communities eagerly embraced steam power, but in London the East India Company, the Admiralty, the Treasury, the Post Office, and the Foreign Office all resisted steamer service as too radical and too costly. Yet the *Hugh Lindsay* caused much excitement among steam enthusiasts in Britain, and they began bombarding newspapers with letters and Parliament and the Court of Directors of the company with petitions. In doing so, they forced the hand of the company and of Parliament.[37]

Under pressure from independent merchants, the British government abolished the company's monopoly on trade in 1833, reducing it to its political role as the administrator of India, and appointed the chairman of the Board of Control that replaced the Court of Directors; de facto, the company was now a branch of the government. But as its profits had vanished along with its monopoly, it became very parsimonious. All new initiatives were now in the hands of the government.

The Euphrates Route

In response to public opinion, the House of Commons appointed a Select Committee on Steam Navigation to India in June 1834.[38] The first witness it called was the novelist and critic Thomas Love Peacock, a rising official in the East India Company. In 1829, when John Loch, the chairman of the company, asked him to look into the question of steam navigation, Peacock wrote a "Memorandum respecting the Application of Steam Navigation to the internal and

external Communications of India."[39] Peacock's "Memorandum"
was entered into evidence before the Select Committee. In it he
argued that steam communications with India should go via the Euphrates River for geopolitical reasons, namely fear of Russia. When
questioned by Sir Charles Grant, the chairman of the committee,
Peacock explained: "The first thing the Russians do when they get
possession of or connexion with any country is to exclude all other
nations from navigating its waters. I think, therefore, it is of great
importance that we should get prior possession of this river." Grant
then asked, "Is it your opinion that the establishment of steam along
the Euphrates would serve in any respect to counteract Russia?" and
Peacock replied, "I think so, by giving us a vested interest and a
right to interfere."[40] At the time, the Russians, far from preparing to
invade Mesopotamia, were worried that British agents then known
to be visiting the khanates of Khiva and Bukhara were the advance
guard of a British invasion of Central Asia. In short, Peacock was
playing the Great Game in which Britain and Russia each used the
other as an excuse to justify their attempts to conquer the lands
lying between them.[41]

The second witness to appear before the Select Committee was
Artillery Captain Francis Rawdon Chesney. From 1829 to 1833, he
had traveled first down the Red Sea, then through Syria and down
the Euphrates to Basra, in order to compare the two routes.[42] When
he returned to England in 1833, he wrote a memorandum on the
Euphrates route that Peacock brought to the attention of King William IV.[43] The king told him of "serious apprehension caused by the
presence of the Russian fleet at Constantinople, as well as by the
gradual advance of that Power toward the Indus."[44] Appearing before
the Select Committee, Chesney recounted his voyages and strongly
endorsed the Euphrates route.[45]

For advice on steamboat technology the Select Committee turned
to Macgregor Laird, the son of the iron founder and shipbuilder William Laird.[46] In 1832 he had built two steamers, one of them with
an iron hull, with which he and his companions explored the lower
Niger River. Upon his return, he was called to testify before the
Select Committee. He described the advantages of iron-hulled river
steamers for explorations; among them were that an iron boat could

be taken apart and the parts packed in the hold of a cargo ship, then reassembled at its destination. Better yet, it could be carried in pieces overland and put together on a river that a wooden boat could not reach. This proved to be important in many exotic locales where rapids interrupted the course of major rivers.[47] The ideas Laird presented were radically new. To be sure, iron steamboats were known since 1822, when the *Aaron Manby* had begun carrying passengers up and down the Seine between Paris and Le Havre. But the British Admiralty clung to its wooden sailing ships, resisted steamers, and adamantly refused to purchase iron ships. For the Select Committee to show an interest in iron steamers demonstrated a technologically adventurous spirit.

Following the advice of Chesney and Peacock, the Select Committee recommended the Euphrates route. Parliament appropriated twenty thousand pounds toward a steamboat expedition using the Euphrates route to be led by Chesney; the East India Company contributed another eight thousand.[48] With that money, Chesney purchased two steamboats from Laird and Sons: the *Euphrates*, a 179-ton boat 105 feet long by 19 wide and drawing three feet of water, and the *Tigris*, a smaller 109-ton, ninety-foot-long boat drawing only eighteen inches of water. Both were powered by Maudslay engines of fifty and forty horsepower, respectively. The boats were armed with cannon, rockets, and muskets for the crew. Many of the crew members were artillerymen trained at Laird's shipyard.[49]

Chesney had the boats and other materiel shipped to Suwaidiyyah in Syria, from where they were to be transported across the desert to Birecik on the upper Euphrates. This part of the Syrian coast was under Egyptian occupation, and obtaining authorization to proceed with the expedition required long negotiations. That 110-mile trek from Suwaidiyyah to the Euphrates was extremely difficult, for the parts—including two seven-ton boilers—had to be carted or carried by camels over hills and through swamps. The expedition had landed in April 1835, but because of the delays, the last of the steamboat parts and equipment did not reach the Euphrates until November.

There its troubles were just beginning. The two steamers were finally launched the following March. The river was full of sandbanks, rocks, and whirlpools and the boats often got stuck. The *Euphrates* was once grounded for almost three weeks. The little *Tigris* sank in a storm with most of its passengers, though Chesney himself survived. The *Euphrates* finally reached Basra on the Persian Gulf in June 1836, fifteen months after landing in Syria. Chesney made an attempt to steam up the river, but had to give up and return to the Mediterranean by way of the ancient desert route to Damascus.[50]

Chesney's expedition demonstrated conclusively that the Red Sea route to India was far better than the Euphrates route, especially as the technology was fast advancing; but then, Chesney had never been interested in transporting passengers and mail between Britain and India. What he did demonstrate was that the Euphrates and Tigris rivers were navigable by steamboats (if just barely) and that Mesopotamia was commercially and strategically inviting and accessible to British penetration.

The Red Sea Route

In 1837, in response to the failure of Chesney's expedition, Parliament appointed a Select Committee "to inquire into the best means of establishing communication by steam with India by way of the Red Sea."[51] The chairman was none other than Lord Bentinck, former governor-general of India and a steam enthusiast in whose opinion "in no part of the world will the power of steam have produced a greater multiplication of the existing means of national wealth and strength as it will in India, as those upon which the maintenance of our political power depends."[52] Once again, the star witness was Thomas Love Peacock, promoted in 1836 to chief examiner of correspondence upon the death of his mentor, James Mill.[53] Peacock, a technological radical, recommended that the company purchase steamers for the Bombay-Suez route, most of them of iron and most built by Lairds. To defray the expense, Parliament voted £37,500 in 1837, £50,000 in 1837–38, £50,000 more in 1839, and another £50,000 in 1840.[54] In short, Parliament, following in the footsteps of the Bombay government and the East India Company, had become a convert to steam navigation via the Red Sea.

By 1837 steam engines and naval architecture had finally made steam power practical on long ocean voyages. The East India Company sent two new steamers, the 620-ton *Atalanta* and the 680-ton *Berenice*, out to India around the Cape to join the *Hugh Lindsay*; the *Atalanta* was the first ship to steam all the way. The next year they were joined by the *Semiramis*, then in 1839 by the *Zenobia* and the *Victoria*, and in 1840 by the *Auckland*, the *Cleopatra*, and the *Sesostris*, some of them built in Bombay with British engines.[55] Between Bombay and Suez, passengers and mail sailed on the new steamers of the Indian Navy. The desert crossing from Suez to Alexandria was in the hands of private companies. On the Mediterranean portion of the route, passengers and mail sailed on Admiralty packets or, after 1840, on the steamers of the Peninsular and Oriental Steam Navigation Company—the P&O—or on French steamers to Marseilles.[56]

Steam was both the means and the incentive for Britain to take over parts of the Middle East. Except for the years 1797–1801 when the French occupied Egypt, the British showed very little interest in the Red Sea until the *Hugh Lindsay* demonstrated its value as a route to India. In 1829, knowing that the *Hugh Lindsay* could not carry enough coal to steam all the way to Suez without refueling, the Bombay government thought of using the port of Aden as a coaling station. But the town of Aden had shrunk to a mere village and the *Hugh Lindsay* had such difficulty finding laborers there that it took several days to load 180 tons of coal.[57] The Bombay Marine briefly occupied the island of Socotra, halfway between Arabia and Somalia, but then abandoned it because it lacked water and good harbors. By 1836 it once again turned its attention to Aden and leased a part of the harbor as a coal depot.

The requirements of steam navigation were necessary, but not sufficient, for Britain to take over Aden. Then came the Egyptian attempt to conquer Yemen in 1836–37.[58] Captain James Mackenzie of the Bengal Light Cavalry wrote:

> I doubt the propriety of our permitting the Pasha [Muhammad Ali, ruler of Egypt] to take Aden. . . . we, so intimately connected with that part of the world in consequence of its being the best and nearest route to India—and so much superior in knowledge, power and civi-

lisation, should not ourselves take and keep possession of Aden whose noble harbours would be of the greatest benefit to the prosecution of our Indian Steam Navigation Plans. Besides giving us commercial advantages in Arabia, Abyssinia and the north coast of Africa . . . it would be the means of extending our knowledge and religion over people now immersed in the profoundest ignorance. It seems to be a law of nature that the civilised nations shall conquer and possess the countries in a state of barbarism and by such means, however unjustifiable it may seem at first sight, extend the blessings of knowledge, industry, and commerce.[59]

In 1836, when a ship from Madras (hence under British protection) foundered off the coast near Aden, the passengers were harassed and their possessions plundered by local tribesmen. This "outrage" led Sir Robert Grant, the governor of Bombay, to write to the governor-general, the Earl of Auckland: "The establishment of a monthly connection by steam with the Red Sea . . . renders it **absolutely necessary** that we should have a station of our own on the coast of Arabia . . . the insult . . . by the Sultan of Aden gives us leave to take possession of the town."[60] When the governor-general demurred, Grant wrote the Secret Committee of the company of his decision "to adopt immediate measure for an attempt to obtain the peaceable possession of Aden, without waiting for instructions from the Governor-General."[61] He then dispatched two warships and eight hundred soldiers to Aden. On January 19, 1839, after a three-hour assault, they seized the city.[62]

Taking Aden was but the first step in Britain's growing involvement in the Middle East; as an old Arab sheikh told Chesney, "The English are like ants; if one finds a bit of meat, a hundred follow."[63] After Egypt seized Palestine and Syria from a faltering Ottoman Empire, it seemed that the Ottoman Empire was about to crumble as the Mughal Empire had a century before. Britain, worried about its new routes to India, decided to shore up the faltering Ottoman Empire by forcing the Egyptians out of Syria and Palestine. In 1840–41 it sent a fleet of warships, including several steamers, to bombard the Egyptian forces in Acre, Sidon, and Beirut. The admiral commanding the fleet, Sir Charles Napier, had larger visions: "Steam navigation, having got to such perfection, Egypt has become almost necessary to England as the half-way house to India, and indeed ought to be an English colony. Now if we wished to weaken Mehemet Ali, with a view, in the event of a breakup of the Turkish Empire, which is not far distant, to have seized Egypt as our share of the spoil, we were perfectly right in our policy."[64] Napier was ahead of his time, but he understood the logic of empire. Britain had the motivation—communications with India—and the means—steam navigation. It was only waiting for the right opportunity to take over Egypt, Palestine, and Mesopotamia.

By 1840 Britain had extended tentacles of steam from India to Burma, Mesopotamia, the Red Sea, and the eastern Mediterranean, following the same impulse that was drawing Americans into the Midwest and beyond. Then came an even more astonishing demonstration of the power of steam: the Opium War.

Britain and China

The first British ships arrived off the coast of China in the 1620s. For the next two centuries, the East India Company and the merchants of Guangzhou (or Canton) carried on a very one-sided trade; the British wanted tea, silk, porcelain, and other fine Chinese products, but offered nothing that interested the Chinese except silver. When a British embassy headed by Lord George Macartney went to China in 1793 to seek more trade and to open diplomatic relations, the Qianlong emperor wrote to King George III: "As your Ambassador can see for himself, we possess all things. I set no value on objects strange or ingenious, and have no use for your country's manufactures."[65]

This one-sided trade began to change after 1800, when the British discovered a product that the Chinese craved in ever-increasing quantities: opium from India. Twice, in 1800 and in 1813, the Chinese government prohibited the import of opium, but the demand was so insistent and the profits so great that the prohibitions only turned honest trade into smuggling lubricated by corruption. In 1816 Britain attempted to make its dealings with China conform to the European pattern of international relations by sending another embassy, but its ambassador, Lord Amherst, the former governor-general of India, was turned away from Beijing without being granted an audience with the Xianfeng emperor. By 1821 China was importing some five thousand chests of opium a year.

Until 1834, trade with China was controlled by the East India Company, which had an interest in keeping the trade flowing smoothly (if illegally). As historian Gerald Graham has explained, "Appeasement in terms of extravagant fees, bribes, and constant self-abnegation proved to be a largely effective, if somewhat humiliating, British operating policy."[66] That year, however, its trading monopoly ended, opening the field to wildcat entrepreneurs who were far less tolerant of Chinese ways or far less willing to accept any humiliation whatsoever. Their attitude was buttressed by the appearance of steamboats. William Jardine, one of the founders of the trading firm Jardine, Matheson & Company, had purchased the 115-ton steamer *Jardine*, hoping to use it as a dispatch boat between Guangzhou and the company's establishment on nearby Lintin Island. The governor of Guangzhou ordered it to leave, however, adding,

> If he presumes obstinately to disobey, I, the acting-governor, have already issued orders to all the forts that when the steamship arrives they are to open a thundering fire and attack her. On the whole, since he has arrived within the boundaries of the Celestial Dynasty, it is right that he should obey the laws of the Celestial Dynasty. I order the said foreigner to ponder this well and act in trembling obedience thereto.[67]

In response, Jardine wrote in the *China Repository*: "Nor indeed should our valuable commerce and revenue both in India and Great Britain be permitted to remain subject to a caprice, which a few gunboats laid alongside this city [Guangzhou] would overrule by the discharge of a few mortars. . . . the results of a war with the Chinese could not be doubted."[68]

The self-righteous arrogance of the Celestial Empire had met its match in the self-righteous arrogance of the Mistress of the Seas. For centuries, the two great powers had been deadlocked by geography.

China's Grand Canal, completed in the sixteenth century to shield its vital north-south trade from pirates, made it impervious to threats from the sea. Britain's men-of-war, meanwhile, only reached as far as the shores of the South China Sea. In 1834, Lord Napier, the British trade commissioner in China, sent two warships, the *Imogene* and the *Andromache*, up the Pearl River to attack the forts protecting Guangzhou, but the only result was that the Chinese temporarily halted the trade.[69]

As more and more Chinese became addicted to opium, illegal imports rose sharply to thirty thousand chests in 1835 and forty thousand in 1838. In 1839 the emperor sent Lin Zexu, a tough-minded official, to Guangzhou with orders to put a stop to the trade. When he arrived, Lin seized and destroyed fifteen hundred tons of raw opium and ordered the house arrest of the Western traders in the city and of the British superintendent of trade, Charles Elliot. The British were outraged. Captain William Hall, a participant in the imminent war, expressed his indignation in these words:

> The harsh and unwarranted measures of the Chinese Commissioner Lin, the imprisonment of Her Majesty's Plenipotentiary and all other English subjects, and his wild career of uncontrolled violence, called imperatively on our part for stronger measures than had yet been resorted to; and such measures were at once adopted by the Court of Directors of the East India Company, as well as by the government of the country, their direct object being the speedy departure of an adequate force for the protection of British trade in China, and to demand proper reparation for the violence and insult offered to Her Majesty's representative.[70]

And so Britain prepared for war. Unlike France, Russia, and the United States, Great Britain was not interested in acquiring more land—it had enough with India and, more recently, Australia. Rather, being the world's dominant manufacturing and mercantile nation, it used its power to impose free trade—that is, trade on its own terms—on the rest of the world.[71] But it was to be a new kind of war, one that would have been impossible just ten years earlier. As Captain William Hall explained, "It was scarcely to be expected that, under these circumstance, hostilities could be altogether avoided; and, as the principal scene of them, if they should occur, would be in rivers, and along the coasts, attention was directed to the fitting out of armed vessels peculiarly adapted for that particular service."[72]

What Hall meant by "armed vessels peculiarly adapted for that particular service" was iron steamboats. Here was the technology that would break the deadlock between Britain and China.

As the Admiralty was opposed to steamers and to iron ships, the task fell upon the East India Company. On Peacock's advice, the Secret Committee of the company ordered six new steamers. Four of them—the *Nemesis, Phlegethon, Ariadne,* and *Medusa*—were from Lairds; the other two—*Pluto* and *Proserpine*—were built in London by Ditchburn & Mare.[73]

The *Nemesis*

The *Nemesis* was the first to be completed. At 660 tons, it was the largest iron steamer yet built. It was 183 feet long by 29 wide, exclusive of the paddle boxes. It was a new kind of vessel, equally able to sail across oceans and steam up rivers and in shallow waters. It had a flat bottom and a draft of only six feet when fully loaded. It was divided by bulkheads into watertight compartments, a design the Chinese had invented centuries before. It was powered by two sixty-horsepower engines. For voyages at sea, it carried masts and sails and two movable keels and a rudder that could be lowered or raised. It was also well armed, with two pivot-mounted thirty-two-pounders, five six-pounders, ten smaller cannon, and a rocket-tube.[74]

In October 1839 the *Nemesis* was assigned to Captain Hall, a veteran naval officer who had served in the West Indies, the Mediterranean, and the Far East. He had also studied steam engineering and worked on steamboats in the United States.[75] He joined the ship in December and recruited a crew. On March 28, 1840, the *Nemesis* left Portsmouth. Before it sailed, Sir John Cam Hobhouse, chairman of the East India Company's Board of Control, informed Foreign Secretary Palmerston, "An Armed iron vessel, called the Nemesis, which has been provided by the Secret Committee of the East

Figure 5.3. The *Nemesis*, British steam-powered gunboat used
in the Opium War (1839–42).

India Company . . . for the service of the Government of India, is
about to proceed to Calcutta. . . . It is desirable that the destination
of the Nemesis, and the authority to which she belongs, should not
be mentioned."[76]

When the governor-general of India heard about the new steamers, he thought "they will be invaluable should that reckless savage
of Ava [i.e., the king of Burma] force us into hostilities with him."[77]
According to Hall, the *Nemesis* "was cleared out for the Russian port
of Odessa, much to the astonishment of every one; but those who
gave themselves time to reflect, hardly believed it possible that such
could be her real destination."[78] Indeed, it was not possible to hide
a heavily armed, privately owned warship from the speculation of
the press. *The Times* guessed that "it can only be against the Chinese;
and for the purpose of smuggling opium she is admirably adapted."[79]
The *Shipping Gazette* opined: "She will, it is said, clear out for Brazil;
but her ultimate destination is conjectured to be the eastern and
Chinese seas."[80] Even Hall did not know the final destination of the
ship he commanded. He had orders to go to Capetown and then
Ceylon; not until the *Nemesis* arrived there on October 6 did Hall
learn that he was to proceed to Guangzhou and place his ship under

the orders of the naval commander in chief in China.[81] Why the secrecy? Certainly not to hide the *Nemesis* from the Chinese, who had no spy in Britain and no way to communicate if they had. Rather, it was to hide from the Admiralty, still so technologically reactionary, the information that the East India Company was going to war with a new and dangerous type of warship.

Before the war broke out, the Chinese had found steamers interesting, but not terribly impressive. Lin called them "cart-wheel ships that put the axles in motion by means of fire, and can move rather fast."[82] Others called them "ships with wind-mills" or "fire-wheel boats" and noted: "Steam vessels are a wonderful invention of foreigners, and are calculated to offer delight to many."[83] The *Nemesis* offered fewer delights. By the time it arrived off Macao in November 1840, war had already broken out. In January 1841, the British decided to attack the forts on the Bogue (or Bocca Tigris) protecting Guangzhou. The Chinese defenses were weak. Before the war began, Commissioner Lin had purchased an armed American merchantman, the *Cambridge*, but kept it locked behind a barrier of rafts, for he had no sailors able to maneuver it. Chinese war junks were one-fifth the size of British men-of-war; their few guns shot cannonballs of less than ten pounds and were difficult to aim.[84] The Chinese also used fire rafts filled with oil-soaked cotton and gunpowder. They relied primarily on the cannon in their forts, some of them cast by Jesuits during the Ming dynasty centuries before and fixed in masonry so they could not be aimed.[85]

With its flat bottom, the *Nemesis* steamed up the shallow channels of the river, pulling fire rafts out of the way, towing larger sailing ships and boats filled with soldiers and destroying Chinese junks with its shells and Congreve rockets. With its help, the British forces were able to destroy the Bogue forts and the defenses on Whampoa Island, opening the way to Guangzhou. In May, Captain Hall wrote to Peacock:

> With respect to the Nemesis I cannot speak too highly in her praise. She does the whole of the advanced work for the Expedition and what with towing Transports, Frigates, large Junks, and carrying Cargoes of provisions, troops, and Sailors, and repeatedly coming into

contact with Sunken Junks—Rocks, Sandbanks, and fishing Stakes in these unknown waters, which we are obliged to navigate by night as well as by day, she must be the strongest of the strong to stand it. . . . as far as fighting goes we have had enough of that being always in advance, and most justly do the Officers as well as the Merchants of Macao say "that she is worth her weight in gold."[86]

John Ochterlony of the Madras Engineers gave another account of the *Nemesis* in action.

After disembarking the whole of the 37th regiment below Chuenpee, she ran alongside the fort, and threw shells into the upper fort, until the advance of the troops compelled the shipping to cease firing, when, taking advantage of her light draught, she ran close up to the sea battery, and poured through the embrasures destructive rounds of grape as she passed; then pushing on over the shallows in Anson's Bay, into the midst of the war junks lying at anchor, she threw Congreve rockets with startling effect, the very first having set fire to the largest of them, which blew up, with all her crew on board. . . . she proceeded on her course across the bay, setting fire to junk after junk, until the whole fleet, eleven in number, was destroyed.[87]

The Chinese did deploy a weapon that surprised the British: paddle-wheelers powered by men walking on treadmills. Some thought "the notion evidently being taken from the steamers," but in fact such boats were a Chinese invention dating back to the T'ang dynasty in the eighth century, if not earlier.[88] The *Nemesis* captured and destroyed several of them.

The arrival of the *Nemesis* and the destruction of the Bogue forts did little to change the relations between China and Britain, however. Even the capture of several ports along the coast did not impress the imperial courtiers in Beijing, for coastal towns, even Guangzhou, were far away on the periphery of their interests.[89] The British residents knew that China was invulnerable from the sea, for its trade was carried on its rivers and on the Grand Canal. To defeat the Celestial Empire, the British had to seize its most vulnerable spot, the junction of the canal and the Yangzi River. This was the strategy that Samuel Ball, tea inspector of the East India Company in China, suggested in a letter that reached Foreign Secretary Lord Palmerston: "The Yang coo kiang [Yangzi] as far as its junction with the Grand Canal ought to be examined and regularly surveyed. This might be done with the aid of a steamer. . . . the island of Kin Shan would be a strong position and enable us to distress the internal commerce greatly by cutting off the communication between the Northern and Southern Provinces by means of the Grand Canal."[90]

This canal crossed the Yangzi River at Zhenjiang, almost two hundred miles upriver from the sea. For this offensive, planned for 1842, the East India Company had brought many more ships. Among them were ten men-of-war, seven other sailing ships, five steam frigates, and six steam-powered gunboats: the *Nemesis* and its sister ships *Phlegethon*, *Proserpine*, *Pluto*, *Medusa*, and *Ariadne*, all designed for river work.[91]

On June 16, 1842, the British fleet entered the Yangzi River and attacked the Chinese forts at Wusong, near Shanghai. The steamers and ships of the line carried, between them, 724 guns. The steamers reconnoitered the river and towed sailing men-of-war into position to bombard the forts.[92] Niu Chien, the governor of the province, described the Chinese defenses.

> We are not neglecting [preparations] to bar them from sailing into the inland waterways. At present, we have sixteen warships. . . . We have also mobilized seventy [merchant and fishing] vessels, large and small, . . . to cruise back and forth in shifts. Skilled artisans have also constructed four water-wheel boats, on which we have mounted guns. . . . If the barbarians should sail into the inland waterways, these vessels can resist them. There is not the slightest worry.[93]

The governor's optimism notwithstanding, the British steamers captured or destroyed the war junks and wheel boats and silenced the guns of the Wusong forts. Steaming and being towed up the river, the fleet captured Zhenjiang on July 21, blocking the Grand Canal, China's main artery of trade, thereby shutting down the transport of rice and other products from the rich central and southern provinces to the capital, Beijing.[94] A month later, China signed the Treaty of Nanjing, granting the island of Hong Kong to Britain, opening five ports to foreign trade, and setting import duties at a

Figure 5.4. East Asia at the time of the Opium War, showing the British
naval offensives against Guangzhou (Canton) and Zhenjiang on the
Yangzi River. Map by Chris Brest.

mere 5 percent. The treaty made China, for the first time, subservi-
ent to a Western power.

The technology that had forced China to submit to Western demands remained effective for a century. The Second Opium War of
1857–60, also known as the Arrow War because the casus belli was
the Chinese arrest of the schooner *Arrow* registered in Hong Kong,
was essentially a replay of the first Opium War. As in the earlier war,
the purpose was not to acquire territory but to "open up" China,
meaning, to force it to admit foreign diplomats and missionaries,
to open eleven more treaty ports and the entire Yangzi Valley to
international commerce, and to legalize the opium trade.[95] This time
the British and their French allies had far more steamers armed with
steel breech-loading cannon that shot explosive shells, while China
was even less prepared than it had been in 1840.[96] By 1857 Rear
Admiral Sir Michael Seymour had assembled a fleet of gunboats
with which to attack Guangzhou. Though this "punitive expedition"
(as the British called their small wars in faraway places) persuaded
the Chinese to concede defeat, renewed British and French demands
led to another expedition, this one up the Yongding River to the
Dagu Forts defending Tianjin and Beijing from the sea. By October
1860, the French and British had taken possession of Beijing and
looted and burned the Summer Palace. China became a quasi-
protectorate of the Western powers, who enforced their hegemony
by stationing gunboats on the Yangzi River, visible signs of China's
humiliation, until the outbreak of World War II. Not until 1867 did
China acquire a steam-powered gunboat of its own.[97] And not until
1947—107 years after the arrival of the *Nemesis*—did the last of the
foreign gunboats leave Chinese waters.

Steamboats on the Niger

Europeans had long been familiar with the coasts of Africa. Yet until
the 1830s they knew as little of the interior of the continent as
they had four hundred years earlier. Three factors account for their
centuries-long ignorance. One was the opposition of coastal Africans who, with the complicity of European traders, profited from the

slave trade and resisted any interference. Another was the prevalence of *falciparum* malaria, a disease so dangerous that it made explorations into the interior almost suicidal. The third was the difficulty of approaching the continent by sea.

Africa is a large table-land with raised edges over which the rivers fall in great cataracts. On some rivers, like the Nile and the Niger, the falls are well inland, allowing navigation in small boats for several hundred miles. On others, like the Congo, the falls are near the coast. The alternating dry and rainy seasons make many rivers navigable only for part of the year. Before the age of steam, the only practicable vessels on African rivers were canoes and, on the Nile, small sailboats. Finally, the lack of natural harbors and offshore sandbars made it hazardous for ships to come too close to the coast.

In spite of these natural obstacles, trade was lucrative, and nowhere more so than in the Oil Rivers of southeastern Nigeria. In this maze of small streams and fetid swamps, independent towns and small kingdoms grew wealthy from the sale of slaves to European traders. After the decline of the slave trade in the early nineteenth century, they turned to palm oil, a product in great and growing demand in the West for the manufacture of soap and the lubrication of industrial machinery; hence the name "Oil Rivers." Ships could not maneuver on the shallow winding channels of the region, nor were outsiders welcome to come and set up shop ashore. Instead, traders lived onboard their ships or on hulks anchored in the channels, from which they conducted their business.

British abolition of the slave trade—though not of slavery itself—in 1807 inaugurated not only a new commercial era on the West African coast but a new military situation as well, for the Royal Navy was entrusted with the task of capturing slave ships and returning the freed slaves to Africa. Navigation in these dangerous waters required information. Hence, in 1822 the Royal Navy sent Captain W.F.W. Owen commanding the *African*, the first steamer in African waters, to survey a thousand-mile stretch of coast from the Ivory Coast to the Bight of Benin.[98]

By then, European interest in Africa was growing, for both commercial reasons—the demand for tropical products and the need for markets for the products of industry—and humanitarian ones—the insistence of abolitionists like William Wilberforce that not only the slave trade but slavery itself should be abolished in the name of Christianity and civilization. Among the topics that aroused great curiosity was the mystery of the Niger River. The English explorer Mungo Park had reached the upper Niger in 1795–97 and again in 1805–6; though he died there, the reports he sent back showed that the river flowed toward the east. In 1821–25 Hugh Clapperton traveled across the Sahara to Lake Chad and on to the Niger, proving that the river did not continue eastward to the Nile, as some had thought. Other expeditions to the upper Niger included those of Alexander Laing and René Caillié, who reached Timbuktu in 1826–28. That left the question: What happened to the Niger after it bent southward below Timbuktu?

The answer came in 1830 when the brothers Richard and John Landers walked north from Lagos to the Bussa Rapids and traveled down the Niger from there to the Bight of Benin, emerging at the Oil Rivers, which we now know as the Niger Delta. The Nigerian historian K. Onwuka Dike called this the "year of destiny . . . when the twin problems of geography and commerce were solved."[99] Knowledge that, with a steamer, it was possible to navigate from the sea to the Niger valley, the source of palm oil, was bound to entice an energetic entrepreneur.

That man was Macgregor Laird. In 1832 he had joined some Liverpool merchants in founding the African Inland Commercial Company "for the commercial development of the recent discoveries of the brothers Lander on the River Niger." Their goal was twofold; in his words:

> . . . those who look upon the opening of Central Africa to the enterprise and capital of British merchants as likely to create new and extensive markets for our manufactured goods, and fresh sources whence to draw our supplies; and those who, viewing mankind as one great family, consider it their duty to raise their fellow creatures from their present degraded, denationalised, and demoralised state, nearer to Him in whose image they were created.[100]

To achieve these goals, Laird built an iron-hulled steamer, the *Alburkah*, that was seventy feet long, displaced fifty-five tons, and was powered by a sixteen-horsepower engine. He also built a larger wooden steamer, the *Quorra*. Both steamers were heavily armed with swivel guns, carronades, and smaller firearms. With these two boats and the sailing brig *Columbine*, Macgregor Laird and Richard Lander set out for the Bight of Benin. It was the first time an iron steamboat had ventured into the ocean, and the plan was greeted with ridicule by more conservative mariners. Yet the expedition reached the Bight of Benin safely. The two steamboats then steamed up the Niger River to the confluence of the Benue River and continued to explore the Niger and the Benue until 1835.[101]

Unfortunately, of the forty-eight men on his expedition, only nine returned alive. Laird himself was ill for years after. The obstacles of ignorance and of river navigation had been overcome, but not the obstacle of disease. Laird thereafter turned his attention to establishing a steamship line between Liverpool and the Niger Delta and to lobbying the East India Company and Parliament, as we saw earlier, to order iron steamboats from his brother's shipyard for service in India and China.

Others, however, were determined to continue in the path that Laird had pioneered, for the 1830s were a decade of increasing trade between Britain and the West African coast. In 1835 John Beecroft, the British consul at Fernando Po (an island in the Bight of Benin) used the *Quorra* that Laird had left behind to explore the Niger Delta and the river up to its confluence with the Benue. In 1840 he did so again with the steamboat *Ethiope*. He used that experience to persuade the British government to fund a major expedition.[102]

The government ordered three new steamers from John Laird's shipyard: the twin ships *Albert* and *Wilberforce*, each of 457 tons with two thirty-five-horsepower engines, and the 249-ton *Soudan*, with one thirty-five-horsepower engine. All three were modeled on the *Nemesis*, with iron hulls, flat bottoms, and retractable keels and rudders. They were also well armed; the larger boats each had a brass twelve-pounder cannon, two twelve-pounder howitzers, and four one-pounder swivel guns, while the *Soudan* had one howitzer and two swivel guns, in addition to small arms. In 1841, these three vessels sailed to the Bight of Benin, then steamed up the Niger to the Benue and then up the Benue. The expedition ended tragically, however, as 53 of the 303 crew members (among them 48 of the 145 Europeans) died, mostly of malaria.[103]

Undeterred by this disaster, the government asked Macgregor Laird to send out another expedition, this one to be led by Beecroft, the Englishman with the best knowledge of the lower Niger. Laird had a 260-ton iron ship, the *Pleiad*, built for the expedition. Rigged as a schooner so it could sail on the open sea, it also had a sixty-horsepower steam engine for river work and—a first in African waters—a propeller instead of paddle wheels. Beecroft died on Fernando Po in 1854 and Dr. William Baikie, the ship's physician, took over the command of the expedition. The *Pleiad* steamed up the Niger and charted the Benue over 250 miles beyond any previous steamer. Baikie made certain that his crew took quinine pills—a known anti-malarial—*before* they reached the Bight of Benin and throughout their stay in Africa. In four months in African waters, not a single crew member died.[104] As we shall see in the next chapter, it was the prophylactic use of quinine, as much as anything else, that opened the doors of tropical Africa to Europeans.

Laird's and the government's persistence in the face of twenty years of reverses were vindicated, for the path into the richest part of Africa now lay open to trade. By the 1850s, Macgregor Laird's African Inland Commercial Company was providing regular mail service between Britain and the Bight of Benin. The lower freight rates that these steamers offered brought increasing competition to the traders in the Niger Delta. In 1857 the British government signed a contract with Laird to maintain a steamer on the river for five years. The first boat to steam upriver under the contract was the *Dayspring* commanded by Dr. Baikie and Lieutenant John Glover.[105] A year later, it was joined by the *Sunbeam* and the *Rainbow*.[106] These were no longer exploratory missions but regular trading ventures that cut deeply into the profits of the African middlemen and their European counterparts on the coast. To the latter, the triumph of steam, quinine, guns, and "civilization" meant competition of a most unfair and unwelcome sort, one they were bound to resist.

From 1857 to the end of the century, steamers had to overcome both the river and the inhabitants of the area. The level of the Niger, like the Missouri, was prone to changing suddenly and unexpectedly.

Snags, floating vegetation, and shifting sandbanks made navigation treacherous, and steamers were often grounded and had to be abandoned until the water rose again the following year. The river people regularly attacked the steamers and the trading posts on the river. At Laird's request, the navy began sending gunboats up the river to escort the cargo steamers. According to the historian of the Royal Niger Company, "although efforts were made to come to a peaceful settlement with the villages responsible for firing on his vessels, these were of no avail. Consequently, the gunboat was left with no option but to bombard those concerned in reprisal as a lesson to all."[107] Starting in 1863, the gunboat H.M.S. *Investigator*, armed with three twelve-pounder howitzers, began patrolling the river. Some inland traders did not wait for naval escorts. The steamer of the trading firm Miller Brothers & Company, the *Sultan of Sokoto* (or *Socotoo*), armed and protected by iron screens, fought its way through the gunfire of the delta towns.

Steamboat technology evolved in parallel with that of ocean steamships. One important innovation was the propeller, first successfully used at sea in 1838. Its application to river work was delayed for several years because it hung below the hull, where it was likely to get tangled in floating vegetation. The boat builder Alfred Yarrow solved this problem in the 1870s when he placed the propeller in a tunnel under the hull, making it possible to build propeller-driven boats with as little as two feet of draft.[108] The British and French adopted high-pressure boilers, long used on the rivers of the American Midwest; once they began to be made of steel, they were much less prone to burst than iron. Other improvements in design included surface condensers that recycled boiler water and did not require scraping and cleaning, and steel hulls that were lighter and resisted puncturing much better than iron.

By the 1870s, four companies ran steamers up and down the Niger. Like the *Sultan of Sokoto*, they were armed and fitted with protective metal screens. In the delta, twenty firms owned a total of sixty trading posts with over two hundred permanent European traders. As competition grew hotter, skirmishes with the delta natives became more frequent. The Royal Navy sent regular punitive expeditions up the river to destroy towns that had fired on passing steamers.

Onitsha was bombarded for three days, then torched. Yamaha, Idah, Aboh, and other towns suffered the same fate. Africans burned down the company stores, and the company and naval gunboats retaliated. In May and June 1876, battles broke out between the *Sultan of Sokoto* and African traders, in which the ship hurled rockets, cannonballs, canister, and chain shot, while the traders used swivel guns. In retaliation, the *Sultan of Sokoto* and the naval gunboats *Oriole, Cygnet, Unicorn,* and *Victoria* burned down the towns of Sabagreia and Agberi, followed a year later by the burning of Emblama.[109] The attacks and counterattacks continued into the 1880s and 1890s. In the 1880s, Sir George Goldie's United African Company ordered twenty gunboats to patrol the Niger and attack recalcitrant towns. In 1887 the British consul Harry Johnson used the gunboat *Goshawk* to kidnap Jaja, the ruler of Opobo and an important palm-oil merchant. In 1894 gunboats participated in the attack on Ebrohimi, in the delta. And in 1897, the gunboat *Ivy* helped transport five hundred soldiers with rockets, machine guns, and rifles in the most spectacular punitive expedition in Nigerian history, the capture and sacking of the royal city of Benin.[110]

Thus was Britain enticed ever more deeply into the conquest of the lower Niger. But no matter who prevailed at any given moment, the British commercial agents and government officials would never pull back, for the French, by the 1880s well ensconced in the Western Sudan and on the upper Niger, would have gladly replaced them if they had. What started out as "free trade imperialism" like the Opium War turned into a series of military campaigns ending in the occupation of Nigeria. The government of China could be coerced into imposing the foreigners' demands upon its own people. In Nigeria, however, there was no central government, and each town and kingdom had to be coerced separately. Not until the entire region had been occupied did the military campaigns cease.

Steamboats and the Scramble for Africa

The lower Niger deserves our emphasis because it reveals so clearly the role of steamboats in the penetration of a commercially and strategically important region of Africa. But it was not unique. Much of Africa is far from any navigable river, but wherever it was possible to make use of steamers, the Europeans did so.

Before 1870 European interest in Africa was largely limited to merchants and missionaries. Steamboats were relatively simple machines employed on rivers easily accessible from the sea. A few examples will suffice. The Gambia had been active in trade for centuries.[111] In 1826 the Royal Navy sent the steamer *African* and the sailing ship *Maidstone* up the river to forestall the French. More steamers began arriving in the 1840s, in particular the *Dover*, a Laird-built riverboat that made thirty-one trips up and down the river in 1849 alone, ending the slave trade and protecting British traders. In 1851 the British seized Lagos, a small town on a lagoon near the Niger Delta, and stationed a small steamer to patrol the lagoon. In the 1850s and early 1860s, they used gunboats to attack coastal and river towns in Sierra Leone. And in 1857, the French governor of Senegal, Louis Faidherbe, sent five hundred soldiers four hundred miles up the Senegal River to Médine to fight al-Hadj Umar's Tokolor warriors.[112]

After 1870 the European penetration of the continent speeded up in a frenzy known as the "Scramble for Africa." Steamers were used wherever navigation was possible. However, many of Africa's important rivers, including the Congo, the upper Niger, and the upper Nile, could not be reached from the sea because of rapids. To be used on such rivers, steamers had to be sent upstream as far as possible, then carried in pieces on the heads of porters and reassembled above the rapids, a difficult job that delayed the penetration of the interior by months or years.

The Congo River basin was closed to access from the sea by several sets of rapids between Matadi, eighty miles from the sea, and Stanley Pool, two hundred miles further upstream. Rain forests and mountain ranges along the coasts added to the difficulty of penetrating this enormous region. Not until 1877 did a European, Henry Morton Stanley, first visit this region, entering from East Africa and navigating down the river to the Atlantic. Meanwhile, King Leopold II of Belgium, intent on acquiring a colonial empire of his own, founded the International Association for the Exploration and Civilization of Africa. When Stanley became Leopold's agent in Africa, he ordered several steamboats to bring the Congo basin under Leopold's sway. To reach the navigable part of the river, he spent a year and employed hundreds of African workers carving a path and carrying the parts over the mountains. The boats then had to be reassembled on the upper Congo. The *En Avant* was finally launched in 1881, two years after it had been unloaded from a freighter on the lower river. In 1881 the missionary George Grenfell had a steamer built in England, shipped to the mouth of the Congo, and carried by porters across 250 miles of mountain trails to Stanley Pool, where it arrived in 1884.[113] In 1886 Leopold wrote to Stanley: "Take good care of our Navy; it sums up at this moment almost all of our governmental force."[114] Providing steamers for the Congo remained horrendously difficult and costly until 1898 when a railroad was completed connecting Matadi with Leopoldville. After that, steamers proliferated on the Congo and its tributaries; by 1898 there were 43, and 103 by 1901.[115] The Congo became a steamboat colony.

A similar situation greeted Europeans who penetrated the Zambezi watershed in southeastern Africa. David Livingstone had two small steamers, the *Pioneer* and the *Lady Nyassa*, built for Lake Nyassa. Like the Congo steamers, they had to steam up the Shire River; they were then taken apart and the pieces carried overland around Murchison's Falls, then reassembled for the lake. The Established Church of Scotland mission that followed in Livingstone's footsteps in 1875 did the same with their steamer, *Ilala*.[116]

Steamers were also used in military operations. As the British were penetrating the lower Niger, the French were installing themselves on the upper reaches of that river, in what is now Mali and Niger. Their plan was to haul gunboats in pieces from Senegal to the upper Niger, then use them to control the trade on the river. In 1880–82 Governor Galliéni of Senegal and Minister of the Navy Jauréguiberry authorized the creation of a gunboat flotilla to patrol the Niger as far as the Bussa Rapids. In 1883 the first of the gunboats, the *Niger*, aided in the capture of Bamako, the first major port on the river. A year later it was patrolling the river between Bamako and Diafarabé. By 1894 the flotilla helped take the fabled city of Timbuktu.[117] At the same time, the French were engaged in the conquest of Dahomey (now the Republic of Benin). They began by blockading the coast in 1890. Two years later a major expedition, led by Colonel Alfred-Amédée Dodds, steamed up the Ouémé River on gunboats named *La Topaze*, *L'Opale*, *L'Emeraude*, *Le Corail*, and *L'Ambre*. They towed canoes, transported soldiers, carried cannon and engineers, and evacuated the wounded and sick. Within a few weeks, they had taken Abomey, the capital of King Behanzin.[118]

Most spectacular of all the gunboat campaigns was the conquest of the Sudan in 1898, so dramatically described by Winston Churchill in his book *The River War*. For that expedition, General Horatio Kitchener assembled 8,200 British and 17,600 Egyptian soldiers, along with 3,524 camels and 3,594 horses, mules, and donkeys. After overcoming great difficulties transporting men, animals, and equipment around the cataracts of the Nile and across the desert, they confronted 50,000 Dervishes, the largest and most warlike military force in Africa up to that time. The British forces were supported by ten gunboats armed with quick-firing cannon and Maxim and Nordenfeldt machine guns. After one battle, Churchill reported, "The loss on the gunboats was limited to the single Soudanese soldier, who died of his wounds, and a few trifling damages. The Arab slaughter is variously estimated, one account rating it at 1,000 men; but half that number would probably be no exaggeration."[119] Approaching Omdurman, the Khalifa Abdullah threw his entire army into the battle in an attempt to push the Anglo-Egyptian force into the river. Not only did the invaders have far more powerful firearms than the Khalifa's troops, but they were backed by the gunboats lining the riverbanks. In Churchill's words:

> But at the critical moment the gunboat arrived on the scene and began suddenly to blaze and flame from Maxim guns, quick-firing guns and rifles. The range was short; the effect tremendous. The terrible machine, floating gracefully on the waters—a beautiful while devil—wreathed itself in smoke. The river slopes of the Kerreri Hills, crowded with the advancing thousands, sprang up into clouds of dust and splinters of rock. The charging Dervishes sank down in tangled heaps. The masses in the rear paused, irresolute. It was too hot even for them.[120]

By the time the battle ended, over nine thousand Dervishes had lost their lives and perhaps twice that many had been wounded, against 48 dead and 428 wounded in the Anglo-Egyptian force. Churchill again: "Thus ended the battle of Omdurman—the most signal triumph even gained by the arms of science over barbarians. Within the space of five hours the strongest and best-armed savage army yet arrayed against a modern European Power had been destroyed and dispersed, with hardly any difficulty, comparatively small risk, and insignificant loss to the victors."[121]

Conclusion

History is not a science. Historians cannot employ the scientific method, changing one variable while keeping others constant in order to test a hypothesis. In every historical event, many variables contribute to the outcome. Seldom can one find situations in which most causes are similar and only a few differ enough for us to draw tentative conclusions. One such case is that of two wars that took place at the same time, instigated by the same protagonists: the Opium War and the First Anglo-Afghan War of 1839–42.

In both wars the motivations were similar, namely the expansionism of the East India Company and its governor-general in India, Lord Auckland, and their desire to impose British civilization and free trade upon recalcitrant Asians. To be sure, the details differed, including the abilities of the military leaders and the support of Great Britain. But most glaring of all was the contrast between the technologies involved. In Afghanistan, the Army of the Indus relied on camels to cross high mountains and barren deserts. Its artillery was weak and its firearms no better than those of the Afghans they confronted. What slight advantages they had over the Afghans were less than those the Mughal leader Babur had enjoyed in his conquest of India three centuries earlier. Of the sixteen thousand British and Indian troops who reached Kabul in 1841, only 161 returned alive. It was the most humiliating defeat in British imperial history since the American War of Independence.[122] Fortunately for

British pride, news of this debacle was overshadowed by the victory in the Opium War.

Steamers, more than anything else, explain the difference between these two conflicts. In China, steamboats breached the ecological barrier of rivers and shallow waters that had long impeded the European advance. Likewise, steamers were invaluable in the Euro-American conquest and settlement of the Midwest and in the European conquest of Burma, the Red Sea, sub-Saharan Africa, and many other places that sailing ships could not reach. They were but the first of the three new technological systems that made the New Imperialism of the late nineteenth century so rapid and easy; the other two were advances in medicine and in weapons, as we shall see in the next two chapters.

Notes

1. Macgregor Laird and R.A.K. Oldfield, *Narrative of an Expedition into the Interior of Africa, by the River Niger, in the Steam-Vessels Quorra and Alburkah, in 1832, 1833, and 1834*, 2 vols. (London: Richard Bentley, 1837), vol. 2, pp. 397–98.

2. Nabil-i-Azam, *The Dawn-Breakers: Nabil's Narrative of the Early Days of the Baha'i Revelation*, trans. and ed. Shogi Effendi (New York: Baha'i Publishing Committee, 1932), p. 131.

3. On the many ideas and projects that never moved a boat, see Philip H. Spratt, *The Birth of the Steamboat* (London: C. Griffin, 1958), pp. 17–20.

4. On the early steamboats, see ibid.; James T. Flexner, *Steamboats Come True: American Inventors in Action* (New York: Viking, 1944); and Michel Mollat, ed., *Les origines de la navigation à vapeur* (Paris: Presses Universitaires de France, 1970).

5. On Fulton, see Kirkpatrick Sale, *The Fire of His Genius: Robert Fulton and the American Dream* (New York: Free Press, 2001).

6. Louis C. Hunter, *Steamboats on the Western Rivers: An Economic and Technological History* (Cambridge, Mass.: Harvard University Press, 1949), pp. 8–12; Carl Daniel Lane, *American Paddle Steamboats* (New York: Coward-McCann, 1943), p. 30.

7. James Hall, *The West: Its Commerce and Navigation* (Cincinnati: H. W. Derby, 1848), pp. 123–26; Sale, *Fire of His Genius*, p. 188; Lane, *American Paddle Steamboats*, pp. 32–33; Flexner, *Steamboats Come True*, pp. 344–45; Hunter, *Steamboats*, pp. 13, 62, 122–33.

8. Sale, *Fire of His Genius*, pp. 188–94.

9. Hunter, *Steamboats*, p. 61. The U.S. Congress became so concerned about the appalling number of exploding steamboat boilers that it passed a bill "to provide for the better security of the lives of passengers on-board of vessels propelled in whole or in part by steam"; see *Congressional Globe*, 25th Cong., 2nd Sess., vol. 6, no. 9 (February 5, 1838).

10. Sale, *Fire of His Genius*, p. 194.

11. Ibid., pp. 188–89.

12. Quoted in ibid., p. 190.

13. Hiram M. Chittenden, *History of Early Steamboat Navigation on the Missouri River: Life and Adventures of Joseph La Barge* (New York: Harper, 1903), vol. 2, pp. 382–83; William John Petersen, *Steamboating on the Upper Mississippi* (Iowa City: State Historical Society of Iowa, 1968), pp. 80–88. See also Hunter, *Steamboats*, p. 552; and Hall, *The West*, p. 128.

14. Hunter, *Steamboats*, p. 552.

15. Letter in the *Niles Register* of May 1819, quoted in Robert G. Athearn, *Forts of the Upper Missouri* (Englewood Cliffs, N.J.: Prentice-Hall, 1967), p. 4.

16. Henry Atkinson, *Wheel Boats on the Missouri: The Journals and Documents of the Atkinson-O'Fallon Expedition, 1824–26*, ed. Richard Jensen and James Hutchins (Helena: Montana Historical Society Press, and Lincoln: Nebraska Historical Society Press, 2001); Petersen, *Steamboating on the Upper Mississippi*, pp. 90–105.

17. Petersen, *Steamboating on the Upper Mississippi*, pp. 175–76.

18. R. G. Robertson, *Rotting Face: Smallpox and the American Indian* (Caldwell, Idaho: Caxton Press, 2001), pp. 240–42.

19. Harry S. Drago, *The Steamboaters* (New York: Dodd, Mead, 1967), p. 120; Chittenden, *Early Steamboat Navigation*, vol. 2, pp. 383–84. See also Carlos A. Schwantes, *Long Day's Journey: The Steamboat and Stagecoach Era in the Northern West* (Seattle: University of Washington Press, 1999), chapters 1–4; and Chittenden, *Early Steamboat Navigation*, vol. 2, pp. 384–85.

20. Joseph M. Hanson, *The Conquest of the Missouri: The Story of the Life and Exploits of Captain Grant Marsh* (Mechanicsburg, Pa.: Stackpole Books, 2003), pp. 61–68.

21. Hanson, *Conquest of the Missouri*, pp. 51–54; Drago, *The Steamboaters*, pp. 121–22.

22. Chittenden, *Early Steamboat Navigation*, vol. 2, pp. 386–93; Hanson, *Conquest of the Missouri*, 301–6.

23. Hanson, *Conquest of the Missouri*, pp. 66, 71. It is amazing to see a work published in 2003 that still refers to the Indians as "hostiles" and "savages."

24. Drago, *The Steamboaters*, p. 122.

25. George Henry Preble, *A Chronological History of the Origin and Development of Steam Navigation*, 2nd ed. (Philadelphia: L. R. Hamersley, 1895), p. 125. Sale (*Fire of His Genius*, p. 196) says that by 1838, sixteen hundred steamboats had been built in the United States, of which five hundred had been lost or worn out. Americans' love of transportation did not end with steamboats; later in the century,

Americans laid more miles of railroad track than the rest of the world put together and, later yet, had more cars and more airplanes than the rest of the world.

26. Henry T. Bernstein, *Steamboats on the Ganges: An Exploration in the History of India's Modernization through Science and Technology* (Bombay: Orient Longmans, 1960), p. 28; Gerald S. Graham, *Great Britain in the Indian Ocean: A Study of Maritime Enterprise, 1810–1850* (Oxford: Clarendon Press, 1968), p. 352; H. A. Gibson-Hill, "The Steamers Employed in Asian Water, 1819–39," *Journal of the Royal Asiatic Society, Malayan Branch* 27, pt. 1 (May 1954), pp. 121–22.

27. Satpal Sangwan, "Technology and Imperialism in the Indian Context: The Case of Steamboats, 1819–1839," in Theresa Meade and Mark Walker, eds., *Science, Medicine and Cultural Imperialism* (New York: St. Martin's, 1991), p. 63.

28. Christopher Lloyd, *Captain Marryat and the Old Navy* (London and New York: Longmans Green, 1939), pp. 211–17; Colonel W.F.B. Laurie, *Our Burmese Wars and Relations with Burma: Being an Abstract of Military and Political Operations, 1824–25–26, and 1852–53* (London: W. H. Allen, 1880), pp. 71–72; D.G.E. Hall, *Europe and Burma: A Study of European Relations with Burma to the Annexation of Thibaw's Kingdom, 1886* (London: Oxford University Press, 1945), p. 115; Gibson-Hill, "Steamers Employed in Asian Water," pp. 127–36; Graham, *Great Britain,* pp. 346–57.

29. On the Ganges steamers, see Bernstein, *Steamboats on the Ganges;* A. J. Bolton, *Progress of Inland Steam-Navigation in North-East India from 1832* (London, 1890, in India Office Library P/T 1220); and J. Johnson, *Inland Navigation on the Gangetic Rivers* (Calcutta: Thacker, Spink, 1947).

30. Jean Fairley, *The Lion River: The Indus* (New York: Allen Lane, 1975), pp. 222–25; V. Nicholas, "The Little Indus (1833–1837)," *Mariner's Mirror* 31 (1945), pp. 210–22; Victor F. Millard, "Ships of India, 1834–1934," *Mariner's Mirror* 30 (1944), pp. 144–45.

31. Halford L. Hoskins, *British Routes to India* (London and New York: Longmans Green, 1928), pp. 86–87.

32. Ghulam Idris Khan, "Attempts at Swift Communication between India and the West before 1830," *Journal of the Asiatic Society of Pakistan* 16, no. 2 (August 1971), pp. 120–21; Sarah Searight, *Steaming East: The Hundred Year Saga of the Struggle to Forge Rail and Steamship Links between Europe and India* (London: Bodley Head, 1991), pp. 22–29; Hoskins, *British Routes to India,* pp. 89–96; Gibson-Hill, "Steamers Employed in Asian Water," pp. 122, 134–35.

33. Quoted in John Rosselli, *Lord William Bentinck: The Making of a Liberal Imperialist, 1774–1839* (Berkeley: University of California Press, 1974), p. 292.

34. Khan, "Attempts at Swift Communication," pp. 121–36.

35. Searight, *Steaming East,* p. 47; Khan, "Attempts at Swift Communication," 139–49; Hoskins, *British Routes to India,* pp. 97–117, 183–85.

36. Thomas Love Peacock, "Memorandum on Steam Navigation in India, and between Europe and India; December 1833," appendix 2 of "Report from the Select Committee on Steam Navigation to India with the Minutes of Evidence, Appendix

and Index," in Great Britain, House of Commons, *Parliamentary Papers* 1834 (478.) XIV, pp. 620–23; Millard, "Ships of India," p. 144; Khan, "Attempts at Swift Communication," 150–57; Hoskins, *British Routes to India*, 101–9; Gibson-Hill, "Steamers Employed in Asian Water," 147–50.

37. Hoskins, *British Routes to India*, pp. 110–25.

38. Great Britain, House of Commons, "Report from the Select Committee on Steam Navigation to India," 369–609.

39. On Peacock's interest in steam navigation, see Herbert Francis Brett-Smith and C. E. Jones, eds., *The Works of Thomas Love Peacock*, 10 vols., vol. 1: *Biographical Introduction and Headlong Hall* (London and New York: Constable, 1924), pp. clviii–clx; Felix Felton, *Thomas Love Peacock* (London: Allen and Unwin, 1973), pp. 229–31; A. B. Young, "Peacock and the Overland Route," *Notes and Queries*, 10th ser., 190 (August 17, 1907), pp. 121–22; and Sylva Norman, "Peacock in Leadenhall Street," in Donald H. Reiman, ed., *Shelley and His Circle* (Cambridge, Mass.: Carl H. Pforzheimer Library, 1973), p. 712.

40. "Report of the Select Committee" (1834), pt. 2, pp. 9–10.

41. Russians were also afraid that Britain's ally, the Ottoman Empire, would close the Bosporus to their grain exports from Odessa; see Searight, *Steaming East*, p. 51.

42. Francis Rawdon Chesney, *Narrative of the Euphrates Expedition carried on by order of the British Government during the years 1835, 1836, and 1837* (London: Longmans Green, 1868), vol. 1, pp. 4–6; Searight, *Steaming East*, pp. 51–55; Brett-Smith and Jones, *Works of Thomas Love Peacock*, vol. 1, pp. clxi–clxi; Hoskins, *British Routes to India*, pp. 150–51.

43. Stanley Lane-Poole, ed., *The Life of the Late General F. R. Chesney Colonel Commandant Royal Artillery D.C.L., F.R.S., F.R.G.S., etc. by His Wife and Daughter* (London: W. H. Allen, 1885), pp. 258–70.

44. Chesney, *Narrative*, vol. 1, pp. 145–46.

45. "Report of the Select Committee" (1834), pt. 2, pp. 16–24.

46. On the history of the Laird family and firm, see Cammell Laird and Company (Shipbuilders and Engineers) Ltd., *Builders of Great Ships* (Birkenhead, 1959), pp. 9–12; and Stanislas Charles Henri Laurent Dupuy de Lôme, *Mémoire sur la construction des bâtiments en fer, adressé à M. le ministre de la marine et des colonies* (Paris: A. Bertrand, 1844), p. 6.

47. "Report of the Select Committee" (1834), pt. 2, pp. 56–67.

48. Great Britain, House of Commons, "An Estimate of the Sum required for the purpose of enabling His Majesty to direct that trial may be made of an Experiment to communicate with India by Steam Navigation. Twenty Thousand Pounds; Clear of Fees and All Other Deductions," in *Parliamentary Papers* 1834 (492.) XLII, 459. See also Hoskins, *British Routes to India*, pp. 159–60; and Searight, *Steaming East*, p. 60.

49. Chesney, *Narrative*, vol. 1, pp. ix, 150–54; Cammell Laird, *Builders of Great Ships*, p. 14; Searight, *Steaming East*, p. 60.

50. For a good brief account of the expedition, see Searight, *Steaming East*, pp. 61–70.

51. Great Britain, House of Commons, "Report from the Select Committee on Steam Navigation to India, with the Minutes of Evidence, Appendix, and Index," in *British Sessional Papers, House of Commons* (1837), vol. 6, pp. 361–617.

52. Letter dated June 8, 1832, to R. Campbell, in *The Correspondence of Lord William Cavendish Bentinck, Governor-General of India, 1828–1835*, ed. C. H. Philips (New York: Oxford University Press, 1977), pp. 831–32.

53. Arthur B. Young, *The Life and Novels of Thomas Love Peacock* (Norwich: A. H. Goose, 1904), p. 28; Henry Cole, ed., *The Works of Thomas Love Peacock*, 3 vols. (London: R. Bentley and Sons, 1875), vol. 1, p. xxxvii; Carl Van Doren, *The Life of Thomas Love Peacock* (London and New York: E. P. Dutton, 1911), pp. 218–19; Hoskins, *British Routes to India*, pp. 210–18; Rosselli, *Lord William Bentinck*, pp. 285–92.

54. Great Britain, House of Commons, "Estimate of the Sum required for an Experiment to Communicate with India by Steam Navigation," *Parliamentary Papers* 1837 (445.), 1837–38 (313.), 1839 (142-IV), and 1840 (179-IV).

55. Hoskins, *British Routes to India*, pp. 193–94, 211–26; Gibson-Hill, "Steamers Employed in Asian Water," p. 135: Millard, "Ships of India," pp. 144–48.

56. On the early years of the P&O, see Hoskins, *British Routes to India, pp. 242–63*; and Daniel Thorner, *Investment in Empire: British Railway and Steam Shipping Enterprise in India, 1825–1849* (Philadelphia: University of Pennsylvania Press, 1950), pp. 32–39.

57. Frederick M. Hunter, *An Account of the British Settlement of Aden* (London: Trübner, 1877), p. 165; Robert L. Playfair, *A History of Arabia Felix or Yemen, from the Commencement of the Christian Era to the Present Time: Including an Account of the British Settlement of Aden* (Byculla: Education Society, 1859), p. 161.

58. Harvey Sicherman, *Aden and British Strategy, 1839–1868* (Philadelphia: Foreign Policy Institute, 1972), pp. 7–9.

59. Thomas E. Marston, *Britain's Imperial Role in the Red Sea Area, 1800–1878* (Hamden, Conn.: Shoe String Press, 1961), pp. 37–38.

60. Sicherman, *Aden and British Strategy*, p. 10; emphasis in the original.

61. Ibid., p. 13.

62. Ibid., pp. 11–13; Hunter, *An Account*, p. 165; Playfair, *History of Arabia Felix or Yemen*, pp. 162–63; Marston, *Britain's Imperial Role*, pp. 55–69.

63. Searight, *Steaming East*, p. 71.

64. Sir Charles Napier, *The War in Syria*, 2 vols. (London: J. W. Parker, 1842), vol. 2, p. 184.

65. E. Backhouse and J.O.P. Bland, *Annals and Memoirs of the Court of Peking* (Boston: Houghton Mifflin, 1914), p. 323.

66. Gerald S. Graham, *The China Station: War and Diplomacy, 1830–1860* (Oxford: Clarendon Press, 1978), p. 8.

67. Preble, *Chronological History*, pp. 143–45. See also Gibson-Hill, "Steamers Employed in Asian Water," pp. 122, 153–56; and Arthur Waley, *The Opium War through Chinese Eyes* (London: Allen and Unwin, 1958), pp. 105–6.

68. K. M. Panikkar, *Asia and Western Dominance* (New York: Collier Books, 1969), p. 97.

69. Hosea Ballou Morse, *The International Relations of the Chinese Empire* (Taipei: Ch'eng Wen Publisher, 1971), vol. 1, p. 135.

70. William H. Hall and William D. Bernard, *The Nemesis in China, comprising a history of the late war in that country, with a complete account of the colony of Hong Kong*, 3rd ed. (London: H. Colburn, 1846), pp. 1–2.

71. John Gallagher and Ronald Robinson, "The Imperialism of Free Trade," *Economic History Review* 6, no. 1 (1955), pp. 1–16.

72. Hall and Bernard, *Nemesis in China*, p. 2.

73. Edith Nicholls, "A Biographical Notice of Thomas Love Peacock, by his granddaughter," in Cole, *Works of Thomas Love Peacock*, vol. 1, pp. xlii–xliii; Brett-Smith, *Works of Thomas Love Peacock*, vol. 1, p. clxxi.

74. For a full description of the *Nemesis*, see Hall and Bernard, *Nemesis in China*, pp. 1–12.

75. "Hall, Sir William Hutcheon," in *Dictionary of National Biography*, vol. 8, pp. 978–79; "William Hutcheon Hall," in William R. O'Byrne, *A Naval Biographical Dictionary: Comprising the Life and Services of Every Living Officer in Her Majesty's Navy, from the Rank of Admiral of the Fleet to That of Lieutenant, Inclusive* (London: J. Murray, 1849), pp. 444–46.

76. Secret draft from Sir John Hobhouse to Viscount Palmerston, February 27, 1840, in India Office Records, L/P&S/3/6, p. 167.

77. Sir George Eden, Lord Auckland, to Sir John Cam Hobhouse, member of the Board of Control of the East India Company, Simla, April 1, 1839, in Broughton Papers, British Museum, Add. MS 36, 473, p. 446.

78. Hall and Bernard, *Nemesis in China*, p. 6.

79. *The Times*, March 30, 1840, p. 7. See also Peter Ward Fay, *The Opium War, 1840–1842: Barbarians in the Celestial Empire in the Early Part of the Nineteenth Century and the War by Which They Forced Her Gates Ajar* (Chapel Hill: University of North Carolina Press, 1975), p. 261.

80. Note in *The Shipping Gazette*, reprinted in *The Nautical Magazine and Naval Chronicle* 9 (1840), pp. 135–36.

81. Letter no. 122 from Auckland to the Secret Committee of the East India Company, Calcutta, November 13, 1840, in India Office Records L/P&S/5/40; Hall and Bernard, *Nemesis in China*, pp. 18, 61.

82. Waley, *The Opium War*, p. 105.

83. Lo Jung-Pang, "China's Paddle-Wheel Boats: Mechanized Craft Used in the Opium War and their Historical Background," *Tsinghua Journal of Chinese Studies*, n.s., no. 2 (1960), p. 191; William Hutton to *Nautical Magazine* 12 (1843), p. 346.

84. G.R.G. Worcester, "The Chinese War-Junk," *Mariner's Mirror* 34 (1948), p. 22.

85. John Lang Rawlinson, *China's Struggle for Naval Development, 1839–1895* (Cambridge, Mass.: Harvard University Press, 1967), pp. 3–17; Jack Beeching, *The Chinese Opium Wars* (New York: Harcourt Brace Jovanovich, 1975), pp. 51–52; Fay, *The Opium War*, pp. 272–73.

86. Captain W. H. Hall to Thomas Love Peacock, Canton, May 1841, in India Office Records, L/P&S/9/7, pp. 59–60. See also pp. 61–82 and Fay, *The Opium War*, pp. 264–90.

87. John Ochterlony, *The Chinese War: An Account of All the Operations of the British Forces from the Commencement to the Treaty of Nanking* (1842; reprint, New York: Praeger, 1970), pp. 98–99.

88. Lo, "China's Paddle-Wheel Boats," pp. 194–200; Rawlinson, *China's Struggle*, pp. 19–21; human-powered wheel boats had been tried on the Mississippi-Missouri rivers before being displaced by steamers.

89. G.R.G. Worcester, "The First Naval Expedition on the Yangtze River, 1842," *Mariner's Mirror* 36, no. 1 (January 1950), p. 2.

90. Samuel Ball to George W. S. Staunton, February 20, 1840, in India Office Records, L/P&S/9/1, p. 519.

91. On the steamers used in the 1842 campaign, see Charles R. Low, *History of the Indian Navy (1613–1863)* (London: R. Bentley, 1877), vol. 2, pp. 140–46; Edgar Charles Smith, *A Short History of Naval and Marine Engineering* (Cambridge: Cambridge University Press, 1938), p. 114; Fay, *The Opium War*, pp. xv–xxi, 313, 341; Gibson-Hill, "Steamers Employed in Asian Water," pp. 121, 128; and Preble, *Chronological History*, p. 190.

92. Hall and Bernard, *Nemesis in China*, pp. 322–26; Lo, "China's Paddle-Wheel Boats," pp. 189–93; Fay, *The Opium War*, 349–50; Worcester, "The Chinese War-Junk," p. 3.

93. Lo, "China's Paddle-Wheel Boats," pp. 189–90.

94. Ochterlony, *The Chinese War*, pp. 331–35; Fay, *The Opium War*, pp. 351–55; Worcester, "The Chinese War-Junk," p. 8; Graham, *China Station*, pp. 214–24.

95. P. J. Cain and Anthony G. Hopkins, *British Imperialism, 1600–2000*, 2nd ed. (Harlow and New York: Longman, 2002), pp. 288, 362–63; Ronald Findlay and Kevin H. O'Rourke, *Power and Plenty: Trade, War, and the World Economy in the Second Millennium* (Princeton: Princeton University Press, 2007), pp. 388–89.

96. Rawlinson, *China's Struggle*, pp. 30–32.

97. Richard N. J. Wright, *The Chinese Steam Navy, 1862–1945* (London: Chatham, 2000), pp. 14, 20; Rawlinson, *China's Struggle*, pp. 32–35.

98. Paul Mmegha Mbaeyi, *British Military and Naval Forces in West African History, 1807–1874* (New York: Nok Publishers, 1978), p. 60; K. Onwuka Dike, *Trade and Politics in the Niger Delta, 1830–1885: An Introduction to the Economic and Political History of Nigeria* (Oxford: Clarendon Press, 1956), p. 15.

99. Dike, *Trade and Politics in the Niger Delta*, p. 18.

100. Laird and Oldfield, *Narrative of an Expedition*, vol. 1, p. vi.

101. Ibid., vol. 1, pp. 2–9. See also Dike, *Trade and Politics in the Niger Delta*, pp. 18, 62–63; Christopher Lloyd, *The Search for the Niger* (London: Collins, 1973), pp. 131–41; and "Laird, Macgregor," in *Dictionary of National Biography*, vol. 11, pp. 407–8.

102. Philip D. Curtin, *The Image of Africa: British Ideas and Actions, 1780–1850* (Madison: University of Wisconsin Press, 1964), pp. 298, 308; Lloyd, *Search for the Niger*, p. 152.

103. William Allen (Captain, R.N.) and T.R.H. Thomson (M.D., R.N.), *A Narrative of the Expedition Sent by Her Majesty's Government to the River Niger, in 1841, under the Command of Captain H. D. Trotter, R.N.* (London: Richard Bentley, 1848), 2 vols.; "Mr Airy, Astronomer-Royal, on the Correction of the Compass in Iron-Built Ships," *United Service Journal and Naval and Military Magazine* (London), pt. 2 (June 1840), pp. 239–41; Lloyd, *Search for the Niger*, p. 150.

104. William Balfour Baikie, *Narrative of an Exploring Voyage up the Rivers Kwóra and Bínue (commonly known as the Niger and Tsádda) in 1854* (London, 1855; reprint, London: Cass, 1966); Thomas J. Hutchinson, *Narrative of the Niger, Tshadda, and Binuë Exploration; Including a Report on the Position and Prospects of Trade up Those Rivers, with Remarks on the Malaria and Fevers of Western Africa* (London, 1855; reprint, London: Cass, 1966), pp. 8–9.

105. John Hawley Glover, *The Voyage of the Dayspring, Being the Journal of the Late Sir John Hawley Glover, R.N., G.C.M.G., Together with Some Account of the Expedition up the Niger River in 1857, by A. C. G. Hastings* (London: J. Lane, 1926); Dike, *Trade and Politics in the Niger Delta*, p. 169; Mbaeyi, *British Military and Naval Forces*, pp. 123–24.

106. Ronald Robinson and John Gallagher with Alice Denny, *Africa and the Victorians: The Climax of Imperialism* (New York: St. Martin's, 1961), p. 37; Lloyd, *Search for the Niger*, pp. 128–30, 199; Glover, *Voyage of the Dayspring*, 16–20.

107. Geoffrey L. Baker, *Trade Winds on the Niger: The Saga of the Royal Niger Company, 1830–1970* (London and New York: Radcliffe Press, 1996), pp. 4–5.

108. Alfred F. Yarrow, "The Screw as a Means of Propulsion for Shallow Draught Vessels," *Transactions of the Institution of Naval Architects* 45 (1903), pp. 106–17.

109. Baker, *Trade Winds on the Niger*, pp. 12–29; Dike, *Trade and Politics in the Niger Delta*, pp. 205–7.

110. Robert V. Kubicek, "The Colonial Steamer and the Occupation of West Africa by the Victorian State, 1840–1900," *Journal of Imperial and Commonwealth History* 18, no. 1 (January 1990), pp. 16–26; Obaro Ikime, *The Fall of Nigeria: The British Conquest* (London: Heinemann, 1977), pp. 105–10; D.J.M. Muffett, *Concerning Brave Captains: Being a History of the British Occupation of Kano and Sokoto and the Last Stand of the Fulani Forces* (London: A. Deutsch, 1964), pp. 284–85; Dike, *Trade and Politics in the Niger Delta*, p. 212.

111. Donald R. Wright, *The World and a Very Small Place in Africa: A History of Globalization in Niumi, the Gambia*, 2nd ed. (Armonk, N.Y.: M. E. Sharpe, 2004), chapters 2–5.

112. Douglas Porch, *Wars of Empire* (London: Cassell, 2000), p. 117; Mbaeyi, *British Military and Naval Forces*, pp. 73, 80–83, 116–17, 133–34; Wright, *The World*, pp. 136–37; Kubicek, "Colonial Steamer," pp. 12–14.

113. André Lederer, *Histoire de la navigation au Congo* (Tervuren: Musée Royal de l'Afrique Centrale, 1965), pp. 11–20; Harry H. Johnston, *George Grenfell and the Congo* (London: Hutchinson, 1908), pp. 97–100.

114. Lederer, *Histoire de la navigation au Congo*, p. 95.

115. Ibid., pp. 130, 137.

116. Richard Thornton, *The Zambezi Papers of Richard Thornton, Geologist to Livingstone's Zambezi Expedition* (London: Chatto and Windus, 1963), pp. 243–44, 296; Alexander J. Hanna, *The Beginnings of Nyasaland and North-Eastern Rhodesia, 1859–1895* (Oxford: Clarendon Press, 1956), pp. 13–14.

117. Alexander S. Kanya-Forstner, *The Conquest of the Western Sudan: A Study in French Military Imperialism* (Cambridge: Cambridge University Press, 1969), pp. 75–135.

118. A. de Salinis, *La marine au Dahomey: Campagne de "La Naïade" (1890–1892)* (Paris: Sanard, 1910), pp. 117–39, 301; Luc Garcia, *Le royaume du Dahomé face à la pénétration coloniale* (Paris: Karthala, 1988), pp. 150–62; David Ross, "Dahomey," in Michael Crowder, ed., *West African Resistance: The Military Response to Colonial Occupation* (London: Hutchinson, 1971), pp. 158–59.

119. Winston S. Churchill, *The River War: An Account of the Reconquest of the Soudan* (1933; New York: Carroll and Graf, 2000), pp. 205–6.

120. Ibid., p. 274.

121. Ibid., p. 300.

122. See James A. Norris, *The First Afghan War, 1838–1842* (Cambridge: Cambridge University Press, 1967); and John H. Waller, *Beyond the Khyber Pass: The Road to British Disaster in the First Afghan War* (New York: Random House, 1990).

✸ Chapter 6 ✸

Health, Medicine, and the
New Imperialism, 1830–1914

We justly celebrate science for its conquest of infectious diseases in the nineteenth and early twentieth centuries. But this achievement was not intended for all humans, for it took place in the context of empire-building. In some cases, the goals of Western medicine and public health were purely domestic and their consequences in the non-Western world were quite fortuitous. In many cases, however, advances in medicine and public health came in response to the needs or consequences of imperialism. Often, these advances made empire-building and colonialism easier and less costly in terms of human lives. In tropical Africa, it gave Europeans the ability to advance into areas that had been off-limits for centuries. Only later—in some cases a century later—did these advances benefit the indigenous peoples of the colonized lands.

In the history of medicine and public health in the nineteenth century, we can distinguish three eras, with a great deal of overlap. The first was the era of traditional European medicine based on the theory of humoral pathology. The second era saw the increasing importance of experimentation in the mid-nineteenth century and the use of statistics to tease out the relations between diseases and the environments in which they occurred. In the third era, in the late nineteenth and early twentieth centuries, a new scientific approach to disease turned medicine and public health into the effective (if much criticized) practices they are today.

Medicine and Africa in the Early Nineteenth Century

Until well into the nineteenth century, medicine was an art rather than a science, and advice on health was a branch of literature. In

The Influence of Tropical Climate . . . on European Constitutions, a popular book first published in 1813, James Johnson advised Britons going to the tropics to wear flannel or woolen clothes, eat a diet high in vegetables, and avoid excessive drink, exercise, and "passions." His recommendations were based entirely on anecdotes, personal opinions, and quotations from Shakespeare and Latin authors.[1]

The theory of humoral pathology, then widely accepted among physicians, taught that illnesses were caused by an imbalance between the bodily fluids, and that the best cure consisted of rebalancing these fluids. For those who succumbed to "fevers," doctors, like the prominent American Dr. Benjamin Rush, prescribed copious bloodletting and heroic doses of calomel, a compound of mercury that caused the patient to salivate.[2] The resulting dehydration and loss of blood killed many who would have survived without any treatment at all.[3]

In the early nineteenth century, medical science did not distinguish clearly among the various "fevers."[4] Fevers were not classified by their causes but by their symptoms; hence an illness that caused frequent bouts of fever was called "remittent," one that recurred every other day was "tertian," and one that lasted two out of every three days was "quartan." The causes of fevers were unknown, but speculation abounded. One of the most popular medical ideas of the early nineteenth century was the miasmatic theory, which blamed diseases on miasmas or putrid vapors emanating from swamps or from rotting matter in the earth that was released when the soil was turned over. Thus malaria got its English name from the Italian *mal'aria*, or "bad air"; the French called it *paludisme* from the Latin word for swamp. Malaria was often confused with typhoid, yellow fever, and other diseases. Attempts to prevent it were equally mistaken. People believed that it could be avoided by staying indoors at night with the windows closed to keep out the night air. Its association with swamps led many to conclude that one could escape it by staying at higher elevations or by quickly moving to the drier parts of Africa away from the coast. No one associated malaria (or any other illness) with mosquitoes.

The opposite theory, contagionism, argued that diseases were spread by contact between humans. Like the miasmatic theory, contagionism had been around for a very long time. Quarantines, used since the Renaissance to isolate the sick during plague and smallpox epidemics, were based on the contagionist idea. Doctors seeking evidence sometimes drew the wrong conclusions. Thus after an outbreak of yellow fever in Barcelona in 1822 showed that some people who had no contact with the sick nonetheless came down with the disease, medical experts discredited contagionism and argued for abolishing quarantines.[5]

The first transformation of medicine from folk art to empirical practice appeared in the early nineteenth century and was closely linked with the beginnings of European imperialism in Africa and with the use of statistics. Medical statistics date to the late seventeenth century, when John Graunt and William Petty collected the Bills of Mortality from London parishes to track the incidence of the plague and other epidemics. But as a tool of scientific research, statistics came into its own in the early nineteenth century.[6]

In the 1820s Europeans learned that the death rate in Africa was much higher than in the Caribbean or elsewhere. The new statistical approach to epidemiology was part of the broader interest in "social physics" in France and Belgium and of the statistical movement in Britain.[7] In 1840 the *United Service Journal and Naval and Military Magazine* published a statistical analysis of disease and death rates among British troops in West Africa and among those who returned from there between 1823 and 1836. The conclusion was startling: 97 percent of those who served in Africa either died or had to be invalided out of the service.[8]

As that knowledge penetrated official circles, Britain began to reconsider its commitments in the region. In 1830 a Select Committee of the House of Commons recommended reducing European troops to a minimum and replacing them with African or black West Indian soldiers. As a result, the number of white Britons in West Africa dropped to two hundred or less, the lowest it had been for a century or would ever be again. Those who remained moved to mountainous areas, believing that altitude protected against fevers.[9] For the same reason, the French reduced their garrison in Senegal.[10] European interest in Africa seemed to be waning.

But it was not to be. Instead, the 1830s and 1840s saw a resurgence of interest in Africa, especially in Britain but also in France. In Britain, the end of the slave trade and an upsurge in religious evangelism opened the door to those who wished to bring Christianity to Africans, such as members of missionary organizations like the Church Missionary Society. Though their death rates were as high as those of merchants and soldiers, many believed, in Philip Curtin's words, "that God would save from harm any who went to Africa to do His work."[11] Commercial interests were also involved. As we saw in chapter 5, Macgregor Laird wrote of "the opening of Central Africa to the enterprise and capital of British merchants" along with "those who . . . consider it their duty to raise their fellow creatures . . . nearer to Him in whose image they were created."[12] Merchants like Laird believed that once the slave trade was gone, both Christianity and a new "legitimate" form of trade could flourish.

The French were less inspired by missionary zeal, but they, too, increased their presence from Senegal to Gabon in competition with the British. And so, four hundred years after Prince Henry the Navigator sent the first boats out into the Atlantic toward the Guinea Coast, there began a new wave of European imperialism directed at Africa.

The motivations that led the British to renew their interest in Africa culminated in the Niger expedition of 1841. Geographically, that expedition was a success when the *Albert* steamed up the Niger and its tributary, the Benue River. Medically, however, it was a disaster. Of the 159 Europeans who entered the Niger, 48 died in the first two months and 55 were dead before the expedition returned to Britain. Most succumbed to malaria or to the treatment inflicted upon those who had a fever.[13] The sad results were published in 1843 by the senior medical officer of the expedition, James Ormiston M'William, under the title *Medical History of the Expedition to the Niger during the years 1841–1842 comprising An Account of the Fever which led to its abrupt Termination.*[14]

The Discovery of Quinine Prophylaxis

By then, a new approach to tropical fevers had appeared: the use of quinine. In the early seventeenth century, Spaniards had learned from South American Indians that the bark of the cinchona tree that grew in the Andes could cure fevers. This knowledge remained largely confined to the Spanish Empire, for cinchona bark was rare and costly, it tasted very bitter, and, as it was imported to Europe by Jesuits, many Britons suspected it of being part of a popish plot against Protestants. By the early nineteenth century, cinchona bark had lost much of its former popularity. James Johnson even recommended against it. In 1820, however, two French pharmacists, Joseph Pelletier and Joseph Caventou, isolated quinine and other anti-malarial alkaloids from the bark. Three years later, the first commercial production of quinine began in the United States, where it was sold under the name "Dr. Sappington's Anti-Fever Pills" and was widely used as a cure for the malaria that was prevalent in the Mississippi Valley.[15]

The French began their long and costly conquest of Algeria in 1830. There, the prevalent form of malaria was caused by the *Plasmodium malariae*, less fatal than the *P. falciparum* prevalent in tropical Africa. The town of Bône (now Annaba) was particularly insalubrious, being surrounded by swamps. In 1833, of 5,500 soldiers stationed there, 4,000 were hospitalized and 1,000 died. During the years 1830–47, the annual death rate among French soldiers was 64 per thousand.[16]

French army doctors at the time were under the influence of Dr. François Broussais, head of the army medical school in Paris, who believed that fevers should be treated with bleedings, leeches, and starvation. In 1834, Dr. François Clément Maillot was assigned to the hospital at Bône. Like his colleagues, he confused malaria, typhoid, and dysentery and practiced a reduced form of bleeding. However, breaking with tradition, he fed the sick and gave them 120 to 200 centigrams of quinine daily from the very onset of fever. As a result, the death rate among the sick dropped from one out of four to one out of twenty. Two years later, his *Traité des fièvres* converted the head of the army medical service in Algeria, Dr. Jean André Antonioni, to the use of quinine.[17]

In sub-Saharan Africa, quinine was used from the 1820s on, but with poor results, because it was administered only *after* a person had fallen ill, and by then it was ineffective against *falciparum* malaria. What worked in Africa was not the drug itself but quinine *prophylaxis*, that is, taking it ahead of time so that one's bloodstream was saturated with quinine before getting infected. The discovery of quinine prophylaxis came through trial and error, not systematic research. T.R.H. Thomson, the naval surgeon onboard the *Soudan* on the 1841–42 Niger expedition, administered up to sixty centigrams of quinine a day to members of the expedition who came down with a fever. He also gave himself up to a gram of quinine before the expedition reached Africa and continued taking it until he returned to England; only then did he come down with a fever.[18] Like other doctors of his day, he classified fevers as "primary," "secondary remittent," "tertian ague," and so on. Yet from his experiments, he drew the conclusion that quinine "produced a most marked and beneficial effect" against fever in general; to avoid the fever prevalent in Africa, he concluded that one had to take large doses of quinine before, during, and after one's stay there.

In Britain, the medical disasters that had befallen previous expeditions did nothing to dampen the enthusiasm for further attempts. In 1847 Dr. Alexander Bryson presented the Admiralty with his *Report on the Climate and Principal Diseases of the African Station.* In it, he attributed the African fevers to "the fever-exciting agencies of the land" as well as "immoderate indulgence in the use of intoxicating liquors," "depressing passions," "laborious duties of surveying," "long continuance on salt provisions," and "exposure to chilling air of the night."[19] Despite his ignorance of the causes of fevers, his solution was clear: quinine. As he explained,

> Cinchona bark and the sulphate of quinine are both extremely useful agents for the prevention of fever, when properly administered on these expeditions; and although it would appear their powers have been considerably underrated, and their administration is apprehended but indifferently understood; still the numerous instances on record in which they have been successfully employed, leave no room to doubt that their more general use upon the station is most urgently required.

Furthermore, he wrote, "it would be advisable not only to administer, daily, one of these febrifuges to men so long as they are exposed to the influence of the land, and the vicissitudes of the weather in open boats, but to continue its use for at least fourteen days after their return on board."[20] In response to this report, the director general of the army's Medical Department sent a circular to the British officials in West Africa, advising them to use quinine prophylactically.[21]

Before the departure of the Niger expedition in 1851, Bryson wrote instructions for the crew; among them were taking six to eight grains (thirty-nine to fifty-two centigrams) daily from the day the ship crossed the bar off the African coast until fourteen days after returning to the ocean.[22] As we saw in chapter 5, the original leader of the expedition, John Beecroft, died before the steamer *Pleiad* reached Fernando Po. When his second-in-command, the young physician Dr. William Balfour Baikie, took over, he saw to it that all European members of the crew took their quinine as prescribed by Dr. Bryson. None of them died.[23] Upon the return of the expedition, Dr. Bryson published an article "On the Prophylactic Influence of Quinine" in the *Medical Times Gazette*, making the results widely known.[24]

The prophylactic use of quinine did not suddenly make Africa healthy for outsiders. Until the 1860s, quinine was extracted from bark imported from South America, where cinchona trees grew wild in the Andean forests. As bark hunters plundered this valuable resource, cinchona trees became scarce, supplies of bark fluctuated widely, and quality varied. For many years, quinine was very expensive; in the early years, an ounce of quinine cost forty shillings in Britain or sixteen dollars in the United States, a fortune at the time. Because of the cost, the French army could not afford to supply all its troops with quinine.[25] Yet by mid-century, the European powers were increasingly involved in colonial expansions: the French in Algeria and Senegal, the British in India, and the Dutch in the East Indies. As their involvements grew, so did their demand for quinine. The only solution to the supply problem, in their view, was to create a source of cinchona bark in their own tropical colonies. This meant sending agents to the Andes to collect cinchona seeds and smuggle them out, for the Andean republics knew the value of their monopoly and forbade the export of seeds. Then the seeds had to be sprouted and transplanted in suitable locations in European colonies. To do so required several scientific expeditions.[26]

The first to venture into the Andes in search of cinchona seeds was the British explorer Hugh Algernon Weddell. In the late 1840s, he sent some *Cinchona calisaya* seeds to the Muséum national d'histoire naturelle in Paris, where they sprouted. A few seedlings were transplanted to Algeria in 1850 but died from the hot, dry winds.[27] The botanical garden of Leiden in the Netherlands grew a seedling from one of Weddell's seeds and sent it to Java, where it grew to produce more plants; unfortunately they were of poor quality.

In 1853–54 the Dutch government sent the botanist Justus Karl Hasskarl to Peru to collect seeds and plants. He returned with cases of *Cinchona pahudiana*, named after the Dutch minister of colonies C. F. Pahud. Eighty plants reached Java. In 1856 another naturalist, F. Junghuhn, imported several other species of cinchona, forming the basis for the experiment station of Tjiniroean at an altitude of 1,566 meters in the mountains of Java.[28]

British interest in cinchonas grew as a result of the Indian Rebellion of 1857, when many European soldiers sent to India came down with malaria. In 1859 the India Office and the Royal Botanic Garden at Kew near London appointed Clements Robert Markham, a clerk at the India Office who had explored the Andes, to lead a cinchona-gathering expedition. Markham and John Weir, a gardener, collected *C. calisaya* seeds in Bolivia and Peru. At the same time, Richard Spruce and Robert Cross, also botanical explorers, went to Ecuador, from which they returned with *C. officinalis* and *C. succirubra* seeds, while G. J. Pritchett brought other varieties from Peru. Markham and Weir's seeds germinated at Kew, but the seedlings died on the voyage to India. Spruce's and Pritchett's seedlings, sent in cooler weather, survived the trip and formed the basis for the government's cinchona plantation at Ootacamund in southern India that supplied the Europeans in India with quinine.[29]

Originally, the India Office had intended that the cinchona plantations produce quinine "for the treatment of the complaints of Europeans." Clements Markham, however, asked: "Did the government undertake Cinchona cultivation in order that the use of quinine, in some form or another, might be extended to the people of India, now entirely debarred from its use; or did they undertake it as a mere speculation?" He went to India to investigate the possi-

bility of developing a febrifuge that Indians could afford. In cinchona bark he found several alkaloids besides quinine that could be
manufactured cheaply enough for mass distribution. In the end, a
compromise was reached. A cheap febrifuge called totaquine was
sold in the post offices of Bengal for a nominal sum, but elsewhere
in India, the production of the government cinchona plantations
was reserved for British personnel.[30]

These early transfers and plantations produced a bark with little
alkaloid content at very great expense. In 1865, however, the English trader Charles Ledger, a resident of Peru, smuggled another
variety out of Bolivia: C. Ledgeriana, also known as C. calisaya, the
same variety that Markham and Weir had smuggled out of South
America but lost between England and India. The British government declined to buy his seeds, but he finally sold a pound of seeds
to the Dutch for twenty-four pounds sterling. Planted at the government's cinchona experiment station of Tjiniroean, it produced bark
with the highest quinine content of all. Dutch scientists worked to
develop more resistant and better-yielding varieties of C. calisaya.
Grown in Java and marketed in Amsterdam from 1872 on, Dutch
quinine soon came to dominate the world market; by 1897, the
Netherlands provided two-thirds of the world's cinchona. The supply rose from 10 tons in 1884 to 516.6 tons in 1913, while the price
dropped from twenty-four pounds per kilogram to between one and
two in 1913. This was the source of the quinine used during the
Scramble for Africa of the late nineteenth century.[31]

Public Health at Mid-Century

After 1850 explorers routinely brought quinine with them, though
they did not always take it regularly. David Livingstone is a case in
point. He read about the use of quinine on the Niger in the 1840s.
On his own expedition across southern Africa up the Zambezi to
Loanda on the Atlantic coast and back in 1853–56, he carried "Livingstone Pills," a mixture of quinine, calomel, rhubarb, and resin of
jalap, as a treatment for malaria. By 1857 he had become convinced
to use quinine as a preventive as well as a remedy. In planning for

another expedition up the Zambezi in 1858, he wrote: "It will be desirable to give quinine wine to all the Europeans before entering and while in the delta,"[32] and had his crew take two grains (thirteen centigrams) a day diluted in sherry. He and his associates were disappointed with its results, for such a low dose did not prevent attacks. At the onset of fever, they increased the dose to ten to thirty grains (sixty-five to two hundred centigrams). Though he and his white companions were often sick, few died. When his medical supplies were stolen during his last expedition (1866–73), Livingstone wrote in his journal: "I felt as if I had now received the sentence of death."[33] Explorers who followed Livingstone—Henry Morton Stanley, Verney Lovett Cameron, Richard Burton, John Speke, Gerhard Rohlfs, and others—had much the same experience: much sickness but few deaths.[34]

Between advances in medical knowledge and their application to public health there were often long delays. Some delays can be attributed to the costs involved, as in the case of quinine. Others resulted from the conservatism of the medical profession. Change often came with the arrival of a new generation of doctors; the use of bloodletting declined in the 1830s and that of calomel in the 1840s and 1850s as a change in medical fashion rather than as a consequence of new knowledge. At other times, the obstacle was the resistance of laymen, both civilian and military, to the advice of their doctors. Thus, only gradually after 1860 did it become customary among Europeans in the tropics to filter water, dispose of sewage, and move to higher ground during the rainy season or at the first signs of an epidemic. These measures were just as important as quinine prophylaxis in maintaining the health of Europeans in Africa.[35]

Especially relevant to the subsequent history of Africa is the impact of quinine prophylaxis on the European troops sent to Africa, both in barracks and on campaign. Philip Curtin has analyzed the medical results of European military actions in the nineteenth century, using the copious statistics gathered by the British and French armies. In the early nineteenth century, the death rates on campaign were higher than even the horrific rates suffered by troops stationed on the coast. Not only did soldiers succumb to malaria and, at certain times, to yellow fever, they also suffered from high rates of gastrointestinal diseases, the result of drinking polluted water and eating contaminated food. In 1824–26, during the first Anglo-Asante War, the death rate was 638 per thousand per year, of which 382 per thousand were from "fevers" and 221 from gastrointestinal diseases.[36]

By the 1860s, military medicine had made such advances that some—but not all—military campaigns could be carried out with remarkably few deaths. Britain conducted two "punitive expeditions" designed not to conquer territory but to overawe Africans and to test new weapons and tactics. The first, in 1867–68, was a six-month attack on Magdala in Abyssinia involving some 68,000 men sent to free a few British officers captured by the Abyssinian ruler Tewodros. On that campaign, the death rate among white troops was only 3.01 per thousand per month (36.12 per thousand per year), compared to 1.18 in barracks in India and .74 in Britain. Some of this was due to luck, as there was no yellow fever or cholera in Abyssinia, and very little malaria or typhoid fever. The rest was the result of careful diet and the availability of distilled water. These precautions only applied to white troops, for the Indian "coolies" who accompanied the expedition suffered much higher disease and death rates.[37]

The Magdala campaign was followed by a war against Asante in 1873–74. This kingdom lay inland from the Gold Coast in a much less salubrious environment than Abyssinia. So unhealthy was this part of Africa that the Danes, the French, and the Dutch had abandoned their forts, leaving only the British. The expedition against Kumasi, the capital of the Asante kingdom, lasted two months and involved 2,500 soldiers. These troops were given five grains (32.4 centigrams) of quinine a day. Water filters were brought along, though some drank unfiltered water. Of the 2,500, 1,503 were treated but only 53 died, 40 of them from disease and 13 from enemy action. The death rate was 8.7 per thousand per month (or 104.4 per thousand per year), of whom 4.57 died from "fevers and sunstroke" and 3.26 from dysentery and diarrhea. This was one-sixth the death rate in the first Anglo-Asante War of 1824–27. The impressive medical results of these two expeditions encouraged those who advocated military advances and led the British public to believe that the army could now cope safely with the African disease environment. Thereby they contributed to the growing European enthusiasm for tropical conquests.[38]

From Empirical to Scientific Medicine

The nineteenth century witnessed tremendous advances in public health and declines in disease and death rates, both in the Western world and among Westerners in the tropics. Popular opinion views these advances as a consequence of discoveries by heroic scientists whose findings were applied by doctors and engineers to solve real-world problems and whose achievements were enthusiastically received by a grateful public. This is clearly a romantic and oversimplified view of the way public health improvements actually took place. Instead, the interactions between science, applications, and results followed a zigzag course, or rather several at once.

One of the complications was the symbiotic relationship between empirical discoveries and scientific explanations. In the course of the nineteenth century, rapid industrialization and urbanization in Europe and North America created both problems and opportunities for experimentation that led to valuable improvements in public health. Likewise, the increased involvement of Western nations in the tropical world also provided opportunities for empirical experiments and the popularization of their results; the introduction of quinine prophylaxis is a good example. Another complication in the history of scientific medicine is that there were always several theories competing to explain diseases, and only gradually did one of them—the germ theory—gain ascendency over the others. And finally, even in cases where empirical practice or laboratory discoveries clearly indicated what needed to be done, there were long delays due to ingrained habits, resistance to new ideas, vested interests, or just the costs involved in making the necessary improvements.[39]

In many instances, empirical advances preceded scientific explanations for diseases by decades. We already saw how the introduction of vaccination by Edward Jenner in the 1790s reinforced the demographic advantage that Europeans had over Indians in

North America. Another famous example of empirical practice preceding a scientific explanation is Dr. John Snow's experiment with the Broad Street pump in 1854. That year, as a cholera epidemic raged in London, Snow noticed a correlation between the incidence of the disease and the supply of water by different companies. When he removed the handle from the pump in Broad Street, forcing the inhabitants go to another neighborhood to fetch water, the number of new cholera cases in the Broad Street neighborhood dropped sharply.[40]

When Dr. Snow described the effects of water supplies on cholera, he was not just recounting anecdotes but presenting numbers that made his argument more persuasive to a public that had learned to trust numbers.[41] As he could not prove the connection between water and cholera, his findings encountered a great deal of resistance from eminent scientists. Nonetheless, through his efforts and those of other sanitary reformers, cities in Europe and the Americas gradually improved their water supply. Improvement, at the time, meant filtering through sand and charcoal to remove suspended solids and unpleasant odors. In tropical settings, this was often done haphazardly with equipment that left dangerous bacteria in the water. During military campaigns thirsty soldiers neglected even that precaution. Yet such practices did contribute to the lowering of the death rates, as military doctors were quick to point out after the expeditions to Magdala and Kumasi.

One of the most dramatic medical advances of the nineteenth century was the discovery in the 1850s by Ignaz Semmelweis, a Hungarian physician, that washing his hands before delivering babies reduced the incidence of puerperal fever among women giving birth, until then the most common cause of death among women. This was followed in the 1860s by the discovery by the English surgeon Joseph Lister that sterilizing instruments and cleaning other surfaces with carbolic acid reduced the rate of infection during surgery.

Another empirical practice that proved beneficial was to remove European troops and other personnel to higher ground at the outbreak of epidemics. This practice dated back to the Middle Ages, when those who could afford it fled the cities during the plague. In the tropical colonies, Europeans moved to hill stations or even, occasionally, back to Europe, in the belief that they were escaping the miasmas or bad air of the lowlands.

Yet what turned medicine into a science was not statistics and epidemiology but a breakthrough in bacteriology. Since the invention of the microscope in the seventeenth century, scientists had been aware of the existence of "animalicules," creatures too small for the naked eye to see. Not until the mid-nineteenth century, however, were microscopes and laboratory techniques able to associate specific microorganisms with particular effects. In the 1860s the Frenchman Louis Pasteur showed that microbes of yeast caused fermentation and that heat killed these microbes and thereby stopped or prevented fermentation and putrefaction. In 1877–78 he identified the bacillus of anthrax. Soon thereafter the German Robert Koch investigated the disease rates in Hamburg and Altona, two neighboring cities that received their water from the same river. Altona's water passed through a slow sand filter covered with a slimy deposit, while Hamburg's water was unfiltered. From this observation, Koch deduced that the sand and the slime effectively removed harmful bacteria from the water. He also developed a method of growing bacteria in a laboratory culture. In 1883, when a cholera epidemic broke out in Egypt, Koch discovered the germ *Vibrio cholerae* and its mode of transmission from sewage to drinking water, thereby providing evidence to back up Snow's hypothesis.[42] Identifying the specific bacteria responsible for a particular disease led quickly to the development of vaccines; by 1893 a vaccine against cholera was available. From that time on, the germ theory of disease gradually replaced the miasmatic theory in medical literature.

Science and Tropical Diseases

Cholera aroused particular interest because it was new to Europe and America in the nineteenth century. The disease is endemic to India, where it flourishes especially during the pilgrimages to Benares (now Varanasi) on the Ganges. In 1820–22 it appeared in Ceylon, Indonesia, China, Japan, and the Middle East. In 1826 it reached Persia and the Ottoman Empire. During the centuries when travel between India and Europe was by sailing ship and voyages took six to nine months, anyone infected with cholera either died or recovered before the ship reached its destination. The introduction of steamers in the 1820s and 1830s shortened travel times and made cholera transportable across oceans. In 1831 the pandemic was carried from Mecca to the countries to which Muslim pilgrims returned, from Morocco to the Philippines; that year it also reached Britain and, a year later, Canada and the United States. Other pandemics followed in the early 1850s (including the one that Dr. Snow encountered in London), in 1863–75, in 1881–94, and from 1899 to the 1920s.[43]

Cholera fascinated physicians and the general public because its effects on the human body were so spectacularly horrible. Some people carried the bacillus without falling ill. Others, seemingly healthy one moment, the next moment lost most of their body fluids through diarrhea and vomiting. As a newspaper reporter described them, "one minute warm, palpitating, human organisms—the next a sort of galvanised corpse, with icy breath, stopped pulse, and blood congealed—blue, shriveled up, convulsed."[44] The discoveries of Snow and Koch therefore led reformers to demand protective measures. Preventing the scourge, however, meant separating sewage from drinking water, which required massive investments in pipes and filtration plants and other urban infrastructures.[45] Only wealthy industrial nations could afford the expenses, and then only piecemeal. In the tropics, colonial officials insisted on segregating Europeans from "native filth and disease" by building separate neighborhoods or cantonments (military housing) at a distance from the native cities.[46]

Quinine prophylaxis was the "magic bullet" of mid-nineteenth-century imperialism, but it was far from the solution to the problem of malaria in the tropics. Death rates from malaria among Europeans in India and tropical Africa dropped until the 1860s, then leveled off, although they continued to drop in Algeria. This method of prevention was limited because quinine was expensive, even after the cinchona plantations of Java and India came into full production. It was too expensive for the French army to protect all its soldiers overseas; a manual of military medicine published in 1875 argued that the prophylactic use of quinine was not necessary in Algeria and should be reserved for epidemics and for especially malarial posts.[47] It was also much too costly for the native inhabitants of malarious regions. Besides, the belief in miasmas died hard and quinine prophylaxis was never fully accepted, even by those who could afford it.

Scientific interest in malaria increased toward the end of the century, especially in Italy, with its malarial swamps near Rome and in the Po valley, and in France and Britain, because of their colonial empires. In 1883 the British physician Patrick Manson of the Imperial Maritime Customs Service in China reported that the *Filiaria bancrofti* worm that caused elephantiasis was transmitted by mosquitoes, raising the possibility that mosquitoes might transmit other diseases.[48] In 1880 the French military doctor Alphonse Laveran identified the *Plasmodium malariae* in the blood of patients infected with malaria, a discovery that triggered a burst of research on the disease in several countries. In 1897, Ronald Ross, a surgeon-major in the Indian Medical Service, showed that bird malaria was transmitted by the *Culex* mosquito. Manson publicized Ross's discovery and got him assigned to full-time research. A year later, Ross in India and three scientists in Italy—Giovanni Battista Grassi, Giuseppe Bastianelli, and Amico Bignami—worked out the complex life cycle of the human malarial *plasmodium* as it spends one part of its life in the gut of the *Anopheles* mosquito, and another infecting humans.[49] The connection between malaria and mosquitoes added two new methods of prevention to the arsenal of public health. Mosquito eradication was attempted, although it was difficult, for the *Anopheles* mosquito could breed out in the country and travel far. Screens and mosquito nets became popular in the United States, but were disliked by the British on the grounds that they interfered with the flow of air. However, netting was used to keep mosquitoes from biting the sick and transmitting the disease to the healthy.[50]

Another disease, typhoid, bore a close connection to imperialism, but indirectly. Until the 1870s, it was often confused with other continued fevers. In the 1870s, just as malaria, cholera, and some gastrointestinal diseases were becoming less of a threat to Europeans overseas, the incidence of typhoid rose in the European garrisons in

India and northern Africa; in India, the number of typhoid-related deaths among white soldiers multiplied fivefold between 1860 and 1900; in Algeria, as in France, it doubled until the 1880s, then declined.

The cause of this disease was the bacteria *Salmonella typhi* carried by water, flies, or dust from the feces of an infected person to the digestive tract of a healthy one. Before antibiotics, the probability of an infected person dying of typhoid was 30 percent, and those who recovered remained carriers of the disease for months after they seemed healthy, some even for many years.

In 1880 Karl Joseph Eberth identified a bacteria that he called the *Bacillus typhosus* (now *Salmonella typhi*). At first, the disease was associated with sewage-tainted water, like cholera. Efforts to combat the disease focused on replacing the "dry system" of night-soil removal in carts with the new "wet system" of flush toilets and sewage pipes that carried the waste out of town. As cities in Europe and America installed waterborne sewage systems that dumped raw sewage into rivers and lakes, typhoid infection rates rose in cities downstream. Prevention required cleaning water by installing slow sand filters and disinfecting it with chlorine or bromine. All these measures were very costly and did not reach the cities in the wealthier industrial countries until well into the twentieth century.

When a British army invaded Egypt in 1882, medical men expected to find dysentery, scurvy, ophthalmia, and other diseases known to be prevalent there, but not typhoid. In fact, during the invasion itself, which lasted from July 17 to October 3, 1882, there were fewer deaths from disease than from combat, a rare event before the twentieth century. However, no sooner were British troops installed in the barracks in Cairo and Alexandria that they had requisitioned from the Egyptian army than typhoid broke out among them. They blamed the filthy conditions of the barracks and the poor quality of the water, which came either directly from the Nile or from wells that tapped underground water that also came from the river. The filters the British army used were ineffective against the disease. Not until many years later was it possible to test water for bacteria and were measures taken to provide soldiers with safe water and effective sewage removal.[51]

Before the twentieth century, yellow fever ranked among the most terrifying of diseases. It was known to cause epidemics in West Africa and the Caribbean, but mainly among newcomers, for the native inhabitants and African slaves seemed immune to it. Though generally confined to the tropics, it was known to break out in port cities in the temperate zone during the summer months. All the ports of North America, from New Orleans to Boston, and the Mississippi Valley suffered periodic epidemics from the seventeenth through the nineteenth centuries; Philadelphia was hit several times, most terribly in 1793.[52] It seemed to follow no known pattern; it was not associated with swamps and "miasmas" like malaria, nor was it propagated from person to person like smallpox. Even after bacteriologists had established the germ theory as the best explanation for so many other diseases, no one could find a bacterium associated with yellow fever, for a good reason: it is caused by a virus, much too small to be seen in the microscopes of the time.

Yet hypotheses abounded. Among the most plausible was the one proposed by the Cuban physician Carlos Finlay. In 1881 he suggested that the disease might be transmitted by the mosquito *Stegomyia fasciata* (now known as *Aedes aegypti*), but he could not prove it. In May 1900, after American soldiers sent to Cuba had proved vulnerable to yellow fever, Surgeon General George Sternberg appointed a Yellow Fever Board under Dr. Walter Reed, who had distinguished himself with his work on typhoid. Though skeptical at first, Reed agreed to work with Finlay and test his hypothesis, using volunteers. These experiments proved conclusively that the disease was caused by the bite of the *A. aegypti* mosquito.[53]

Health and Empire at the Turn of the Century

The connection between improvements in the health of Europeans and the New Imperialism in the late nineteenth century has not been lost on historians. As Philip Curtin has written, "While the medical reforms were not a direct cause of the later scramble for Africa, they were clearly a technological leap forward. As such, they were necessarily an important permissive factor."[54]

During the Scramble for Africa, not all European military campaigns were as fortunate as the British expeditions to Magdala and Kumasi. The French forces in Senegal suffered from much higher death rates than did the British. In the French campaigns in the Western Sudan in 1883–88, the death rate from malaria was 97.74 per thousand per year, from typhoid fever 24.24, and from gastrointestinal ailments 60.79, contributing to a total of 200.24 per thousand per year, twice the rate of that of the British Kumasi campaign. In later years, the French reduced the death rate by moving white troops to the coast and sending the officers home during the rainy season.[55]

We can compare British and French military medicine in the two major campaigns of the end of the century: the Anglo-Egyptian campaign in the Sudan in 1898 and the French conquest of Madagascar in 1894–95. Among British troops going up the Nile to Omdurman in 1898, the death rate was 8.599 per thousand per month (an annual rate of 103 per thousand), two-thirds of them from "enteric fevers," mainly typhoid from drinking unfiltered Nile water. Among the French troops in Madagascar, in contrast, the death rate per thousand was 44.67 per month (or 536 per year), three-quarters of them from malaria. The cause was the low dosage of quinine given the soldiers for reasons of economy: 40 to 80 centigrams per week, compared to 136 centigrams per week for British troops during the Asante campaign or 227 centigrams per week in the 1890s.[56] Evidently the French army and the French public tolerated higher death rates among their troops than did the British.[57] Whatever the differences, the overall trend is clear. Until the 1860s, death rates in tropical Africa were horrendous, even suicidal, for Europeans. But after that decade, scientific and technological advances had made sub-Saharan Africa if not a healthy place, at least tolerable for the Europeans sent there.

When the United States declared war on Spain in April 1898, army medical men were worried about the diseases they expected to find in Cuba: malaria, dysentery, and especially yellow fever. They did not know what caused yellow fever nor how to prevent it; all they knew was that it was at its most dangerous during the summer rainy season, and they urged President McKinley to postpone the planned invasion of Cuba until the fall. Meanwhile, lured by the prospect of an exciting and victorious war, scores of volunteers joined the state militias. The generals of the regular army, wishing to seize the glory of victory before the volunteers were ready for combat, lobbied to invade Cuba in June, against the advice of their medical officers.[58]

While the regular army was achieving a quick victory in Cuba, the volunteers were still training in five camps scattered throughout the southern United States. Medical officers knew the dangers of typhoid and how to prevent it with adequate latrines and sewers. The line officers who commanded the camps, however, would brook no interference with their authority and considered the presence of human excrement a normal part of military life. The result was what the medical officers had predicted: an epidemic of typhoid swept through the camps. In some regiments, 90 percent of the volunteers came down with typhoid. There were 20,738 hospital admissions and 1,590 deaths. Overall, of every thousand recruits in the camps, 192 fell ill and fifteen died.[59]

Meanwhile in Cuba, the U.S. Fifth Army was going from victory to victory. One reason for its success was that the Spanish army was decimated by disease. From 1895 to 1898, 16,000 Spanish soldiers had died from yellow fever; when the Americans invaded, only 55,000 out of a Spanish army of 230,000 were available to fight.[60] But soon the Americans encountered much more dangerous enemies than the bedraggled Spanish soldiers. At first, the most serious problems were malaria and dysentery. The troops were not protected against the *falciparum* malaria prevalent on the island because John Guiteras, head of the Cuban Sanitary Department and an expert on tropical diseases, did not think quinine could prevent malaria and recommended it only "when the individual is subjected to extraordinary depressing influences."[61] Then on July 6, the first case of yellow fever appeared in the Fifth Army; by July 13 there were a hundred cases. On August 2, General Shafter recommended that the Fifth Army withdraw from the island. During the fall of that year, when it was evacuated to Camp Wikoff in eastern Long Island, 80 percent of the troops were to some degree ill.[62] The flight from Cuba ranks among the most ignominious retreats in American history.

American schoolchildren are taught that the campaign of 1898 was a "splendid little war." True, it was over quickly, and the United States gained a vast overseas empire with little effort. But the cost was borne by the 2,565 soldiers who died of diseases while only 345 died in combat, a ratio of almost eight to one. (In comparison, during the Civil War, known for its horrendous health conditions, the ratio of deaths from disease to deaths in combat was two to one.) And most of the illnesses and deaths were preventable. While the army's troops were falling ill and dying, the Marines had good sanitation and the navy, which used the ocean as a sewer, remained healthy.[63]

After the war, there was much soul-searching among the military doctors. In August 1898 Surgeon General Sternberg appointed a Typhoid Board under Major Walter Reed to investigate the disaster. After reviewing the evidence, the board's report identified typhoid as caused by a specific germ, different from malaria and other fevers. It also found that it was transmitted not only by water but also by flies that carried the germ from feces to food.[64]

As for yellow fever, the findings of Doctors Finlay and Reed were almost immediately beneficial. Even before their results were published, the American forces occupying Cuba were hard at work cleaning the Cuban cities of filth and decaying animal and human corpses and fumigating houses, partly in the belief that disgusting things bred disease, and partly from the moral superiority complex common among conquering whites in an age of racial and ethnic arrogance. When Reed and his associates demonstrated that one particular mosquito caused yellow fever, the campaign became much more focused, for the *Aedes aegypti* mosquito was known as a domestic creature that could breed in the tiniest amount of stagnant water and rarely flew more than a few hundred feet from its place of birth. In 1901 Colonel William Gorgas, the medical officer of the American forces occupying Havana, began a campaign to rid the city of mosquitoes by fumigating houses, emptying or sealing all open water containers and puddles that might harbor mosquito larvae, and installing mosquito nets and screens. The results were astonishing. The number of cases of yellow fever in Havana dropped from fourteen hundred in 1900 to thirty-seven in 1901 and to zero in 1902.[65]

Other subtropical cities like New Orleans and Rio de Janeiro quickly followed suit.

Before the Anglo-Boer War of 1899–1902, South Africa was known as a very healthy environment; British troops stationed there were no more likely to get sick and die than in Britain itself. When the war broke out in late 1899, medical men knew what caused most diseases and how to prevent them. The problem was not a lack of knowledge but the resistance of the military authorities to the idea that their responsibilities included health and sanitation. Almroth Wright, who had developed an anti-typhoid vaccine in the mid-1890s, proposed that all troops sent overseas should be immunized. But the army left it up to the soldiers, and fewer than 4 percent volunteered. Although the vaccine was not perfect, the death rate among the vaccinated was less than one-tenth what it was among the unvaccinated.

When the British army besieged Ladysmith between November 1899 and March 1900, an epidemic broke out almost immediately; 177 per thousand contracted typhoid or other continued fevers, and another 186 per thousand came down with a gastrointestinal infection. Altogether, 465 men died. In Bloemfontein, which the British occupied from March to July 1900, there were 8,568 cases of typhoid, of whom 964 died, and 2,121 of gastrointestinal diseases, causing another 81 deaths. Altogether, of a British army of half a million men, twenty thousand died, 70 percent of them from diseases.

Much worse was to happen to the civilians caught in the fighting. At first, most of these were refugees trying to escape the fighting in the countryside. But after the British victories of 1900, the Boers switched to guerrilla warfare and the British responded by herding civilians into concentration camps. By the end of the war, 115,700 Africans were incarcerated in sixty-six camps, where the death rates per year were as high as 446 per thousand, mostly from respiratory diseases. Among whites, 116,000, or well over half the Boer population—mainly women and children—were incarcerated; of these, over 27,900 died, including 81 percent of the children under the age of sixteen.

Experts argued over the causes of the epidemics, with most blaming the infected water supply, the lack of filters, and the bad habits of soldiers, while others also blamed flies, dust laden with fecal particles, and infected clothing and blankets. Line officers were faulted for leaving water filters, ambulances, and medical supplies behind when they moved their troops. A Royal Commission, appointed in 1901, ruled the Royal Army Medical Corps blameless. Not until the eve of World War I did Western armies take sanitation seriously enough to prevent such epidemics.[66]

The most spectacular achievement of the new approach to tropical disease and public health was its role in the construction of the Panama Canal. A French company headed by Ferdinand de Lesseps, the builder of the Suez Canal, had attempted to build a canal across the Isthmus of Panama in the 1880s. Many problems bedeviled this project, not the least of them the diseases that decimated the workforce and the engineers who directed it. Yellow fever and malaria were especially severe in the humid forests of Panama, but typhoid, smallpox, pneumonia, dysentery, and other ailments also afflicted the workers. In 1884, of 19,000 workers, 6,000 were sick. Of those who got yellow fever, half died. Of those admitted to the hospitals, three-quarters died. All told, an estimated 21,000 Frenchmen and 24,000 Jamaicans and Haitians perished. Finally, overwhelmed by intractable engineering challenges and riddled with financial scandals, the company went bankrupt in 1899.

After the Spanish-American War, the United States suddenly found itself in possession of Puerto Rico, Hawaii, and the Philippines and in control of Cuba. Under the influence of Theodore Roosevelt, a naval enthusiast and disciple of Captain Alfred Mahan, the United States began building a navy to rival that of Great Britain. With possessions on both sides of the globe, it seemed imperative that it should possess a canal to link the two oceans. In 1903, in a classic case of informal imperialism, the United States engineered the secession of the province of Panama from the republic of Colombia to which it belonged. In exchange, the French chief engineer Philippe Bunau-Varilla, representing the breakaway province, signed a treaty granting the United States control over a slice of the new nation through which a canal was to be built.

The prospective canal builders faced the same disease ecology that had thwarted the French attempt. In 1905 John Stevens, the chief American engineer, gave Major General Gorgas his full support and sufficient funding to rid the cities at the two ends of the canal project, Colón and Panama City, of yellow fever. Within a year and a half, using the same draconian measures he had imposed on Havana, Gorgas had achieved that goal. Though other diseases did not disappear, the number of deaths among the workers dropped from 1,273 in 1907–8 (of which 205 were from malaria) to 414 in 1913–14 (14 from malaria). The average death rate that year was 7.92 per thousand (2.06 per thousand among whites and 8.23 per thousand among black workers), much lower than in the United States.[67]

Conclusion

The advances in medicine and public health over the course of the nineteenth and early twentieth centuries were closely related to the role played by the industrial nations of Europe and North America in the rest of the world. As this was the high point of Western imperialism, it is not surprising to find imperialism and medicine closely linked.[68] The connections, however, were multiple and complex.

Some of the advances resulted from a two-way interaction between the imperialists' urge to explore, penetrate, and conquer and the medical advances needed to achieve their goals; in other words, power over people required power over nature. Although malaria was present in Europe, the discovery of quinine and its prophylactic use were associated with the French invasion of Algeria and the British explorations of the Niger River. Likewise, the transfer of cinchona to India and Java was deliberately encouraged by the Dutch and British colonial ministries. And toward the end of the nineteenth century, Alphonse Laveran in Algeria and Ronald Ross in India were responsible for elucidating the role of the *Anopheles* mosquito in transmitting malaria. We find the same direct connection between disease and empire in the American conquest of Cuba and the discovery by Carlos Finlay and Walter Reed of the role of the *Aedes aegypti* mosquito in transmitting yellow fever.

In other cases, the connections are more remote. Cholera was a permanent part of life in India long before it came to the attention of Europeans. Only starting in the 1830s, when steamships shortened travel times, could the disease reach Europe and North America; and only then did doctors, scientists, and public health officials in the West pay it much attention.

Smallpox and typhoid represent yet another kind of interaction between imperialism and disease. We already saw, in chapter 3, the role that smallpox played in the conquest of the Americas in the early modern period. The story did not end there. The introduction of smallpox inoculation, and later of vaccination, once again favored people of European origin over Native Americans. Typhoid appeared late in history. Though the disease itself favored no one people over another, its dramatic appearance in Egypt in the 1880s, in the Spanish-American War of 1898, and in the Anglo-Boer War of 1899–1902 forced the imperial powers Britain and the United States to confront it and take measures such as immunizations, water filtration, and sewage disposal in order to protect their soldiers from this disease.

Overall, the results of the advances in the health of Europeans in the tropics were astonishing. The most spectacular improvements occurred in West Africa. The average death rate among European soldiers in Sierra Leone in the years 1817–38 was 483 per thousand; by 1909–13, their death rate in British West Africa had dropped to 5.56–6.65 per thousand, a decline of over 98 percent. Elsewhere in the tropics, the declines were less dramatic but nonetheless significant: 96.24 percent in the Netherlands East Indies, 95.96 in French West Africa, 92.13 percent in Ceylon, 75.23 percent in South Africa, and 71.91 percent in Jamaica.[69]

None of these advances was designed to improve the health of the inhabitants of the places under colonial domination. In some cases, improvements in health occurred as an unintended byproduct of measures taken to protect the citizens of the imperial powers. Thus, by eradicating yellow fever from Havana and the Panama Canal Zone, the U.S. Army made those places much healthier for their inhabitants. Likewise, totaquine was made cheaply avail-

able to the people of Bengal, although nowhere else was an anti-
malarial drug within reach for the majority of those who suffered
from malaria.[70] In situations where prevention was costly, such as
providing quinine or installing water filtration plants and waterborne sewage disposal systems, the policy was to segregate the Europeans or Americans in their own neighborhoods or cantonments, or
to remove them entirely during epidemics and ignore the native
quarters. Worse yet, in many places, as the health of whites in the
tropics improved, that of the indigenous inhabitants declined as soldiers, traders, and porters spread syphilis, gonorrhea, trypanosomiasis, cholera, and other diseases.[71] In short, public health, like other
technologies, was an economic good and a costly one. Like all other
costly economic goods, it was not designed for human welfare, but
for the welfare of some humans.

Notes

1. James Johnson, *The Influence of Tropical Climate, more especially the climate of
India, on European Constitutions: The principal effects and diseases thereby induced,
their prevention and removal, and the means of preserving health in hot climates, rendered
obvious to Europeans of every capacity* (London: J. J. Stockdale, 1813), pp. 415ff. See
also Paul F. Russell, *Man's Mastery of Malaria* (London: Oxford University Press,
1955), pp. 98–99.

2. Wesley W. Spink, *Infectious Diseases: Prevention and Treatment in the Nineteenth
and Twentieth Centuries* (Minneapolis: University of Minnesota Press, 1978), p. 12;
Dennis G. Carlson, *African Fever: A Study of British Science, Technology, and Politics
in West Africa, 1787–1864* (Carlton, Mass.: Science History Publications, 1984),
pp. 43–44.

3. Philip D. Curtin, *Disease and Empire: The Health of European Troops in the
Conquest of Africa* (New York: Cambridge University Press, 1998), pp. 24–25; Curtin, *The Image of Africa: British Ideas and Action, 1780–1850* (Madison: University
of Wisconsin Press, 1964), pp. 192–93.

4. On the confusions of medical "science" in the early nineteenth century, see
Carlson, *African Fever*, chapter 3: "Theoretical Chaos."

5. William H. McNeill, *Plagues and Peoples* (Garden City, N.Y.: Doubleday),
p. 266.

6. On the origin of statistics, see Daniel R. Headrick, *When Information Came of
Age: Technologies of Knowledge in the Age of Reason and Revolution, 1700–1850* (New

York: Oxford University Press, 2000), chapter 3: "Transforming Information: The Origin of Statistics," especially pp. 84–89. On medical statistics, see Philip D. Curtin, *Death by Migration: Europe's Encounter with the Tropical World in the Nineteenth Century* (New York: Cambridge University Press, 1989), pp. 162–222: "Appendix: Statistical Tables."

7. Headrick, *When Information Came of Age*, pp. 81–89. See also Carlson, *African Fever*, chapter 5: "Changing Analysis."

8. "Western Africa and Its Effects on the Health of Troops," *United Service Journal and Naval and Military Magazine* 12, no. 2 (August 1840), pp. 509–19.

9. Curtin, *Disease and Empire*, pp. 16, 20.

10. Ibid., pp. 12–18.

11. Ibid., p. 20.

12. Macgregor Laird and R.A.K. Oldfield, *Narrative of an Expedition into the Interior of Africa, by the River Niger, in the Steam-Vessels Quorra and Alburkah, in 1832, 1833, and 1834*, 2 vols. (London: Richard Bentley, 1837), vol. 1, p. vi, and vol. 2, pp. 397–98.

13. Carlson, *African Fever*, pp. 14–16, 51–52.

14. James Ormiston M'William, *Medical History of the Expedition to the Niger during the years 1841–1842 comprising An Account of the Fever which led to its abrupt Termination* (London: John Churchill, 1843).

15. Jaime Jaramillo-Arango, *The Conquest of Malaria* (London: Heinemann, 1950), p. 87; Russell, *Man's Mastery of Malaria*, pp. 105, 133.

16. Curtin, *Death by Migration*, p. 5.

17. René Brignon, *La contribution de la France à l'étude des maladies coloniales* (Lyon: E. Vitte, 1942), pp. 20–21; A. Darbon, J.-F. Dulac, and A. Portal, "La pathologie médicale en Algérie pendant la Conquête et la Pacification," in *Regards sur la France: Le Service de Santé des Armées en Algérie, 1830–1958* (*Numéro spécial réservé au Corps Médical*) 2, no. 7 (Paris: October–November 1958), pp. 32–38; Général Jaulmes and Lieutenant-Colonel Bénitte, "Les grands noms du Service de Santé des Armées en Algérie" in ibid., pp. 100–103.

18. T.R.H. Thomson, "On the Value of Quinine in African Remittent Fever," *The Lancet* (February 28, 1846), pp. 244–45. See also Curtin, *Death by Migration*, p. 63; Curtin, *Disease and Empire*, pp. 21–23.

19. Dr. Alexander Bryson, *Report on the Climate and Principal Diseases of the African Station* (London: William Clowes, 1847), pp. 195–96, 210–17.

20. Ibid., pp. 218–19.

21. Philip Curtin, "'The White Man's Grave': Image and Reality, 1780–1850," *Journal of British Studies* 1 (1961), pp. 105–23.

22. Thomas Joseph Hutchinson, *Narrative of the Niger, Tshadda, and Binuë Exploration; Including a Report on the Position and Prospects of Trade up Those Rivers, with Remarks on the Malaria and Fevers of Western Africa* (London, 1855; reprint, London: Cass, 1966), pp. 211–21.

23. Carlson, *African Fever*, pp. 86–87.

24. Dr. Alexander Bryson, "On the Prophylactic Influence of Quinine," *Medical Times Gazette* (London: January 7, 1854), p. 7.

25. Carlson, *African Fever*, p. 48; Darbon, Dulac, and Portal, "La pathologie médicale en Algérie," p. 33.

26. On the cinchona transfer, see Daniel R. Headrick, *The Tentacles of Progress: Technology Transfer in the Age of Imperialism, 1850–1940* (New York: Oxford University Press, 1988), pp. 231–37.

27. Julius Heinrich Albert Dronke, *Die Verpflanzung des Fieberrindbaumes aus seiner südamerikanischen Heimat nach Asian und anderen Ländern* (Vienna: R. Lechner, 1902), p. 13.

28. Fiammetta Rocco, *Quinine: Malaria and the Quest for a Cure That Changed the World* (New York: Perennial, 2004), pp. 206–49; Norman Taylor, *Cinchona in Java: The Story of Quinine* (New York: Greenberg, 1945), pp. 39–45; Pieter Honig, "Chapters in the History of Cinchona. I. A Short Introductory Review," in Pieter Honig and Frans Verdoorn, *Science and Scientists in the Netherlands East Indies* (New York: Board for the Netherlands Indies, Surinam and Curaçao, 1945), pp. 181–82; K. W. van Gorkum, "The Introduction of Cinchona into Java," in ibid., pp. 182–90; P. van Leersum, "Junghuhn and Cinchona Cultivation," in ibid., pp. 190–96; Dronke, *Die Verpflanzung des Fieberrindbaumes*, pp. 14–15.

29. Lucile H. Brockway, *Science and Colonial Expansion: The Role of the British Botanic Gardens* (New York: Academic Press, 1972), chapter 6. On the cinchona transfer expeditions, see Clements Markham, *Travels in Peru and India while Superintending the Collection of Cinchona Plants and Seeds in South America, and Their Introduction into India* (London: John Murray, 1862); Markham, *Peruvian Bark: A Popular Account of the Introduction of Cinchona Cultivation into British India, 1860–1880* (London: John Murray, 1880); Donovan Williams, "Clements Robert Markham and the Introduction of the Cinchona Tree into British India," *Geographical Journal* 128 (1962), pp. 431–42; M.R.D. Seaward and S.M.D. Fitzgerald, eds., *Richard Spruce (1817–1893): Botanist and Explorer* (Kew: Royal Botanic Garden, 1996); Gabriele Gramiccia, *The Life of Charles Ledger (1818–1905): Alpacas and Quinine* (Basingstoke: Macmillan, 1988); and Dronke, *Die Verpflanzung des Fieberrindbaumes*, pp. 28–30.

30. Williams, "Clements Robert Markham," pp. 438–39; Brockway, *Science and Colonial Expansion*, pp. 120–33.

31. Taylor, *Cinchona in Java*, pp. 45–55; Dronke, *Die Verpflanzung des Fieberrindbaumes*, pp. 17–23. On the cinchona and quinine commerce, see Emile Perrot, *Quinquina et quinine* (Paris: Presses Universitaires de France, 1926), pp. 46–49; and M. Kerbosch, "Some Notes on Cinchona Cultivation and the World Consumption of Quinine," *Bulletin of the Colonial Institute of Amsterdam* 3, no. 1 (December 1939), pp. 36–51.

32. Michael Gelfand, *Livingstone the Doctor, His Life and Travels: A Study in Medical History* (Oxford: Blackwell, 1957), p. 127.

33. Horace Waller, *The Last Journals of David Livingstone* (New York: Harper and Brothers, 1875), vol. 1, p. 177.

34. Robert I. Rotberg, *Africa and Its Explorers: Motives, Methods and Impact* (Cambridge, Mass.: Harvard University Press, 1970) passim.

35. Curtin, "'White Man's Grave,'" pp. 106–7; Curtin, *Death by Migration*, pp. 61–66, 160.

36. Curtin, *Disease and Empire*, pp. 5, 15–18, 49.

37. Ibid., pp. 44–46.

38. Ibid., pp. 67–69, 229.

39. Such resistance is not unique to the nineteenth century; witness the resistance to stem-cell research, the teaching of evolution, and AIDS prevention, and the continued belief in paranormal phenomena in our own day.

40. John Snow, *On the Mode of Communication of Cholera*, 2nd ed. (London: J. Churchill, 1855). See also Sandra Hempel, *The Strange Case of the Broad Street Pump: John Snow and the Mystery of Cholera* (Berkeley: University of California Press, 2007).

41. See Theodore Porter, *Trust in Numbers: The Pursuit of Objectivity in Science and Public Life* (Princeton: Princeton University Press, 1995).

42. Spink, *Infectious Diseases*, pp. 19–21; Curtin, *Death by Migration*, p. 117.

43. McNeill, *Plagues and Peoples*, pp. 261–64; Curtin, *Death by Migration*, pp. 71–73, 145–49; Spink, *Infectious Diseases*, pp. 163–65.

44. Quoted in Steven Shapin, "Sick City: Maps and Mortality in the Time of the Cholera," *New Yorker* (November 6, 2006), p. 110.

45. The case of Chicago is especially instructive. To provide proper drainage, many houses had to be raised by several feet and the course of the Chicago River, which had always flowed into Lake Michigan, was reversed in order to keep the lake water clean and make the city's sewage flow through a canal to the Illinois and Mississippi rivers; see Louis P. Cain, "Raising and Watering a City: Ellis Sylvester Chesborough and Chicago's First Sanitation System," *Technology and Culture* 13 (1972), pp. 355–72.

46. Headrick, *Tentacles of Progress*, chapter 5: "Cities, Sanitation, and Segregation"; Curtin, *Death by Migration*, pp. 108–9.

47. Curtin, *Death by Migration*, pp. 132–35.

48. Michael Worboys, "Manson, Ross and Colonial Medical Policy: Tropical Medicine in London and Liverpool," in Roy Macleod and Milton Lewis, eds., *Disease, Medicine, and Empire: Perspectives on Western Medicine and the Experience of European Expansion* (London and New York: Routledge, 1988), p. 23.

49. Douglas M. Haynes, *Imperial Medicine: Patrick Manson and the Conquest of Tropical Disease* (Philadelphia: University of Pennsylvania Press, 2001), pp. 86–88; Michael Colbourne, *Malaria in Africa* (London: Oxford University Press, 1966), p. 6; Rocco, *Quinine*, pp. 251–80; Spink, *Infectious Diseases*, pp. 366–69; Worboys, "Manson, Ross and Colonial Medical Policy," pp. 23–24.

50. Curtin, *Death by Migration*, pp. 136–40.

51. Curtin, *Disease and Empire*, pp. 117–35, 157–67; Curtin, *Death by Migration*, pp. 112, 150–52.

52. J. H. Powell, *Bring Out Your Dead: The Great Plague of Yellow Fever in Philadelphia in 1793* (Philadelphia: University of Pennsylvania Press, 1949).

53. Vincent J. Cirillo, *Bullets and Bacilli: The Spanish-American War and Military Medicine* (New Brunswick, N.J.: Rutgers University Press, 2004), pp. 113–16; John R. Pierce, *Yellow Jack: How Yellow Fever Ravaged America and Walter Reed Discovered Its Deadly Secrets* (Hoboken, N.J.: Wiley, 2005), pp. 148–88; Spink, *Infectious Diseases*, pp. 155–56.

54. Curtin, "'White Man's Grave,'" p. 110.

55. Curtin, *Disease and Empire*, p. 87.

56. Ibid., pp. 187–98.

57. Ibid., pp. 26, 84–105.

58. Pierce, *Yellow Jack*, pp. 103–4.

59. Cirillo, *Bullets and Bacilli*, pp. 57–72; Curtin, *Disease and Empire*, p. 124.

60. Pierce, *Yellow Jack*, p. 103.

61. Mary C. Gillett, *The Army Medical Department, 1865–1917* (Washington, D.C.: Center for Military History, U.S. Army, 1995), pp. 129–30.

62. Pierce, *Yellow Jack*, pp. 104–9; Gillett, *Army Medical Department*, p. 186.

63. Pierce, *Yellow Jack*, pp. 109–10; Cirillo, *Bullets and Bacilli*, pp. 1–3, 72.

64. Cirillo, *Bullets and Bacilli*, pp. 72–75; Gillett, *Army Medical Department*, pp. 173–95; Pierce, *Yellow Jack*, pp. 110–14.

65. David G. McCullough, *The Path between the Seas: The Creation of the Panama Canal, 1870–1914* (New York: Simon and Schuster, 1977), p. 413; Pierce, *Yellow Jack*, pp. 113–15; Cirillo, *Bullets and Bacilli*, p. 118.

66. Curtin, *Disease and Empire*, pp. 123–25, 208–19. See also Cirillo, *Bullets and Bacilli*, pp. 136–52.

67. McCullough, *Path between the Seas*, pp. 140, 171–73, 415–26, 465–67, 581–82. See also Cirillo, *Bullets and Bacilli*, pp. 118–19.

68. On the many connections between science and imperialism, see, for example, Deepak Kumar, *Science and Empire* (Delhi: Anamika Prakashan, 1991), and Kumar, *Science and the Raj, 1857–1905* (Delhi: Oxford University Press, 1995), as well as Teresa Meade and Mark Walker, eds., *Science, Medicine, and Cultural Imperialism* (New York: St. Martin's, 1991).

69. Curtin, *Death by Migration*, pp. 7–10. See also Philip Curtin, Steven Feierman, Leonard Thompson, and Jan Vansina, *African History* (Boston: Little Brown, 1978), p. 446.

70. That is still true today. One hundred fifty years after the discovery of quinine prophylaxis, malaria still kills more people than any other disease.

71. Rita Headrick, *Colonialism, Health and Illness in French Equatorial Africa, 1885–1935* (Atlanta: African Studies Association Press, 1994), chapter 2: "The

Medical History of French Equatorial Africa to 1914." See also Radhika Ramasub-
ban, "Imperial Health in British India, 1857–1900," pp. 38–60; Anne Marcovich,
"French Colonial Medicine and Colonial Rule: Algeria and Indochina," pp. 103–
17; and Maryinez Lyons, "Sleeping Sickness, Colonial Medicine and Imperialism:
Some Connections in the Belgian Congo," pp. 242–56, all in Macleod and Lewis,
Disease, Medicine, and Empire.

Chapter 7

Weapons and Colonial Wars, 1830–1914

The nineteenth century was, after the sixteenth, the century that witnessed the most rapid and dramatic expansion of European power in the world. Yet it did not start out that way. During the first four decades, Africa (except for the Cape of Good Hope) was still off-limits to Europeans; European Americans occupied less than a quarter of North and South America; and Asia, with the exception of Java and half of India, was still under Asian rule. Attempts to expand beyond these limits were met with great difficulties, and sometimes failures, as we saw in chapter 4. Then, beginning in the 1830s and 1840s, came a new era in which industrialization and advances in science made it increasingly easy, cheap, and tempting for Europeans to engage in wars of conquest. We have already reviewed two examples of this transformation: steamboats and tropical medicine. Yet steamboats were restricted to shallow waters and advances in medicine and pharmacology made survival possible but did not ensure victory. It is a third technological change—the invention of new firearms and their uneven diffusion around the world—that made the "New Imperialism" of the late nineteenth century so rapid and dramatic.

The Gun Revolution

The nineteenth century saw more innovations in firearms than any period before or since. Innovations that increased the ease of loading, the rapidity of fire, and the accuracy and range of bullets gave those who possessed new weapons the ability to dominate and coerce those who did not. The causes of these innovations were three: rivalries between the European nations and wars in the United

States; a culture that exalted and rewarded inventors; and the industrialization of the Western world that provided the machine tools and materials with which to manufacture powerful new firearms.

Until the 1840s, the standard military gun was the muzzle-loading smoothbore musket. The English version was called the Brown Bess, a gun that remained essentially unchanged from the Battle of Blenheim in 1704 through Waterloo in 1815 and that was still issued to soldiers in the Crimean War in 1854. It shot a round bullet accurately at fifty yards, but seldom more than eighty yards; hence the admonition to soldiers to hold their fire until they could see the whites of their enemies' eyes. Loading such a gun was a complex operation that had to be performed standing up and often took a minute; only highly trained soldiers could load and fire three times a minute. Since the powder in the firing-pan was ignited by a spark from the flint, such guns were very vulnerable to moisture and were not expected to fire more than six or seven times out of ten. It was said that soldiers shot away their weight in lead for every enemy they killed.[1]

Rifling, that is, cutting a spiral groove inside the barrel that made the bullet spin and thereby fly straighter and more accurately, had been known and practiced since the sixteenth century. Late eighteenth-century rifles, like the American Pennsylvania-Kentucky rifles or the German Jägers on which they were modeled, had a range of two to three hundred yards, five or six times further than smoothbore muskets. However, they were only effective if the bullet fit tightly inside the barrel, which meant loading could take up to four minutes, more than twice as long as muskets. For that reason, rifles were mainly used for hunting. Armies issued them only to skirmishers and sharpshooters. Napoleon called them "the worst weapon that could be got into the hands of a soldier." The British Army did not introduce a rifle, the Baker, until 1800. In Algeria, only the elite Chasseurs d'Orléans were equipped with rifles.[2]

The long period of technological conservatism in gun making came to an end after the Napoleonic Wars. The first significant innovation was percussion ignition. In 1807 Alexander Forsyth, a Scottish clergyman and amateur chemist, patented the use of potassium chlorate as a detonator that exploded when hit by a hard object; when placed in the firing-pan and struck with a hammer, it ignited the powder without the need for a match or a flint. Nine years later, Joshua Shaw of Philadelphia introduced copper percussion caps that worked even when wet. Percussion guns misfired only 4.5 times per thousand rounds, a great advance over flintlocks that misfired 411 times per thousand. Percussion was of special interest to the armies of western Europe, where it often rained. The British Army experimented with percussion caps in 1831. In 1836 it equipped the Guards regiments with Brunswick rifles with percussion locks, and in 1839 it began converting all infantry guns from flintlocks to percussion. The French and American armies followed suit in the 1840s.[3]

The next invention was the needle-gun, patented by the German Johann Nicholas Dreyse in 1836 and adopted by the Prussian Army in 1842. This was the first mass-produced breech-loading rifle, a gun that could be loaded and fired five to seven times a minute and from a prone position instead of standing up. This gave soldiers a great advantage on the battlefield, as a Prussian army demonstrated in 1866 at the Battle of Sadowa against the muzzle-loading Austrians. It had the disadvantage, however, of leaking hot gases from the breech when fired, causing soldiers to fire from the hip, reducing the accuracy to two hundred yards.[4]

While the Prussians were developing a military breechloader, the French were working on the problem of finding a bullet that would slide easily into the muzzle of a rifle, yet grip the rifling on the way out. In 1848 Captain Minié of the French army introduced a cylindrical bullet with a conical head and a hollow base that expanded when fired. The bullet came with a pre-measured amount of powder in a paper cartridge. With this bullet, rifles were as easy and quick to load as muskets. In the early 1850s, the French and British armies began replacing their muskets with rifles using Minié bullets. The results were astonishing. At a hundred yards, a Minié rifle hit the target 94.5 percent of the time whereas a musket scored 74.5 percent of the time; at four hundred yards, Minié rifles hit the target 52.5 percent of the time, compared with 4.5 percent for muskets. In short, infantry guns had become dangerous long-distance weapons.[5]

In the arms race, the United States was not far behind. Though its army was small, its citizens had a voracious demand for firearms with which to hunt and fight Indians on the frontier. Experimental breech-loading rifles were introduced in the 1820s, but the first really successful gun of this type was the one patented by Christian Sharps in 1844 and manufactured from 1848 on. Sharps's single-shot rifles and carbines (shorter and lighter rifles) were much used from the Mexican-American War (1846–48) to the Civil War (1861–65) and in the frontier fighting of the period. At first Sharps rifles used paper cartridges that leaked gases, but when the manufacturer switched to sturdier linen cartridges, they quickly became the most popular firearm in America.[6]

One more weapon needs to be mentioned: the Colt revolver. In the 1830s, while a sailor on a ship traveling between Boston and Calcutta, Samuel Colt invented a pistol with a revolving chamber to hold cartridges and bullets. He patented it in 1836 and offered it to the U.S. Army, but the army rejected it because it was too costly. Colt's company went bankrupt and he would have given up entirely if it had not been discovered by the Texas Rangers in 1839. When the Mexican War broke out in 1846, Captain Walker of the Rangers ordered a thousand .44-caliber six-shooters, starting a fashion that has not yet ended.[7]

The growing demand for firearms and the development of machine tools that could produce parts in great numbers led the arms industry toward the goal of interchangeable parts, an idea that had been proposed during the French Revolution but was not fully realized until the 1870s.[8] During the Great Exhibition of London in 1851, British officials were so impressed by the American firearms exhibited there that they sent three artillery officers and an engineer to the Springfield Arsenal in Massachusetts. As a result of their visit, the British government built a new arsenal at Enfield to mass-produce rifles on the "American system." The muzzle-loading Enfield rifle, introduced in 1853, had an effective range of eight hundred yards. The powder was supplied in pre-measured paper cartridges that were greased to make them waterproof. To load the gun, the soldier was to bite off the end of the cartridge and pour the powder into the muzzle. The rumor that Enfield cartridges were greased with pig or cow tallow caused the sepoys in the Indian Army—both Hindus and Muslims—to rise up in rebellion in 1857.[9]

The American Civil War pitted against each other two huge armies hastily recruited and armed with any weapons that soldiers could find. The sudden demand outstripped the ability of the government arsenals and gave private manufacturers, some no more than small craft shops, an incentive to produce guns in large numbers. None of the guns used during that war survived long into the postwar period, but the demand produced a flurry of innovations that bore fruit in the 1870s. Among them were the first repeating rifles, the Henry and the Spencer. The latter could fire seven times faster than the muzzle-loaders used by the Confederate Army and by many Union soldiers as well. It was light and durable, though dangerous to handle, for the magazine that held the bullets sometimes exploded.[10]

Meanwhile, the arms race in Europe accelerated when France introduced its answer to the Prussian Dreyse needle-gun, the Chassepot breech-loading rifle, in 1866. It was more accurate than the needle-gun, but it leaked gas from the breech and fouled easily. At the same time, the British Army began converting its Enfield rifles to breechloaders, using a breechblock patented by Jacob Snider. The resulting hybrid, called Snider-Enfield, was adopted in 1867 and used in the punitive expedition of 1868 to Magdala, among others.

When U.S. government orders dried up after the Civil War, many American firearms manufacturers went bankrupt or were bought up by others. The Remington Company survived by introducing a rifle with a "rolling-block" breech that was simple, reliable, and quick firing and that proved superior to European guns. During the Cuban revolution of 1868–78, Spain bought 300,000 of them. When the Franco-Prussian War broke out in 1870, France purchased 145,000. Egypt, Chile, Mexico, Argentina, and other countries also ordered Remingtons, along with millions of rounds of ammunition.[11]

Rapid-firing breechloaders and repeating rifles could only operate if the primer, powder, and bullet were all contained in a cartridge. Paper cartridges, as in the Dreyse and Enfield, were too delicate and allowed gases to escape during firing, causing soldiers to keep their guns away from their face and thus reducing the accuracy of their aim. In response, gunsmiths invented a variety of metal cartridges. The first was the Frenchman Houllier's cartridge of 1846 that was ignited by hitting a protruding pin that made it dangerous to handle. Many of the rifles used in the American Civil War accepted the safer rim-fire cartridges made of copper or brass foil. In 1867 both Colonel Edward Boxer of the Woolwich Arsenal in England and Colonel Hiram Berdan of the U.S. Army developed effective brass center-fire cartridges of the kind that have been used in all rifles ever since.[12]

The pace of innovations continued into the 1870s. After lengthy trials, the British Army adopted a new rifle, the Martini-Henry, to replace the Snider-Enfield. It had a smaller bore than its predecessor, which meant that soldiers could carry more ammunition. It used brass cartridges that were much more reliable than the earlier paper cartridges. It proved both dependable and accurate to eight hundred yards. It remained in service, with many modifications, until World War I, and was widely used in Britain's colonial wars in Africa.[13] In 1874 the French army replaced its Chassepot with the Gras, a bolt-action single-shot breechloader that used metal cartridges. The new German army equipped its infantrymen with the Mauser 1871. The U.S. Army, no longer pressured to innovate after the Civil War and faced with a drastically reduced budget, converted its Civil War–era Springfield rifles to breech-loading rather than replacing them with newer guns.[14]

Just as the armies of the Western powers were upgrading to single-shot, metal-cartridge rifles, inventors forced the pace of obsolescence by introducing repeating rifles that outperformed the experimental models used in the American Civil War. The Winchester Repeating Arms Company improved on the repeating mechanism of the Henry to produce first the Model 1866 and then the Model 1873, the most popular guns used on the Western frontier. Though their range was less than two hundred yards, they could fire fifteen rounds in a few seconds.[15] To save money, the U.S. Army delayed switching to repeating rifles until 1892. Meanwhile, in 1879–80 France turned its Gras into a repeating rifle, the Gras-Kropatschek, by adding a tubular magazine. Germany followed suit with a repeating Mauser in 1884.[16] Another innovation was the box maga-

zine, invented by the Scottish watchmaker James Lee. Unlike the
Winchester tube magazine that had to be filled by hand, this magazine, when empty, could be replaced quickly with a full one.

Technological advance did not stop there. Three innovations
once again forced armies to replace their obsolete rifles with newer
models. One was the substitution of steel for iron, the result of the
Bessemer, Siemens-Martin, and Gilchrist-Thomas steel-making pro-
cesses and the rise of a steel industry. Steel was much stronger and
more durable than iron for barrels and for the many parts in the lock
mechanism, but it required larger and more complex machine tools.
With the introduction of steel, the era of small workshops making
guns by hand came to an end, and blacksmiths could no longer make
gun parts; instead, gunsmiths simply replaced broken parts with
factory-made parts that needed little or no fitting.

Gunpowder, used since the fourteenth century, emitted black
fumes that gave away the shooter's position and left a residue that
fouled gun barrels. In 1885 the French chemist Paul Vieille invented
a mixture of celluloid and ethyl alcohol that not only exploded without smoke and left no residue but was much more powerful and
produced much higher muzzle velocities than gunpowder. The
French "Poudre B" was quickly followed by German, British, and
American versions.[17] Smokeless powder was so powerful that it could
not simply be substituted for gunpowder in existing rifles. Its higher
velocity increased the range and accuracy of fire and also allowed
smaller calibers; the military rifles of the 1890s had a caliber of .236
to .300, one-third that of a Brown Bess.[18] The speed of the bullet
compensated for its small diameter. As one historian wrote, the new
bullets cause "horrible wounds, the entry hole is barely visible, while
the exit looks like a funnel . . . flesh is reduced to mush."[19]

One final, especially sinister "improvement" was the invention of
hollow or flat-head bullets that mushroomed out upon hitting flesh,
creating a fist-sized hole. As such bullets were invented by Captain
Bertie-Clay of the Indian ammunition works at Dum Dum near Calcutta, they became known as "dum-dum" bullets. They were especially designed for colonial warfare on the grounds that "savage
tribes, with whom we were always conducting wars, refused to be
sufficiently impressed by the Mark II bullet; in fact, they often ig-

nored it altogether, and, having been hit in four or five places, came on to unpleasantly close quarters."[20]

The major armies tried to adapt their existing rifles to smokeless powder, but in the competitive atmosphere of the late nineteenth century, they soon realized that they needed all new weapons. France replaced its Gras-Kropatscheks with the new Lebel 1886 with an eight-cartridge magazine, the standard military rifle until after World War I. Britain adopted the Lee-Metford in 1888, followed by the Lee-Enfield. Germany's Mauser produced a series of magazine repeaters designed for smokeless powder. And in the United States, where military purchases lagged behind the demands of civilians, Remington introduced the Remington-Lee magazine repeater.[21]

With the new weapons, soldiers had firepower undreamed of in earlier times. A soldier could shoot lying down or hidden behind a rock or a tree. He could keep shooting as fast as he could pull the bolt and then the trigger and as long as his ammunition held out. A good marksman could hit a target a kilometer away. With that, the evolution of guns stabilized. Many of the rifles designed in the late 1880s and 1890s were used, with minor improvements, in World War I and even as late as World War II. The difference between the rifles that infantrymen carry today and those of the 1890s are minor compared with the extraordinary changes that took place between 1840 and 1890.

Generals were interested in many other weapons besides the guns carried by lowly foot soldiers. Artillery had always carried prestige, even before the reign of the great artilleryman Napoleon. Colonial armies operating in difficult terrain used light mortars and mountain guns that shot explosive shells and could be carried on the backs of mules or camels. Innovations in heavy artillery—and there were many—seldom played a part in colonial campaigns unless the targets were within reach of naval guns on warships.[22]

More important in colonial settings were machine guns. The first of these was the Gatling, a product of the American Civil War. This multi-barreled monster could fire up to three thousand rounds per minute, but it often fouled or overheated. Nonetheless, it was adopted by the British Army in 1871 and saw action in Africa from the 1870s through the 1890s, while the French preferred the similar

Montigny developed in the 1860s. These were followed by the Nordenfeldt 1877, a four-barreled gun that fired 216 rounds per minute, and the five-barreled 1892 model that fired 600 rounds per minute; these were so heavy that they were more often used on gunboats and torpedo boats than on land.

The first machine gun that proved practical in colonial warfare was the Maxim, patented by the American Hiram Maxim in 1884. It was a single-barreled automatic rifle powered by the gases that escaped from the cartridge upon firing and could fire eleven rounds per second. The inventor could not find enough buyers in the United States, so he moved to Britain. When he demonstrated his gun in London in 1884, the Chinese ambassador who witnessed the demonstration declared that his country could not afford the cost of ammunition, at £5 (or $25) per minute. Field Marshal Lord Wolseley, however, "exhibited the most lively interest in the gun and its inventor: and, thinking of the practical purposes to which the gun might be put, especially in colonial warfare, made several suggestions to Mr. Maxim."[23]

Steel, smokeless powder, brass cartridges, and the many precision parts firearms were made of required industrial manufactures. Only industrialized nations had the wherewithal to make such weapons. Only the larger European countries and the United States had the steel mills, chemical plants, and arms factories they needed to arm themselves. The Latin American republics imported what they could afford. The rest of the world fell far behind.

Guns in Africa

The firearms revolution of the nineteenth century greatly increased the firepower of those who possessed the new weapons. But acquiring new weapons did not in and of itself confer an advantage over other people. Only where the gap between the technologies possessed by two belligerents was very large could it make up for disadvantages in manpower or local knowledge. Colonel Charles E. Callwell, the foremost theoretician of colonial wars, described the gap in these terms in his book *Small Wars*, first published in 1896.

> In small wars of the old musket days, it was not unusual to find the
> enemy in possession of fire arms effective up to longer ranges than
> those of the regular troops and as efficient in their use. . . . But in the
> present day it is safe to assume that the enemy from the nature of his
> weapons, want of training and so forth, is almost invariably far infe-
> rior to the trained infantry as regards the efficacy of musketry.[24]

Colonel Callwell was a man who had much experience in colonial
warfare and a profound knowledge of weapons and their uses. Nonetheless, like so many who wrote about colonial warfare in his day,
Callwell attributed the victories of the Europeans less to their weapons and tactics than to the morale, zeal, resolution, daring, discipline, and valor of their troops and to the strategy of their generals.
Like others of his time, he called the people whom Europeans confronted savages, fanatics, and barbarians, and he referred to their
armies as hordes. Even the Egyptians and Chinese were, in his eyes,
semi-civilized.

Such contempt for non-Western peoples was characteristic of an
age of racism and social Darwinism. Charles Wallis, a district commissioner in Sierra Leone and the author of a book of advice to
British officers posted to West Africa, made similar comments, call-
ing Africans "savage and lawless races," "fanatical" and "cunning
savages," "like the wild animals of his own forests." Like Callwell,
he acknowledged the power of the new firearms but attributed the
British successes to "the stern discipline and enthusiastic *esprit de
corps* of the British army," the bravery, endurance, dash, and vigor
of the soldiers and, especially, to "that indispensable factor in the
machine of West African warfare—the British officer."[25] Surprisingly, such attitudes still persist in more recent writings.[26] To avoid
attributing such simplistic generalizations to entire peoples, we need
to consider the role of weapons in the events of the period.

Since the sixteenth century, European traders had sold guns to
Africans in exchange for slaves. These weapons, known as Dane
guns, cost one-fifth as much as European military weapons. The demand for them was huge and growing until the early nineteenth
century. Between 1796 and 1805, the last decades of the slave trade,
Britain exported 150,000 to 200,000 guns a year to West Africa,

along with 847,075 pounds of gunpowder and 200,000 pounds of lead and shot, while other European countries exported at least as many guns. These were the most important trade goods sold to Africans in exchange for slaves, at a ratio of 2.5 to 6.2 guns per slave.[27] The trade continued after the end of the slave trade, especially when steamboats began traveling up the Niger. By the 1860s, the gun manufacturers of Birmingham were exporting 100,000 to 150,000 guns a year to Africa, and the gun makers of Liège almost as many.[28] Later in the century, French, German, and Portuguese traders exported guns to East Africa, and Arab traders carried guns across the Sahara to the Sudan.[29] In southern Africa beginning in the 1870s, diamond mine owners found that to recruit workers, they needed to offer them guns.[30]

Dane guns were cheaply made flintlock or percussion-lock muskets, with rolled iron barrels and loose tolerances. They had a limited range and accuracy. Soldiers poured gunpowder and pieces of lead or iron down the barrel, then waited until they were a few feet from the enemy before firing. Such guns were unreliable, especially in wet weather, and slow to load. If packed with too much gunpowder—a common occurrence—they were likely to burst.[31] Dane guns had one advantage, however: unlike rifles and breechloaders with their intricate parts, they could be repaired by African blacksmiths, though they could not make entire guns, for their furnaces were not hot enough to melt iron.[32] Gunpowder was mostly imported; what was made locally using imported sulfur was uncorned and readily absorbed moisture. Instead of bullets, which were costly, Africans often shot pebbles or bits of iron.[33]

Firearms were very unevenly distributed in Africa before the colonial period. The states of the West African coast had large supplies of guns and ammunition that they used for hunting, capturing slaves, and dominating the trade of the interior. Some coastal states even imported swivel guns and other light artillery pieces to arm their war canoes.[34] In South Africa until the 1860s, Afrikaner gunsmiths assembled guns from imported barrels and parts and locally made stocks.[35]

The interior of Africa was much more bereft of guns. To the first European visitors, the armies of the Western Sudan seemed medieval, with their horses, suits of armor, swords, spears, and battle-axes, as did the towns with their walls and moats.[36] Firearms were known but rarely used for lack of powder, shot, and parts. The explorer Hugh Clapperton, who reached the Sokoto Caliphate in 1826, counted forty-two muskets among Caliph Bello's fifty thousand soldiers.[37] During the 1840s and 1850s, Wadai, also in the central Sudan, had three hundred guns; by the 1870s, there were four thousand flintlock muskets.[38] In parts of central Africa where horses could not survive, states and stateless peoples still used bows and arrows or javelins and shields.[39]

When European armies rearmed with breech-loading rifles, their old guns, made obsolete by the ongoing arms race, made their way to sub-Saharan Africa, albeit with considerable delay. When the French army introduced the Gras rifles in 1874, it sold its obsolete Chassepots to gun makers in Liège, who sold them to traders; by 1890 Chassepots could be bought in African trading posts.[40] But imported rifles were expensive. In Ibadan in southern Nigeria in the 1870s, rifles cost ten to fifteen pounds sterling apiece. At the diamond mines of Kimberley, South Africa, where muskets cost four pounds (or three months' wages for a miner), breechloaders cost twenty-five pounds, six times as much.[41] In the central Sudan toward the end of the century, a rifle cost as much as fifteen to thirty slaves or five or six camels. Moreover, their ammunition was an industrial product that had to be imported at great expense.[42]

Europeans were of two minds on the issue of selling modern military rifles to Africans. Whenever armies replaced their rifles with a newer model, traders bought the obsolete castoffs and shipped them abroad. Governments viewed this practice with misgivings. In the seventeenth and eighteenth centuries, the Dutch East India Company prohibited the sale of guns to Africans in the Cape Colony, but white farmers often evaded this rule. As early as 1830, the French closed down the gun trade across the Sahara. In 1854 Britain tried to stop sales of firearms to Africans.[43] In South Africa the situation varied by location. The two Boer republics, the Orange Free State and Transvaal, reserved gun ownership for whites. In the Cape Colony there was much debate about free trade, the loyalty of Africans, and their skills with guns until 1878, when the Cape parliament passed a law regulating gun ownership, leading to the de facto disarmament of Africans.[44] Elsewhere, the gun traffic boomed. In 1888 the British consul in Zanzibar wrote: "Unless some steps are taken to check this immense import of arms into East Africa the development and pacification of this great continent will have to be carried out in the face of an enormous population, the majority of whom will probably be armed with first-class breech-loading rifles."[45] Two years later, the European powers engaged in conquering Africa signed a treaty in Brussels permitting free trade in muskets but outlawing the export of breechloaders to Africa between twenty degrees north and twenty degrees south. This act was renewed in 1899.[46] Although some traders evaded the prohibition, it unquestionably helped tip the balance of power against Africans during the colonial wars of the late nineteenth century.

The Scramble for Africa

Explorers were the vanguard of the Scramble for Africa. Once steamboats and quinine prophylaxis had made the penetration of the continent less suicidal, missionaries, traders, and adventurers began crisscrossing the continent, abetted by an insatiable craving for information and the dissemination of adventure stories in the popular press of Europe and North America.

All explorers brought with them the latest firearms, but they put them to use in very different ways. The most famous of all, David Livingstone, who made several trips through south-central Africa between 1849 and 1861, used his guns for hunting and, rarely, for self-defense.[47] Samuel White Baker, a wealthy big-game hunter who traveled to the upper Nile in 1863–64, brought breechloaders and several years' worth of ammunition and used them not only for hunting but also to frighten the Africans he encountered; so did Verney Lovett Cameron, who explored Angola in the early 1870s.[48] In the central Sudan in 1865–67, Gerhard Rohlfs used his rifles and carbines to overawe the local inhabitants who "showed an inclination to oppose forcefully our camping there . . . a few blind shots made them see reason."[49]

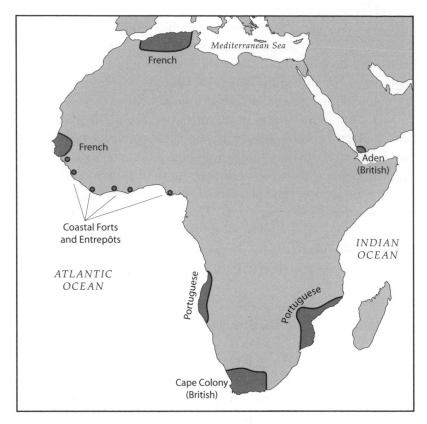

Figure 7.1. Africa in the mid-nineteenth century, showing the limits of European control before the Scramble for Africa. Map by Chris Brest.

No one, however, blurred the line between exploration and military conquest quite as blatantly as Henry Morton Stanley. He was a violent and brutal man whose actions the British consul in Zanzibar described as "unequalled in the annals of African discovery for the reckless use of the power that modern weapons placed in his hands over natives who never before had heard a gun fired."[50] He took with him far more equipment than any other explorer, thanks to the financial support of the *New York Herald* and later of King Leopold of Belgium. On his first expedition in 1871–73 in search of Livingstone, he brought six tons of materiel and 190 men, most of them porters. On his expedition to Lake Tanganyika and the Congo in 1879–84, he had a steamer, the *Lady Alice*, and eight tons of equipment, including dozens of rifles and elephant guns and even a Krupp cannon. On his final expedition in 1886–88 to rescue Emin Pasha (né Eduard Schnitzer) on the upper Nile, he brought with him hundreds of Remington and Winchester rifles, 100,000 rounds of ammunition, 350,000 percussion caps, two tons of gunpowder, and a machine gun, which he said was "of valuable service in helping civilisation to overcome barbarism."[51] Nor did he hesitate to use all the firepower at his disposal. When he encountered resistance or was greeted with less than enthusiasm by Africans, he would pursue them "up to their villages; I skirmish in their streets, drive them pell-mell into the woods beyond, and level their ivory temples; with frantic haste I fire their huts, and end the scene by towing the canoes into mid-stream and setting them adrift."[52] In one encounter on Lake Tanganyika, "the beach was crowded with infuriates and mockers . . . we perceived we were followed by several canoes in some of which we saw spears shaken at us. . . . I opened on them with the Winchester Repeating Rifle. Six shots and four deaths were sufficient to quiet the mocking."[53]

By Stanley's day, the era of explorations had morphed into a frenzy of conquests historians call the Scramble for Africa. The success of the colonialists—the conquest of a continent in less than four decades—was due in large part to the weapons they used, or rather, to the gap between their weapons and those of the Africans.

In the early nineteenth century, the gap was much narrower and the successes much fewer. In southern Africa, the only area of significant European penetration, the various societies, including the Boers, were fairly evenly matched. In the Xhosa-Boer War of 1799–1802, the Xhosa, along with some Khoisan servants who had deserted their Boer masters bringing guns and horses with them, were able to hold off the white advance into their lands.[54] In the first war between the British and the Asante in the interior of the Gold Coast in 1823–31, the British escaped defeat by using Congreve rockets that did more to frighten than to wound the Asante soldiers.[55]

By the 1860s the introduction of breechloaders widened the gap. In their war with the Orange Free State in 1865–68, the Sotho used horses and flintlock and percussion-lock muskets against the Boers' breech-loading rifles and steel cannon. The Boers killed several thousand Sotho and seized their best lands, at a loss of a hundred whites and a hundred of their African allies.[56]

The gap widened further in the 1870s. In the Second Anglo-Asante War of 1873–74, Sir Garnet Wolseley's forces were equipped with new Snider-Enfield rifles and a million rounds of ammunition. They also brought two Gatlings, the first machine guns to be used in Africa, but found them too hard to transport and maintain, and abandoned them. The Asante, meanwhile, had Dane guns. In Wolseley's "punitive expedition" to the Asante capital Kumasi, the British worried more about diseases than enemy bullets.[57]

The Zulu War of 1878–79 saw less of a disparity because the Zulu were skilled warriors by tradition and had developed very effective tactics that had defeated their African rivals on many earlier occasions. Their victory over a British column at Isandhlwana in early 1879 has become famous because it was such a rare event and was much criticized in the British press. Six months later, the British, though burdened with a slow baggage train, reversed the defeat, in part by using Gatling guns and the new Martini-Henry breechloaders against the Zulus' spears.[58]

Repeating rifles arrived in Africa in the 1880s. The Gras-Kropatschek, first used in West Africa in 1885, could fire eight rounds in a few seconds, using cartridges with smokeless powder. In the French campaign against Mahmadou Lamine in Senegal in 1886–87, the force under Lieutenant Colonel Henri-Nicholas Frey was armed with these rifles, as well as Gras single-shot breechloaders and field cannon to shell towns. Lebel rifles, using cartridges with smokeless powder, appeared in Africa in the early 1890s.[59]

Colonial warfare grew fiercer in the 1890s, partly because some of the African states that still held out were deeper in the interior, and partly because they, too, were beginning to acquire breechloaders, albeit with a decade's delay behind the imperial forces. By the time the French decided to conquer the kingdom of Dahomey (today the Republic of Benin), its army had acquired six Krupp cannon, five machine guns, and seventeen hundred rifles of various sorts, including Chassepots, Winchesters, Sniders, and Spencers. Its soldiers, however, were not trained in the use of these weapons; accustomed to Dane guns, they tended to aim too high. In a decisive battle near the capital, Abomey, several thousand men in the Dahomey army were killed or wounded, while the French lost ten officers and sixty-seven soldiers.[60]

Likewise, in their war with the British in 1893–96, the Ndebele had acquired Sniders and Martini-Henrys. But when thousands of warriors massed together attacked the British, they were mowed down by machine guns, including the new and much more powerful Nordenfeldts and Maxims.[61] Later writers on colonial warfare approved of machine guns, especially Maxims. According to Colonel Callwell, "Against rushes of Zulus, ghazis, or other fanatics the effect of such weapons is tremendous as long as their fire is well maintained." And Charles Wallis maintained that "the rapidity of fire of the Maxim always has some moral effect upon savages, who may never have heard or seen one before."[62]

Colonial battles were lopsided in numbers as well as armaments. Dahomey was defeated by 2,000 troops, Ijebu in Nigeria by 1,000, the Sokoto Caliphate by 1,100, and Rabah in the central Sudan by 320. In most colonial units, half the men were porters and most of the rest were Africans; only the officers and NCOs were white. To conquer Algeria, France needed 100,000 soldiers led by generals and field marshals; but in the Western Sudan its garrison never numbered more than 4,000 led by majors and colonels.[63]

African armies, meanwhile, numbered in the tens of thousands. Their tactics, developed over years of warfare with other Africans, did not change quickly enough in the face of the invaders. In forests, tall grasslands, or broken terrain, ambushes were common but seldom decisive in the face of heavily armed colonial forces. On the open savannas of the Sudanic belt, they employed cavalry armed with spears or muskets, easy targets for soldiers with rifles. The African states that had rifles seldom had enough ammunition to train their soldiers in firing or tactical movements. The Zulu, the most proficient warriors in southern Africa, used a frontal assault with two horns to encircle their enemies, a tactic that worked brilliantly at Isandhlwana, but failed them thereafter.[64]

In response to these tactics, colonial forces, whose strategy was always offensive, that is, aimed at conquering, resorted to the favorite defensive tactic of earlier centuries: the square, last used at Waterloo. As several experts on "small wars" have noted, such a tight formation would have been too vulnerable against an enemy armed with modern rifles, but against spears and Dane guns it proved very effective. Colonial forces even on occasion resorted to firing in volleys, causing a powerful psychological effect.[65]

Two major encounters between colonial and African forces—the French against Samori Toure in the Western Sudan and the British against the Mahdists in the eastern Sudan—illustrate the variety of experiences at the height of the Scramble and the roles that weapons and tactics played in them.

Samori Toure was a self-appointed military leader among the Dyula people of the upper Niger region. During the 1870s he gradually built up a small elite corps. His was the first African force equipped exclusively with firearms. At first, these were mostly percussion-cap muskets, but he also purchased some modern rifles from the British in Freetown, Sierra Leone. By 1887 his army had approximately fifty breechloaders and thirty-six repeating rifles. As the French moved into his territory in the early 1890s, he moved eastward into the upper Volta region and continued to purchase or capture Gras and Mauser rifles and Winchester repeaters. By 1894 his army may have possessed as many as six thousand rifles before the French cut off his access to his suppliers in Sierra Leone.

Samori had also sent some blacksmiths to take a course at the French arsenal at Saint-Louis, in Senegal. These and other blacksmiths were able to repair muskets and even make new ones using imported barrels, while their wives made gunpowder with a mortar and pestle. Cartridges for his Gras breechloaders were reloaded using spent cartridge cases picked up after battles. His blacksmiths were even able to make a few imitation Gras-Kropatschek repeaters.

His tactics were also innovative. Instead of fighting pitched battles, he engaged in guerrilla-style skirmishes. The French expected to defeat him in a few weeks, but his scorched-earth policy delayed the French advance for several years. When he was finally captured in 1898, his army still had four thousand rifles and even a small cannon, but was running out of ammunition.[66]

The best-known battle in which a European-led force armed with modern weapons confronted an army of traditional African warriors took place at Omdurman on the upper Nile. In the 1890s the eastern Sudan was ruled by Khalifa Abdallahi, the successor to the Mahdi, a religious leader who had led a revolt against Egyptian rule a decade earlier. Meanwhile, the British occupied Egypt to the north while a French expedition was approaching from the west and Belgians from the south. In 1896 Britain sent a large expedition under General Horatio Kitchener to conquer the Sudan. The Egyptian troops under his command carried breechloaders, while his British soldiers had newer Lee-Metford repeating rifles, Maxim guns, and field cannon. Supporting them (as we saw in chapter 5) were gunboats on the Nile armed with cannon with explosive shells.[67] Opposing them was a Mahdist army numbering in the tens of thousands armed with spears, swords, and muskets, whose tactic was the massive frontal assault.[68]

The two forces clashed at Omdurman near Khartoum on September 2, 1898. In memorable passages, Winston Churchill, then a war correspondent, described the battle.

> The infantry fired steadily and stolidly, without hurry or excitement, for the enemy were far away and the officers careful. Besides, the soldiers were interested in the work and took great pains. . . . And all the time out on the plain on the other side bullets were shearing through flesh, smashing and splintering bone; blood spouted from terrible wounds; valiant men were struggling on through a hell of whistling metal, exploding shells, and spurting dust—suffering, despairing, dying. Such was the first phase of the battle of Omdurman.[69]

After the battle, the British counted eleven thousand Dervish dead, with a loss of twenty-eight British and twenty Egyptians. Churchill called it "the most signal triumph ever gained by the arms of science against barbarians. Within the space of five hours the strongest and best-armed savage army yet arrayed against a modern European Power had been destroyed and dispersed, with hardly any difficulty, comparatively small risk, and insignificant loss to the victors."[70]

Figure 7.2. The Battle of Omdurman, September 2, 1898. Major General Sir
Horatio Kitchener, the Sirdar of Egypt, commanding British and Egyptian army,
defeats Mahdist forces led by Abdullah and Osman Digna. Colored lithograph
by A. Sutherland. Courtesy of the Art Archive/Eileen Tweedy. Note the British
troops using rifles against the attacking Sudanese.

Yet, even as Churchill was witnessing the battle of Omdurman,
the easy victories of "the arms of science against barbarians" were
coming to a close, and colonial warfare was becoming more difficult,
costly, and unpredictable.

North America

Many will be surprised—and a few will be offended—to see the westward expansion of the United States conflated with European imperialism in Africa and Asia. After all, the inhabitants of the western
states are now just as American as those of the original Atlantic
seaboard settlements, that is to say, of predominantly European origin. The consequences are completely different from those of the

New Imperialism, which lasted a century and ended with the collapse of colonial empires.

Yet here, as so often in history, hindsight distorts the past. To the people living at the time, the expansion of European Americans in the nineteenth century was of a piece with European conquests in other parts of the world. In 1856 Jefferson Davis, secretary of war under President Franklin Pierce, wrote:

> The occupation of Algeria by the French presents a case having much parallelism to that of our western frontier, and affords an opportunity of profiting by their experience. Their practice, as far as understood by me, is to leave the desert region to the possession of the nomadic tribes; their outposts, having strong garrisons, are established near the limits of the cultivated region, and their services performed by large detachments making expeditions into the desert regions as required.[71]

There was a good reason for Davis to find similarities between the French in Algeria and the United States in its wars with the Indians. In both cases, the conquest was slow, difficult, and costly, requiring strong garrisons and large detachments.

Far from being the triumphal advance into the wilderness depicted in old-fashioned textbooks, the advance of the Europeans into North America had been remarkably slow until Davis's time. Historical atlases tell the tale of the westward expansion as follows. The first European settlements in what is now the United States were Saint Augustine, Florida, in 1565, Jamestown, Virginia, in 1607, and the Plymouth Colony in Massachusetts in 1620. After the American War of Independence, the United States was made up of thirteen colonies east of the Appalachian mountains and north of Georgia, while Britain claimed Canada, France the center of the continent, and Spain the west. By 1802 the United States had expanded to the Mississippi River. That year, the Louisiana Purchase gave all the territory from the Mississippi River to the Rocky Mountains to the United States. In 1845 the United States annexed Texas, and in 1846–48 it acquired the Oregon Territory in the Northwest and took the Southwest from Mexico. Finally, Alaska was purchased from Russia in 1867.[72] Such maps are very misleading.

Most of the lands that France and Russia sold to the United States did not belong to France or Russia; most of what Mexico lost was not Mexico's to lose; and the Oregon Territory was not Britain's to negotiate away. It was all Indian territory.

Sometimes, historical atlases will include smaller but more honest maps showing areas of settlement.[73] These little maps tell a very different story from the large maps, for they show actual possession and control. For two hundred years, the frontier between Europeans and Indians had moved very slowly. In 1775, Europeans controlled the East Coast and the eastern slope of the Appalachians, along with a few isolated spots such as New Orleans and Santa Fe. By 1802, they had taken possession of the western slope of the Appalachians and little else.[74]

The contrast with the Russian expansion into Siberia is striking. In the 1590s, Russia was confined to the west of the Ural Mountains. By 1646, Russian explorers and fur traders had reached the eastern edge of Siberia and had founded Okhotsk of the sea of that name and Anadyrsk in northeastern Asia. By 1689—after only a hundred years—Russia controlled almost all of Siberia to the Pacific Ocean, 3,500 miles from European Russia. In 1802, while Anglo-Americans were still confined to the eastern quarter of the continent, Russians had founded settlements in Alaska and down the west coast of North America to northern California.

The westward expansion of European Americans accelerated slowly at first. By 1820 they had taken over the Ohio Valley and much of the old South. By 1850 their settlements reached three hundred miles west of the Mississippi River and a few places in Utah, California, and Oregon. At the time Jefferson Davis wrote the words quoted above, European Americans and Canadians possessed less than a third of North America north of the Rio Grande. Then came the rush, and within thirty years, white Americans and Canadians had taken possession of the other two-thirds of the continent.

Many factors account for the slow process of European expansion during the first two hundred years, the gradual acceleration in the first half of the nineteenth century, and the sudden wave of conquest in the second half. Demographic pressure is the most important factor, as European migration swelled to a flood in the nineteenth cen-

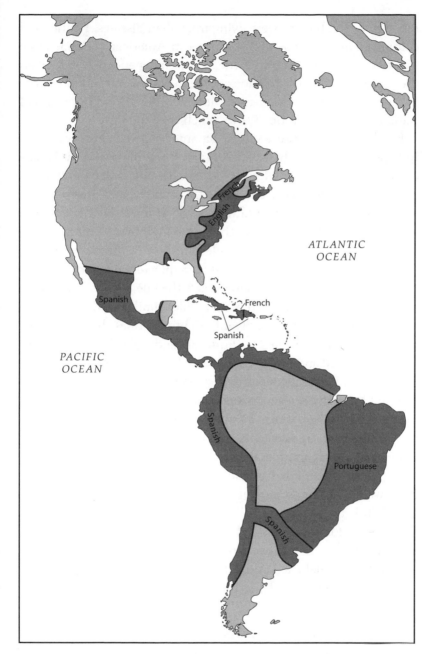

Figure 7.3. Map of the Americas showing the extent of European control
circa 1800. Map by Chris Brest.

tury, just as the Indians were dying from alien diseases. Politics and culture are important as well, as European Americans became more assertive and demanding after the War of Independence. But given the increasingly hostile relations between whites and Indians in the nineteenth century, we must take into account the technologies of warfare and the environments in which they were used.

For the first two centuries, the encounters between Europeans and Indians took place in the wooded eastern third of the North American continent. Here, the Europeans had the advantage of numbers, but the Indians had a better knowledge of the terrain. Like Europeans, Indians had firearms and horses, albeit fewer of them. In the eighteenth and early nineteenth centuries, Indians obtained firearms from English and French traders in the east, from the Hudson Bay Company in the northeast, and from the French at New Orleans (but not from Mexico, for Spain forbade the sale of guns to Indians). Woodlands Indians switched to long rifles similar to those the settlers had, such as the Pennsylvania-Kentucky rifles that were especially suited to hunting and to skirmishes and ambushes. Plains Indians, who hunted and fought on horseback, also acquired lightweight flintlock muskets called Northwest guns whenever they could.[75] Well into the 1870s, gun makers continued to make flintlock muskets for the Indian trade, for flint—unlike percussion caps—was readily available in many places. But the Indians also used their traditional knives and tomahawks, long spears and rawhide shields, and especially their bows and quivers of arrows.[76] Out on the Plains, bows were superior to muskets, as Josiah Gregg, a trader, pointed out: "While the musketeer will load and fire once, the bowman will discharge a dozen arrows, and that, at distances under fifty yards, with an accuracy nearly equal to the rifle."[77]

In 1882 Colonel Richard Dodge, a frontiersman who had fought and lived among the Plains Indians for decades, described their armament.

> Before the Plains Indians obtained firearms, they were armed with bow and lance, and with these weapons were truly formidable, the fighting necessarily being almost hand-to-hand. But the Indian likes this close contest as little as any one, and whenever he could procure a gun his more dangerous arms were discarded.

Thirty years ago, the rifle was little used by mounted Indians, as it could not be reloaded on horseback, but many of them were armed with guns of the most nondescript character, old Tower muskets, and smooth-bores of every antique pattern. Powder and lead were easily obtained from the traders. The former was carried in a horn, the latter was cut into pieces, which were roughly hammered into spherical form. These bullets were purposely made so much smaller than the bore of the gun as to run down when dropped into the muzzle. When going into a fight, the Indian filled his mouth with bullets. After firing he reloaded in full career, by turning up the powder-horn, pouring into his gun an unknown quantity of powder, and then spitting a bullet into the muzzle. There was very little danger to be apprehended from such weapons, so loaded, and the troops did not hesitate, even with the sable alone, to rush on any odds of Indians.[78]

Few U.S. cavalrymen were trained to fight with lances and swords, however. The Texas Rangers, who confronted the Comanche Indians, were armed with long rifles, which could only be loaded standing up. They also carried two single-shot pistols. In the minute it took a Ranger to reload his guns, a Comanche could ride three hundred yards and shoot twenty arrows. If the Rangers did not prevail in the first few minutes of an encounter, their only hope of escape was to have faster horses.[79] The Plains Indians were still too powerful to be defeated in battle.

The balance of firepower began to shift against the Indians in the 1840s with the introduction of the Colt revolvers, which the Texas Rangers were the first to buy. The first encounter between Rangers and Comanches that involved Colts took place on the Pedernales River in May 1844. In a letter to Samuel Colt, Captain Samuel Walker described this skirmish.

In the Summer of 1844 Col. J. C. Hays with 15 men fought about 80 Comanche Indians, boldly attacking them upon their own ground, killing & wounding about half their number. Up to this time these daring Indians had always supposed themselves superior to us, man to man, on horse—at that time they were threatening a descent upon our Frontier Settlements—the result of this engagement was such as to intimidate them and enable us to treat with them.[80]

Figure 7.4. "U.S. Army—Cavalry Pursuing Indians. 1876." Print from *The United States Army and Navy*. Akron, Ohio: Werner Company, 1899. Chromolithograph. Ca. 9 1/4 x 13 1/4. Courtesy of The Philadelphia Print Shop, Ltd.

The Civil War only slowed down the westward movement against the Indians. The war stimulated both the design and the production of firearms. When it ended, the United States was awash in military rifles. To stay in business, gun manufacturers hastened to introduce new and more powerful models and market them to pioneers heading west. Soldiers on the western frontier carried muzzle-loading single-shot Springfield rifles or carbines. These were replaced in 1867 by Springfields converted to breech-loading, rugged, and powerful guns, accurate at well over two hundred yards, well suited to rough treatment in the field, but slow to load. Hunters, settlers, and army scouts generally preferred the Winchester repeating rifle.[81]

Indians also acquired war-surplus Spencers, Henrys, and Sharps whenever they could.[82] Again, Colonel Dodge: "Every male Indian who can buy, beg, borrow, or steal them, has now firearms of some kind. They are connoisseurs of these articles, and have the very best

that their means or opportunities permit."[83] And, further: "It remained, however, for the breech-loading rifle and metallic cartridges
to transform the Plains Indian from an insignificant, scarcely dangerous adversary into as magnificent a soldier as the world can show.
Already a perfect horseman, and accustomed all his life to the use
of arms on horseback, all he needed was an accurate weapon, which
could be easily and rapidly loaded while at full speed."[84]

During the 1860s and 1870s, Plains Indians and European Americans were more evenly matched than they had ever been. As individuals, Indians were better warriors, with greater skills and more
practice. However, they had difficulty obtaining ammunition for
their rifles. Pioneers and army troops were better organized and had
more weapons and ammunition. The Indians won some of the battles, for instance those that took place around Fort Kearney, Nebraska, in 1866 and 1867, but lost others.[85]

Though skilled warriors, the Plains Indians had a weakness that
European Americans were soon to take advantage of: they depended
on buffalo for their food and many of their possessions. To whites,
buffalo hides made fine rugs and blankets, as well as belts for industrial machines. From the 1820s, if not earlier, Plains Indians had
sold buffalo hides to the American Fur Company and others. After
1832, many were shipped on steamboats down the Missouri and Mississippi. By the 1860s, they were selling a hundred thousand hides
a year.[86] After 1872, when the Union Pacific, Kansas Pacific, and
Atchison Topeka & Santa Fe railroads reached buffalo country west
of the Mississippi, white buffalo hunters descended upon the region,
killing the animals for their hides. The U.S. Army, as part of its
policy of subduing the Indians, supplied free cartridges to profes-
sional hunters. A year later, in Colonel Dodge's words, "Where there
were myriads of buffalo the year before, there were now myriads of
carcasses. The air was foul with the sickening stench, and the vast
plains, which only a twelvemonth before teemed with animal life,
was a dead, solitary, putrid desert." The effect on the Plains Indians
was dramatic: "Ten years ago the Plains Indians had an ample supply
of food, and could support life comfortably without the assistance of
the government. Now everything is gone, and they are reduced to

the condition of paupers, without food, shelter, clothing, or any of those necessities of life which came from the buffalo."[87]

Before the end, the Plains Indians had one last glorious victory, the Battle of Little Bighorn in Montana. On June 25, 1876, some seven hundred cavalrymen led by General Armstrong Custer encountered several thousand Cheyenne and Lakota warriors. Not only were their numbers unequal, so were their firearms. Custer's soldiers were armed with the Springfield Model 1873, the standard army rifle, a single-shot breechloader.[88] Before entering Indian territory, Custer had left behind his Gatling machine guns because they were cumbersome. He expected to encounter no more than eight hundred warriors and he was confident that he could defeat them with rifles and pistols alone. Many of the Indians, meanwhile, had Sharps, Winchester, Henry, and Remington repeating rifles and were better marksmen than the U.S. Cavalry. At long range, the Springfields were more accurate, but when the two sides closed, the greater firepower of the repeating rifles gave the advantage to the Indians. The result was a victory for chiefs Sitting Bull and Crazy Horse and a disaster for Custer, who lost his life and half his men.[89]

The Battle of Little Bighorn is deservedly famous as a victory for the Indians. But it was only a minor setback for the juggernaut of Anglo-American conquest. The following year, when the army confronted the resistance of the Nez Perce Indians led by Chief Joseph, the Indians were armed as the Cheyenne and Lakota had been the year before, but the army used Gatlings and howitzers. In 1890 the last Indian resistance was crushed at the Battle of Wounded Knee by soldiers using Hotchkiss machine guns.[90] By then, continental imperialism was over, the frontier was closed, and the attention of the United States was increasingly turning overseas.

Argentina and Chile

In Argentina and Chile, as in North America, the frontier between Europeans and Indians hardly moved from the sixteenth to the midnineteenth century. Then, quite suddenly, Indian resistance collapsed before the pressure of European armies and colonists, and for much the same reason as in North America and Africa.

Until the mid-nineteenth century, the frontier between Europeans and Indians in Argentina followed the Rio Salado, a hundred miles south and west of Buenos Aires; it had barely moved since the 1590s. The Pampas Indians knew the land. Their source of food and wealth was the herds of cattle and horses raised by the Europeans on huge ranches along the frontier. They periodically attacked these outlying estancias and stole their livestock, which they drove over the Andes to Chile and traded for alcohol, tobacco, blankets, metal goods, and other necessities they could not produce themselves. Their tactics consisted of surprise attacks using lances, lassos, and *bolas*; unlike the North American Indians, they did not use bows and arrows on horseback nor did they acquire firearms. Though their horses were of the same mediocre stock as the Europeans', they cared for them and trained them better than did the Argentine cavalry.

In 1833–34 Juan Manuel de Rosas, the governor of Buenos Aires province, led an expedition to the "Desert"—the Europeans' term for Indian territory—that reached the Rio Negro, 450 miles to the south. But the Argentinean hold on the Pampas was very tenuous and the government had to bribe the Indians not to attack the few European settlements in the frontier zone. After Rosas was overthrown in 1852, his soldiers were withdrawn from the frontier to fight in the decade-long civil wars between Buenos Aires and the provinces, each side led by ambitious warlords. The Indians, led by the Araucanian chief Calfucurá, took advantage of the chaos to plunder the frontier settlements and estancias. Between 1854 and 1857, they took 400,000 animals and four hundred captives and reduced Buenos Aires province by 25,000 square miles.[91] From 1865 to 1870, the Argentines joined Brazil and Uruguay in the War of the Triple Alliance against the Paraguayan dictator Francisco Solano López. During this time, the Indians allied themselves with one faction or another and periodically raided Argentinean territory. The frontier remained in the same place and was just as insecure as it had been fifty years earlier.

In 1868 Domingo Faustino Sarmiento, the most celebrated of Argentina's political leaders, was elected president. A former schoolteacher and the author of the best-known Argentinean novel, *Facundo*, he dreamed of transforming Argentina into a modern country. His administration encouraged immigration from Spain and Italy, as well as British investments in railroads, harbors, and other public works. But Sarmiento was also a military leader, as all Argentinean politicians had to be during those uncertain times. He had first visited the United States as a young man during the 1840s. While there, he was very impressed with the removal of the Cherokee and other Indians from the southeast to Oklahoma. Like many of his white contemporaries, he was inspired by Roman notions of civilization and barbarism, and referred to Indians as "American bedouins."[92] He returned to the United States as Argentine ambassador from 1866 to 1868. From Washington, he kept his government informed about the latest American weapons and other military equipment such as waterproof coats, food for soldiers, and torpedo boats.

During his presidency, Sarmiento founded a naval academy and acquired the river steamers *Transporte* and *Choele-Choel*. Two years later he purchased the corvettes *Uruguay* and *Paraná* and four other steamers from Britain. He also acquired Gatling machine guns, a Krupp 7.5-centimeter cannon, and a large number of Remington rifles.[93] As he put it, "No Paraguayan or guerrilla or Indian can stand against our repeating rifles."[94] His goal was to establish a new frontier 75 to 150 miles beyond the old line. It was to be guarded by a line of forts and six thousand troops, stretching fourteen hundred miles from Bahía Blanca on the Atlantic to San Rafael in the Andes. Though they could not prevent Indian raids on white-owned ranches, they forced the Indians to travel much further there and back, giving time for the Argentine troops to catch up with them and recover the booty. In 1872 Calfucurá decided to meet the Argentines in battle; he lost three hundred of his three thousand warriors and most of his horses. After his death in June 1873, Indians continued to raid across the frontier, but they were increasingly defeated in skirmishes with the army. Not only were the whites becoming more skilled at fighting and more familiar with the terrain, they were now armed with the Remingtons that Sarmiento had bought.[95]

In 1874, when Nicolás Avellaneda succeeded Sarmiento as president, the Argentine congress voted funds to strengthen existing forts and build new ones along the frontier and to connect them with telegraph lines. The strategy pursued by Avellaneda and his minister of war, Adolfo Ansina, was almost purely defensive, however.[96] In 1878 General Julio Roca replaced Alsina's defensive strategy with a far more aggressive one. President Avellaneda sent a message to the Argentine congress requesting funds for a war against the Indians: "Today the Nation has powerful means compared to those that the Viceroyalty possessed and even to those that the Congress could count on in 1867. . . . We have 6,000 soldiers armed with the latest modern inventions of war, to fight against 2,000 Indians who have no defense besides dispersion, nor weapons besides primitive lances."[97]

The next year Roca launched an offensive with six thousand men armed with Remington rifles. His objective was not only to cut off contact between the Pampas Indians and their allies in Chile but also to locate and occupy the oases or fertile places in the dry western Pampas where the Indians lived and fed their livestock. Within three months he had pushed the frontier back to the Rio Negro, five hundred miles from Buenos Aires. While the *Uruguay* transported troops and supplies down the coast, the river steamers *Triunfo*, *Rio Neuquén*, and *Rio Negro* steamed up the Colorado and Negro rivers to support the army that exterminated the Indians.[98] In this "Conquest of the Desert," sixteen hundred Indians were killed and over ten thousand made prisoners.[99] During the next six years, with Roca as president, the Argentine army continued the Conquest of the Desert with a campaign that pushed the frontier another 450 miles south to the Rio Deseado, almost to Tierra del Fuego.[100]

Today, Argentine historians depict the Conquest of the Desert as one of the glorious pages in their military history. For Juan Carlos Walther, for example, it was a struggle between Christian civilization and barbaric savages. Most of the rhetoric is about General Julio Argentino Roca and his brilliant officers, some about brave soldiers, and very little about the weapons they used. Yet Europeans had been fighting Indians, with little success, for three hundred years. It is not until Sarmiento and Avellaneda purchased modern weapons that the frontier broke.

The expansion of Chile into Indian territory paralleled that of Argentina, with some notable differences. For several centuries, Chileans and Araucanians had lived side by side with long stretches of peaceful interaction punctuated by wars and raids. Unlike the Indians of the Argentine Pampas who were nomadic hunters and herders, their close relatives the Araucanians were mostly agriculturalists who were partially integrated with European Chileans through trade, religion, and culture. Yet by the mid-nineteenth century, the Europeans began referring to them as savages, barbarians, and lazy, vicious, immoral drunks, and other such derogatory terms. Behind the rhetoric was a desire to seize the lands of the Araucanians lying between central Chile and the region of European settlement around Valdivia and the island of Chiloé to the south.[101]

The Chilean advance into Araucanian territory took place in two phases. The first phase did not begin until the 1860s because European Chileans were involved in protracted conflicts and occasional civil wars between a conservative landowning oligarchy and a liberal urban merchant class. In 1861 the Araucanians began to fight back against land speculators and settlers who were moving south of the Bío-Bío River, the longstanding frontier line. In retaliation, a Chilean army led by Colonel Cornelio Saavedra entered Araucanian territory and built a fort at Angol, a few miles south of the river. The coastal Indians were easily cowed, but those who lived in the mountains, allied to the Argentinean Indians under Calfucurá, put up a fierce resistance. The army engaged in a brutal war, burning crops and houses, stealing cattle, capturing Indian women and children, and putting male prisoners to death. They were aided by a smallpox epidemic that decimated the Indians in the winter of 1869.[102] Despite the actions of the Chilean army, the Araucanians still retained control of the mountain valleys.

The remaining Araucanians were given a reprieve when Chile became embroiled in the War of the Pacific against Bolivia and Peru over the coastal zone between seventeen and twenty-five degrees south. In this war, the victorious Chileans seized the mineral-rich provinces of Antofagasta from Bolivia and Tarapacá from Peru. By

the time the war ended, the Argentine army had conquered the
Pampas up to the Andes and pushed the Indians across the mountains into Araucania. In 1881, Chilean and Argentinean Indians
united in one last effort to push the settlers back north of the Bío-
Bío River and destroy the forts. The Chilean press demanded revenge. This time, a Chilean army of over two thousand men, back
from its victory against Bolivia and Chile and armed with American
rifles and other modern weapons, engaged in a war of extermination.[103] As historian Ravest Mora explains,

> If the native reaction could not prevent a gradual advance, it could
> even less resist a violent and radical invasion like that of the Argen-
> tines. Far from me to interpret this ineffectual result as a weakening
> of the warlike resolve of the Mapuche people, the product, according
> to some, of a secular contact and commerce along the frontier. It was
> rather, in my opinion, a consequence of the brutal disequilibrium
> between the weapons of the contestants: lances, sabers, lariats, and
> slings against repeating rifles, cannon, Gatling machine guns, with,
> finally, the help of railroads and telegraphs.[104]

By 1882, the Araucanians who had preserved their independence
for three and a half centuries had been exterminated or forced onto
reservations.

Ethiopia

Not all wars between Europeans and indigenous peoples were so
lopsided or victories so resounding as the events recounted here.
At the end of the nineteenth century, at the very time when the
Mahdist army was being massacred at Omdurman, other Africans
were acquiring modern weapons and learning to use them effectively. In the Western Sudan, Samori Toure succeeded in keeping
the colonial forces at bay for several years; in Ethiopia, Menelik
actually defeated them.

Ethiopia is always singled out as the great exception to the Scramble for Africa, the one African state that managed to preserve its
independence. It succeeded in doing so for many reasons: a more

cohesive population, a better-organized state, the leadership of Emperor Menelik, and the errors of the Italians who attempted to conquer it. In this story, weapons and tactics feature prominently.

In the early nineteenth century, the rulers of Ethiopia (then called Abyssinia) and its provinces tried to obtain muskets and cannon from Europe, and even to manufacture some domestically, but were unsuccessful. What few firearms Ethiopians had obtained under Emperor Tewodros II (1855–68) were destroyed by the British during their punitive expedition to Magdala in 1868. Emperor Yohannes IV, who ruled from 1872 to 1889, was more successful. In exchange for his support against Tewodros the British gave Yohannes six mortars, six howitzers, 725 muskets, 130 rifles, and ammunition and gunpowder. Yohannes trained his soldiers well. With those weapons, they were able to expel an Egyptian invasion in 1875–76 and capture twenty thousand Remington rifles and twenty-five to thirty cannon, as well as animals and supplies.[105] The British military advisor to Yohannes reported: "The Abyssinians are better armed with firearms than is commonly supposed. They have matchlocks . . . , short guns, Brown Besses—part of the English present to King John—Snyders, now Remingtons, and a variety of other fire-arms."[106]

Nonetheless, Ethiopia was still far behind the Europeans in the quantity and quality of their arms. In the late 1880s, Menelik, ruler of the province of Shoa, was aware of the Scramble going on all around his country, with Italians and French on the Red Sea and British in the Sudan and Kenya. Even before he became emperor, the Italians had attempted to buy his support against Yohannes with a gift of several thousand Remingtons and Vetterlis (a Swiss rifle) and hundreds of thousands of cartridges. When Menelik succeeded Yohannes in 1889, Ethiopian forces were armed with breechloaders and used them to conquer neighboring peoples not yet under European rule, thereby joining the Scramble for Africa.[107]

In 1889 Italy and Ethiopia signed the treaty of Wichali, which the Italians interpreted as conceding them a protectorate over Ethiopia. As a bribe to Menelik, they gave him thirty-nine thousand rifles and twenty-eight cannon. In addition, Menelik got ten thousand rifles from Russia and purchased machine guns and field artillery from private traders.

In early 1896 Italy invaded Ethiopia with an army of 10,596 Italian and 7,100 Eritrean soldiers armed with rifles and fifty-six cannon. Though General Baratiere had no reliable maps or knowledge of the terrain, he overconfidently divided his force into three columns several kilometers apart. On March 1, 1896, at Adowa in the mountains of northern Ethiopia, he encountered Menelik's army of one hundred thousand men armed with rifles and some forty cannon. Of the 10,596 Italians, almost six thousand were killed, went missing, or were taken prisoner.[108] Menelik's victory was as decisive as that of the British at Omdurman two and a half years later. Not only did it assure the independence of Ethiopia for the next forty years, it also demonstrated, for the first time, that Africans could fight on European terms. Contrary to what many Europeans believed at the time, the lopsided nature of the Scramble was not due to differences in race or "civilization" but to purely military factors that could shift in a generation.

Conclusion

What can we make of these disparate stories? What is clear is that they are not disparate at all, though they are usually told in different chapters or in different books. What unites them is a story of radical advances in the weapons available to Europeans and to European Americans and of the weapons gap that opened in the mid-nineteenth century. This explains the astonishing parallels between the Scramble for Africa, the westward expansion of the United States, the Argentinean Conquest of the Desert, and the Chilean conquest of Araucania.

Before the mid-nineteenth century, Europeans had been able to impose their will on highly organized and urbanized societies: the Incas of Peru, the Aztecs and other peoples of Mexico, the Mughals and their successors in India, and finally even China. Less urbanized and organized peoples, especially horseback-riding hunters and herders and the inhabitants of deserts and mountains, were much more successful in keeping Europeans at bay. This was partly due to environmental factors, such as the diseases of tropical Africa or the difficulty of transporting and supplying armies in desert and mountain areas like Afghanistan, Algeria, and the Caucasus. But it was also because the European armies and weapons of the early modern Military Revolution were not able to overcome the resistance of non-Western peoples in those difficult environments.

After many centuries in which Europeans were thwarted or—as in Algeria—had great difficulty advancing into these lands, the sudden sweep of their advance in the late nineteenth century surely cannot be attributed to a sudden increase in motivation or demand for the land and other natural resources of those regions. Rather it is the means that industrialization placed at their disposal—the steamboats and medical advances we saw in earlier chapters and the firearms described here—that transformed the relations between the peoples of the industrialized nations and those who were still without access to the products of modern industry.

Yet technology always changes and industrial technology changed especially fast. Two factors determined the relations between Europeans and European Americans on the one hand, and Africans, Native Americans, and other non-Western peoples on the other. One was the increasing Western power over nature through advances in technology; the other was non-Western peoples' access to the products of modern industrial technology. As the Ethiopian case demonstrates, already by 1896 some Africans had turned the tables on would-be imperialists.[109] Both of these factors changed quickly in the course of the twentieth century, with results that surprised both sides.

Notes

1. David F. Butler, *United States Firearms: The First Century, 1776–1875* (New York: Winchester Press, 1975), pp. 15–41; William Wellington Greener, *The Gun and Its Development*, 9th ed. (New York: Bonanza, 1910), pp. 119–22, 624; Robert Held, with Nancy Jenkins, *The Age of Firearms: A Pictorial History*, 2nd ed. (New York: Harper, 1978), pp. 171–82; "Small Arms, Military," in *Encyclopaedia Britannica*, 14th ed. (Chicago, 1973), vol. 20, p. 668; Walter H. B. Smith, *Small Arms of the World*, 4th ed. (Harrisburg, Pa.: Military Service Publishing, 1953), p. 37.

2. H. Ommundsen and Ernest H. Robinson, *Rifles and Ammunition* (London: Cassell, 1915), p. 18; G.W.P. Swenson, *Pictorial History of the Rifle* (New York: Drake, 1972), p. 16; Butler, *United States Firearms*, pp. 61–93; Greener, *The Gun and Its Development*, p. 627; "Small Arms," pp. 668–70.

3. J.F.C. Fuller, *Armament and History: A Study of the Influence of Armament on History from the Dawn of Classical Warfare to the Second World War* (New York: Scribner's, 1945), p. 110; William Young Carman, *A History of Firearms from Earliest Times to 1914* (London: Routledge and Kegan Paul, 1955), pp. 104, 113, 178; James E. Hicks, *Notes on French Ordnance* (photoprinted, Mount Vernon, N.Y., 1938), p. 21; Colonel Jean Martin, *Armes à feu de l'armée française: 1860 à 1940, historique des évolutions précédentes, comparaison avec les armes étrangères* (Paris: Crépin-Leblond, 1974), pp. 58–64; J. Margerand, *Armement et équipement de l'infanterie française du XVIᵉ au XXᵉ siècle* (Paris: Editions militaires illustrées, 1945), p. 114; Steven T. Ross, *From Flintlock to Rifle: Infantry Tactics, 1740–1866* (Rutherford, N.J.: Fairleigh Dickinson University Press, 1979), pp. 161–62; Held, *Age of Firearms*, pp. 171–75; Greener, *The Gun and Its Development*, pp. 112–17, 624; "Small Arms," p. 668; Swenson, *Pictorial History of the Rifle*, pp. 19–20.

4. Carman, *History of Firearms*, p. 121; Martin, *Armes à feu*, pp. 124–31; Greener, *The Gun and Its Development*, p. 706; Fuller, *Armament and History*, p. 116; Ross, *From Flintlock to Rifle*, pp. 175–76.

5. Fuller, *Armament and History*, pp. 110, 128–29; Ross, *From Flintlock to Rifle*, 164; Ommundsen and Robinson, *Rifles and Ammunition*, p. 65; Held, *Age of Firearms*, p. 183; Margerand, *Armement et équipement*, p. 116; "Small Arms," p. 669.

6. Louis A. Garavaglia and Charles G. Worman, *Firearms of the American West, 1803–1865* (Albuquerque: University of New Mexico Press, 1984), pp. 135–36; Joseph G. Rosa, *Age of the Gunfighter: Men and Weapons on the Frontier, 1840–1900* (Norman: University of Oklahoma Press, 1995), pp. 34–35; Waldo E. Rosebush, *American Firearms and the Changing Frontier* (Spokane: Eastern Washington State Historical Society, 1962), pp. 48–49.

7. Walter Prescott Webb, *The Great Plains* (Boston: Ginn, 1931), pp. 171–77; Garavaglia and Worman, *Firearms of the American West*, pp. 141–42; Rosebush, *American Firearms*, pp. 37–42.

8. Ken Alder, *Engineering the Revolution: Arms and Enlightenment in France, 1763–1815* (Princeton: Princeton University Press, 1997).

9. Russell L. Fries, "British Response to the American System: The Case of the Small-Arms Industry after 1850," *Technology and Culture* 16 (July 1975), pp. 377–403; Roger A. Beaumont, *Sword of the Raj: The British Army in India, 1747–1947* (Indianapolis: Bobbs-Merrill, 1977), p. 68; Carman, *History of Firearms*, p. 112; Greener, *The Gun and Its Development*, pp. 283–84, 621–25; Ommundsen and Robinson, *Rifles and Ammunition*, pp. 78–79; Swenson, *Pictorial History of the Rifle*, pp. 24–25.

10. Lee B. Kennett, *The Gun in America: The Origins of a National Dilemma* (Westport, Conn.: Greenwood, 1975), p. 92; Greener, *The Gun and Its Development*,

pp. 716–17; Swenson, *Pictorial History of the Rifle*, pp. 28–29; Butler, *United States Firearms*, pp. 222–23.

11. Robert Gardner, *Small Arms Makers: A Directory of Fabricators of Firearms, Edged Weapons, Crossbows, and Polearms* (New York: Crown, 1963), p. 160; Donald Featherstone, *Colonial Small Wars, 1837–1901* (Newton Abbot: David and Charles, 1973), p. 23; Alden Hatch, *Remington Arms in American History* (New York: Rinehart, 1956), pp. 136–42; Richard L. Hill, *Egypt in the Sudan, 1820–1881* (London: Oxford University Press, 1959), p. 109; Kennett, *The Gun in America*, pp. 92–94, 280n37; Margerand, *Armement et équipement*, p. 117; Swenson, *Pictorial History of the Rifle*, pp. 27–32; Greener, *The Gun and Its Development*, pp. 702–12.

12. Hatch, *Remington Arms*, pp. 110–14; "Small Arms," pp. 671–72; Rosebush, *American Firearms*, pp. 48–56; Held, *Age of Firearms*, p. 184; Carman, *History of Firearms*, p. 178.

13. Paul Scarlata, "The British Martini-Henry Rifle," *Shotgun News* (December 6, 2004), pp. 36–40. (I am grateful to Alex Koch for bringing this article to my attention.)

14. Hicks, *Notes on French Ordnance*, p. 27; Martin, *Armes à feu*, pp. 247–61; Ommundsen and Robinson, *Rifles and Ammunition*, pp. 72, 90; Margerand, *Armement et équipement*, p. 117; Rosebush, *American Firearms*, p. 62; Swenson, *Pictorial History of the Rifle*, p. 33.

15. Robert Lawrence Wilson, *Winchester: An American Legend: The Official History of Winchester Firearms and Ammunition from 1849 to the Present* (New York: Random House, 1991), pp. 22, 41; Alarico Gattia, *Cosí sparavano i "nostri": Uomini e armi del vecchio West* (Genoa: Stringa, 1966), pp. 118–19; Douglas C. McChristian, *The U.S. Army in the West, 1870–1880: Uniforms, Weapons, and Equipment* (Norman: University of Oklahoma Press, 1995), p. 113; Rosa, *Age of the Gunfighter*, pp. 138–39; Rosebush, *American Firearms*, p. 52; Butler, *United States Firearms*, pp. 235–47.

16. Martin, *Armes à feu*, pp. 307–13; Swenson, *Pictorial History of the Rifle*, p. 33; Ommundsen and Robinson, *Rifles and Ammunition*, p. 101.

17. Hatch, *Remington Arms*, pp. 190–91; Martin, *Armes à feu*, pp. 317–20.

18. Greener, *The Gun and Its Development*, pp. 727–28.

19. Luc Garcia, *Le royaume du Dahomé face à la pénétration coloniale* (Paris: Karthala, 1988), pp. 159–60.

20. Ommundsen and Robinson, *Rifles and Ammunition*, p. 118.

21. Swenson, *Pictorial History of the Rifle*, pp. 34–35; Martin, *Armes à feu*, pp. 322–28; Hatch, *Remington Arms*, pp. 177–78, 191; Greener, *The Gun and Its Development*, pp. 701, 731; Margerand, *Armement et équipement*, p. 118; Featherstone, *Colonial Small Wars*, p. 24; Hicks, *Notes on French Ordnance*, pp. 27–28.

22. On nineteenth-century artillery, see Ian V. Hogg, *A History of Artillery* (Feltham, England: Hamlyn, 1974), pp. 56–65.

23. Graham Seton Hutchison, *Machine Guns, Their History and Tactical Employment (Being Also a History of the Machine Gun Corps, 1916–1922)* (London: Macmil-

lan, 1938), pp. 31–55; Carman, *History of Firearms*, p. 85. See also John Ellis, *The Social History of the Machine Gun* (New York: Random House, 1975), pp. 64–70, 86–90.

24. Colonel Charles E. Callwell, *Small Wars, Their Principles and Practice*, 3rd ed. (London: General Staff, War Office, 1906), p. 398. This book has gone through many editions and reprintings, most recently in 1996.

25. Charles Braithwaite Wallis, *West African Warfare* (London: Harrison, 1906) passim.

26. See, for example, Byron Farwell, *Queen Victoria's Little Wars* (New York: Harper and Row, 1972), and Robert Giddings, *Imperial Echoes: Eye-Witness Accounts of Victoria's Little Wars* (London: Lee Cooper, 1996), as well as articles in *Shotgun News*, *Soldier of Fortune*, and similar magazines.

27. Joseph E. Inikori, "The Import of Firearms into West Africa, 1750 to 1807: A Quantitative Analysis," *Journal of African History* 18 (1977), pp. 339–68.

28. R. A. Kea, "Firearms and Warfare in the Gold and Slave Coasts from the Sixteenth to the Nineteenth Centuries," *Journal of African History* 12 (1971), pp. 2002–4; Gavin White, "Firearms in Africa: An Introduction," *Journal of African History* 12 (1971), pp. 179–81.

29. R. W. Beachey, "The Arms Trade in East Africa," *Journal of African History* 3 (1962), pp. 454–63; Humphrey J. Fisher and Virginia Rowland, "Firearms in the Central Sudan," *Journal of African History* 12 (1971), pp. 222–24.

30. William Storey, "Guns, Race, and Skill in Nineteenth-Century Southern Africa," *Technology and Culture* 45 (October 2004), p. 693n14; Anthony Atmore, J. M. Chirenje, and S. I. Mudenge, "Firearms in South Central Africa," *Journal of African History* 12 (1971), p. 550; J. J. Guy, "A Note on Firearms in the Zulu Kingdom with Special Reference to the Anglo-Zulu War, 1879," *Journal of African History* 12 (1971), p. 559; Sue Miers, "Notes on the Arms Trade and Government Policy in Southern Africa between 1870 and 1890," *Journal of African History* 12 (1971), pp. 571–72.

31. Robin Law, "Horses, Firearms and Political Power," *Past and Present* 72 (1976), pp. 122–23; Paul M. Mbaeyi, *British Military and Naval Forces in West African History, 1807–1874* (New York: NOK Publishers, 1978), p. 28; Beachey, "The Arms Trade in East Africa," 451–52; Miers, "Notes on the Arms Trade," p. 572.

32. On African iron-making, see Jack Goody, *Technology, Tradition, and the State in Africa* (London: Oxford University Press, 1971), pp. 28–29, and Walter Cline, *Mining and Metallurgy in Negro Africa* (Menasha, Wisc.: George Banta, 1937) passim; Kea, "Firearms and Warfare," p. 205; White, "Firearms in Africa," p. 181.

33. John D. Goodman, "The Birmingham Gun Trade," in Samuel Timmins, ed., *The Resources, Products, and Industrial History of Birmingham and the Midland Hardware District* (London: Cass, 1967), pp. 388, 426; Myron J. Echenberg, "Late Nineteenth-Century Military Technology in Upper Volta," *Journal of African History* 12 (1971), pp. 251–52; H. A. Gemery and Jan S. Hogendorn, "Technological Change, Slavery and the Slave Trade," in Clive Dewey and Antony G. Hopkins,

eds., *The Imperial Impact: Studies in the Economic History of India and Africa* (London: Athlone Press, 1978), pp. 248–50; White, "Firearms in Africa," pp. 173–84; Goody, *Technology*, p. 52.

34. K. Onwuka Dike, *Trade and Politics in the Niger Delta, 1830–1885: An Introduction to the Economic and Political History of Nigeria* (Oxford: Clarendon Press, 1956), p. 107.

35. Storey, "Guns, Race, and Skill," p. 695.

36. D.J.M. Muffett, "Nigeria-Sokoto Caliphate," in Michael Crowder, ed., *West African Resistance: The Military Response to Colonial Occupation* (London: Hutchinson, 1971), pp. 268–99; Echenberg, "Late Nineteenth-Century Military Technology in Upper Volta," pp. 245–53; Fisher and Rowland, "Firearms in the Central Sudan," pp. 215–39.

37. Joseph P. Smaldone, "The Firearms Trade in the Central Sudan in the Nineteenth Century," in Daniel F. McCall and Norman R. Bennett, eds., *Aspects of West African Islam*, Boston University Papers on Africa, vol. 5 (Boston: African Studies Center, Boston University, 1971), p. 155.

38. Joseph P. Smaldone, "Firearms in the Central Sudan: A Reevaluation," *Journal of African History* 13 (1972), p. 594; Fisher and Rowland, "Firearms in the Central Sudan," p. 223n60.

39. Goody, *Technology*, pp. 47–55.

40. Yves Person, *Samori: Une Révolution Dyula*, 3 vols. (Dakar: IFAN, 1968), vol. 2, p. 908.

41. Storey, "Guns, Race, and Skill," p. 693n14.

42. Beachey, "The Arms Trade in East Africa," p. 464; Smaldone, "Firearms Trade," pp. 162–70.

43. Anthony Atmore and Peter Sanders, "Sotho Arms and Ammunition in the Nineteenth Century," *Journal of African History* 12 (1971), p. 539; Smaldone, "Firearms Trade," pp. 155–56.

44. Storey, "Guns, Race, and Skill," pp. 702, 707–8.

45. Beachey, "The Arms Trade in East Africa," p. 453.

46. Shula Marks and Anthony Atmore, "Firearms in Southern Africa: A Survey," *Journal of African History* 12 (1971), pp. 524–28; Beachey, "The Arms Trade in East Africa," pp. 455–57; Smaldone, "Firearms Trade," p. 170.

47. Michael Gelfand, *Livingstone the Doctor: His Life and Travels* (Oxford: Blackwell, 1957), pp. 170–73, 192.

48. Roy C. Bridges, "John Hanning Speke: Negotiating a Way to the Nile," p. 107; James R. Hooker, "Verney Lovett Cameron: A Sailor in Central Africa," pp. 265, 274; and Robert O. Collins, "Samuel White Baker: Prospero in Purgatory," pp. 141–50, 171; all in Robert I. Rotberg, ed., *Africa and Its Explorers: Motives, Methods and Impact* (Cambridge, Mass.: Harvard University Press, 1970).

49. Wolfe W. Schmokel, "Gerhard Rohlfs: The Lonely Explorer," in Robert I. Rotberg, ed., *Africa and Its Explorers: Motives, Methods and Impact* (Cambridge, Mass.: Harvard University Press, 1970), p. 208.

50. Eric Halladay, "Henry Morton Stanley: The Opening Up of the Congo Basin," in Robert I. Rotberg, ed., *Africa and Its Explorers: Motives, Methods and Impact* (Cambridge, Mass.: Harvard University Press, 1970), p. 242.

51. Henry Morton Stanley, *In Darkest Africa, or the Quest, Rescue, and Retreat of Emin, Governor of Equatoria*, 2 vols. (New York: Scribner's, 1890), vol. 1, pp. 37–39. See also his *The Congo and the Founding of Its Free State* (New York: Harper, 1885), pp. 63–64, and Adam Hochschild, *King Leopold's Ghost: A Story of Greed, Terror, and Heroism in Colonial Africa* (Boston: Houghton Mifflin, 1998), pp. 30, 47–49, 97.

52. Quoted in Halladay, "Henry Morton Stanley," p. 244.

53. Henry Morton Stanley, *The Exploration Diaries of H. M. Stanley*, ed. Richard Stanley and Alan Neame (New York: Vanguard Press, 1961), p. 125, quoted in Hochschild, *King Leopold's Ghost*, p. 49.

54. Marks and Atmore, "Firearms in Southern Africa," pp. 519–22.

55. J. K. Fynn, "Ghana-Asante (Ashanti)," in Michael Crowder, ed., *West African Resistance: The Military Response to Colonial Occupation* (London: Hutchinson, 1971), p. 32; Mbaeyi, *British Military and Naval Forces*, pp. 34–35.

56. Atmore and Sanders, "Sotho Arms and Ammunition," pp. 537–41.

57. John Keegan, "The Ashanti Campaign, 1873–74," in Brian Bond, ed., *Victorian Military Campaigns* (London: Hutchinson, 1967), pp. 161–98; Philip Curtin, *Disease and Empire: The Health of European Troops in the Conquest of Africa* (New York: Cambridge University Press, 1998), p. 58; Hutchison, *Machine Guns*, p. 38; Kea, "Firearms and Warfare," p. 201; Fynn, "Ghana-Asante (Ashanti)," p. 40; Mbaeyi, *British Military and Naval Forces*, pp. 34–35.

58. Donald R. Morris, *The Washing of the Spears: A History of the Rise of the Zulu Nation under Shaka and Its Fall in the Zulu War of 1879* (New York: Simon and Schuster, 1964), pp. 352–88, 545–75; Ellis, *Social History of the Machine Gun*, p. 82; Hutchison, *Machine Guns*, p. 39; Scarlata, "The British Martini-Henry Rifle," p. 36.

59. Henri-Nicholas Frey, *Campagne dans le Haut Sénégal et dans le Haut Niger (1885–1886)* (Paris: Plon, 1888), pp. 60–62; B. Olatunji Oloruntimehin, "Senegambia-Mahmadou Lamine," in Michael Crowder, ed., *West African Resistance: The Military Response to Colonial Occupation* (London: Hutchinson, 1971), pp. 93–94; Person, *Samori*, vol. 2, p. 907.

60. Garcia, *Le royaume du Dahomé*, pp. 163–67; David Ross, "Dahomey," in Michael Crowder, ed., *West African Resistance: The Military Response to Colonial Occupation* (London: Hutchinson, 1971), pp. 158–61; Callwell, *Small Wars*, p. 260; Kea, "Firearms and Warfare," p. 213.

61. Atmore, Chirenje, and Mutenge, "Firearms in South Central Africa," pp. 553–54; Ellis, *Social History of the Machine Gun*, p. 90; Hutchison, *Machine Guns*, p. 63.

62. Callwell, *Small Wars*, pp. 440–42; Wallis, *West African Warfare*, p. 56.

63. Pierre Gentil, *La conquête du Tchad*, 2 vols. (Vincennes: Etat-major de l'Armée de terre, Service historique, 1971), vol. 1, p. 99; Alexander S. Kanya-Forstner, *The Conquest of the Western Sudan: A Study in French Military Imperialism* (Cambridge: Cambridge University Press, 1969), p. 10; Michael Crowder, introduction, pp. 6–7, and Muffett, "Nigeria-Sokoto Caliphate," pp. 284–85, both in Michael Crowder, ed., *West African Resistance: The Military Response to Colonial Occupation* (London: Hutchinson, 1971).

64. Guy, "A Note on Firearms," pp. 560–62; Marks and Atmore, "Firearms in Southern Africa," pp. 519–27; Fisher and Rowland, "Firearms in the Central Sudan," pp. 229–30; Muffett, "Nigeria-Sokoto Caliphate," p. 287; Crowder, introduction, p. 11.

65. Cyril Falls, *A Hundred Years of War* (London: Duckworth, 1953), pp. 118–19; Callwell, *Small Wars*, pp. 30–31, 75–76; Wallis, *West African Warfare*, pp. 44–45; Brian Bond, introduction to Brian Bond, ed., *Victorian Military Campaigns* (London: Hutchinson, 1967), p. 25.

66. M. Legassick, "Firearms, Horses and Samorian Army Organization, 1870–1898," *Journal of African History* 7 (1966), pp. 95–115; Person, *Samori*, vol. 2, pp. 905–12. See also Person, "Guinea-Samori," in Michael Crowder, ed., *West African Resistance: The Military Response to Colonial Occupation* (London: Hutchinson, 1971), pp. 111–43.

67. Callwell, *Small Wars*, pp. 389, 438–39; Hutchison, *Machine Guns*, pp. 67–70.

68. It is well to remember that massive quasi-suicidal frontal assaults by infantrymen against rifle and machine-gun fire was not unique to Africans. It was, in fact, the very tactic used by European soldiers going "over the top" of the trenches in World War I, with just as deadly results. The difference was that in Europe the killing went on for four years and cost many times more lives than the conquest of Africa.

69. Winston S. Churchill, *The River War: An Account of the Reconquest of the Sudan* (1933; New York: Carroll and Graf, 2000), pp. 274, 279.

70. Ibid., p. 300.

71. Quoted in Webb, *The Great Plains*, p. 195.

72. *The Times Atlas of World History*, rev. ed. by Geoffrey Barraclough (London: Times Books, 1979), pp. 160, 164, 220. See also *Historical Atlas of the World* (Skokie, Ill.: Rand McNally, 1997), pp. 58–59, and *Atlas of American History* (Skokie, Ill.: Rand McNally, 1993), pp. 12–18, 22–29.

73. For example, *Times Atlas*, p. 220, and *Historical Atlas of the World*, p. 59.

74. See the map in Michael Williams, *Deforesting the Earth: From Prehistory to Global Crisis: An Abridgment* (Chicago: University of Chicago Press, 2006), p. 205.

75. Frank R. Secoy, *Changing Military Patterns on the Great Plains (17th Century through Early 19th Century)* (Locust Valley, N.Y.: Augustin, 1953), pp. 5, 39, 95–97, 104–5; Garavaglia and Worman, *Firearms of the American West*, pp. 344–60;

Paul Russell, *Man's Mastery of Malaria* (London: Oxford University Press, 1955), pp. 103–30.

76. On traditional Indian weapons, see Colin F. Taylor, *Native American Weapons* (Norman: University of Oklahoma Press, 2001).

77. Josiah Gregg, *Commerce on the Prairies*, ed. Max L. Moorhead (Norman: University of Oklahoma Press, 1954), pp. 416–17.

78. Colonel Richard Irving Dodge, *Our Wild Indians: Thirty-Three Years' Personal Experience among the Red Men of the Great West* (Hartford, Conn.: Worthington, 1882; reprint, New York: Archer House, 1960), pp. 449–50.

79. R. G. Robertson, *Rotting Face: Smallpox and the American Indian* (Caldwell, Idaho: Caxton Press, 2001), p. 245; Webb, *The Great Plains*, pp. 167–69. On military rifles before the Civil War, see Russell, pp. 173–91.

80. John E. Parsons, ed., *Sam'l Colt's Own Record* (Hartford: Connecticut Historical Society, 1949), p. 10.

81. Robert Wooster, *The Military and United States Indian Policy, 1865–1903* (New Haven: Yale University Press, 1988), p. 33; McChristian, *U.S. Army in the West*, p. 113; Butler, *United States Firearms*, p. 190; Gattia, *Cosí sparavano i "nostri,"* p. 106.

82. Rosebush, *American Firearms*, pp. 65–67; Gattia, *Cosí sparavano i "nostri,"* p. 114.

83. Dodge, *Our Wild Indians*, p. 422.

84. Ibid., p. 450.

85. Hatch, *Remington Arms*, pp. 140–41; Gattia, *Cosí sparavano i "nostri,"* pp. 105–6; Rosebush, *American Firearms*, pp. 62–64; Dodge, *Our Wild Indians*, p. 423.

86. Robertson, *Rotting Face*, pp. 242–46.

87. Dodge, *Our Wild Indians*, pp. 295–96.

88. Daniel O. Magnussen, *Peter Thompson's Narrative of the Little Bighorn Campaign of 1876: A Critical Analysis of an Eyewitness Account of the Custer Debacle* (Glendale, Calif.: Arthur H. Clark, 1974), pp. 151–52, and Rosebush, *American Firearms*, p. 71, claim that the cartridges used in these rifles tended to jam and were hard to extract. Archaeological evidence at the site does not back up this claim, however; see McChristian, *U.S. Army in the West*, p. 114.

89. Gregory F. Michno, *Lakota Noon: The Indian Narrative of Custer's Defeat* (Missoula, Mont.: Mountain Press, 1997), pp. 49, 110, 192; Bruce A. Rosenberg, *Custer and the Epic of Defeat* (University Park: Pennsylvania State University Press, 1974), pp. 27, 68–70; Wooster, *Indian Policy*, p. 33; Hatch, *Remington Arms*, pp. 154–55; Rosebush, *American Firearms*, pp. 69–70; McChristian, *U.S. Army in the West*, p. 115.

90. Anthony Smith, *Machine Gun: The Story of the Men and the Weapon That Changed the Face of War* (New York: St. Martin's, 2002), pp. 74, 106; Rosebush, *American Firearms*, pp. 72–74.

91. Kristine L. Jones, "Civilization and Barbarism and Sarmiento's Indian Policy," in Joseph T. Criscenti, ed., *Sarmiento and His Argentina* (Boulder, Colo.: Lynne

Rienner, 1993), pp. 37–38; Richard O. Perry, "Warfare on the Pampas in the 1870s," *Military Affairs* 36, no. 2 (April 1972), pp. 52–58.

92. Jones, "Civilization and Barbarism," pp. 35–37.

93. Nora Siegrist de Gentile and María Haydee Martín, *Geopolítica, ciencia y técnica a través de la Campaña del Desierto* (Buenos Aires: Editorial Universitaria de Buenos Aires, 1981), pp. 92–96.

94. David Viñas, *Indio, ejército y frontera* (Mexico: Siglo Veintiuno Editores, 1982), p. 90.

95. Perry, "Warfare on the Pampas," pp. 54–56; Siegrist de Gentile and Haydee Martín, *Geopolítica*, pp. 96–97.

96. Orlando Mario Punzi, *Historia del desierto: La conquista del desierto pampeano-patagónico: La conquista del Chaco* (Buenos Aires: Ediciones Corregidor, 1983), pp. 48–49; Jones, "Civilization and Barbarism," p. 41; Siegrist de Gentile and Haydee Martín, *Geopolítica*, p. 106.

97. Lobodón Garra [Liborio Justo], *A sangre y lanza, o el último combate del capitanejo Nehuén: Tragedia e infortunio de la Epopeya del Desierto* (Buenos Aires: Ediciones Anaconda, 1969), pp. 428–29.

98. Enrique González Lonzième, *La Armada en la Conquista del Desierto* (Buenos Aires: Editorial Universitaria de Buenos Aires, 1977), pp. 57–96; Ricardo Capdevila, "La corbeta 'Uruguay': Su participación en la Conquista del Desierto y las tierras australes argentinas," in *Congreso Nacional de Historia sobre la Conquista del Desierto celebrado en la ciudad de Gral. Roca del 6 al 10 de noviembre de 1979* (Buenos Aires: Academia Nacional de la Historia, 1980), pp. 259–68.

99. Felix Best, *Historia de las guerras argentinas, de la independencia, internacionales, civiles y con el indio* (Buenos Aires: Ediciones Penser, 1960), pp. 387–91; Juan Carlos Walther, *La conquista del desierto* (Buenos Aires: Círculo Militar, 1964), p. 752; Argentine Republic, Ministerio del Interior, *Campaña del Desierto (1878–1884)* (Buenos Aires: Archivo General de la Nación, 1969), prologue; Roberto Levillier, *Historia de Argentina* (Buenos Aires: Plaza y Janés, 1968), vol. 4, pp. 2956–60; Punzi, *Historia del desierto*, pp. 53–55; Perry, "Warfare on the Pampas," pp. 56–57.

100. *Acción y presencia del Ejército en el sur del país* (N.p.: Comando del Vto Cuerpo de Ejército "Teniente General D. Julio Argentino Roca," 1997), pp. 28–31.

101. Jean-Pierre Blancpain, *Les Araucans et la frontière dans l'histoire du Chili des origines au XIXe siècle: Une épopée américaine* (Frankfurt: Vervuert Verlag, 1990), pp. 134–35; Luis Carlos Parentini and Patricia Herrera, "Araucanía maldita: Su imágen a través de la prensa, 1820–1860," in Leonardo León et al., eds., *Araucanía: La frontera mestiza, siglo XIX* (Santiago: Ediciones UCSH, 2003), pp. 63–100.

102. Patricia Cerda-Hegerl, *Fronteras del sur: La región del Bío-Bío y la Araucanía chilena, 1604–1883* (Temuco, Chile: Universidad de la Frontera, 1990), pp. 131–32, 142; Manuel Ravest Mora, *Ocupación militar de la Araucanía (1861–1883)* (Santiago: Licanray, 1997), pp. 9–10; Blancpain, *Les Araucans*, pp. 140–43.

103. José Bengoa, *Historia del pueblo mapuche: Siglos XIX y XX*, 6th ed. (Santiago: LOM Ediciones, 2000), p. 275; Blancpain, *Les Araucans*, pp. 141–45.

104. Ravest Mora, *Ocupación militar de la Araucanía*, p. 11.

105. R. A. Caulk, "Firearms and Princely Power in Ethiopia in the Nineteenth Century," *Journal of African History* 13 (1972), pp. 610–13; Richard Pankhurst, "Guns in Ethiopia," *Transition* 20 (1965), pp. 20–29; Jonathan Grant, *Rulers, Guns, and Money: The Global Arms Trade in the Age of Imperialism* (Cambridge, Mass.: Harvard University Press, 2007), p. 47.

106. Pankhurst, "Guns in Ethiopia," pp. 29–30.

107. Ibid., p. 30; Caulk, "Firearms and Princely Power," pp. 621–22; Grant, *Rulers, Guns, and Money*, pp. 47–63.

108. Romain H. Rainero, "The Battle of Adowa, on 1st March 1896: A Reappraisal," in J. A. De Moor and H. L. Wesserling, eds., *Imperialism and War: Essays on Colonial Wars in Asia and Africa* (Leiden: E. J. Brill, 1989), pp. 193–97; Pankhurst, "Guns in Ethiopia," pp. 31–32.

109. The other famous case is that of Japan, which defeated Russia in 1905. I have not dealt with Japan in this book because it was by then already an industrial and imperialist nation in its own right rather than an object of Western imperialism.

❋ Chapter 8 ❋

The Age of Air Control, 1911–1936

The tremendous advantage that modern infantry weapons had once given the industrial nations began to dissipate by the end of the nineteenth century, as some non-Western societies acquired similar weapons and adapted their tactics to them. This trend was first revealed in the Ethiopian victory over Italy in 1896 and in the Japanese defeat of Russia in 1905. After World War I, the victorious powers encountered unexpected resistance in the Middle East and Asia fueled by a rising tide of nationalism and by the surplus military weapons that flooded the world.

Just when the Western powers began to face increasing challenges to their domination, an entirely new technology—aircraft—promised to restore the advantage that was slipping away. As early as 1911, the French general Henri-Nicolas Frey had noted the connection between the diffusion of modern weapons and the value of aircraft in colonial settings.

> [I]t has been possible for some of those races to fight, under certain circumstances, with almost equal weapons against Europeans by obtaining rapid-firing rifles by contraband or with the complicity of neighboring countries or by recruiting former sharpshooters trained by us in the use of these weapons. . . . Command of the air allows Europeans 1) to exercise easy, rapid, and continuous police surveillance on barbarian nomadic tribes and on numerous and civilized populations that are, by nature, suspicious, hostile, and quick to revolt; 2) to intervene, with the speed of a bird of prey, in threatened or troubled places by bringing to bear, if necessary, formidable engines of destruction against which the imperfect weapons, the cunning, and the ingenious tricks to which the so-called "inferior" races resort

are reduced to impotence. . . ; 3) and thus to lay the most solid foundation for their domination of those tribes and races.[1]

The Beginnings of Aviation

Few technologies have undergone such rapid change—and none has received so much publicity—as aviation. Flying through the sky and dominating the earth below have long been dreams of mankind. For centuries inventors tried various methods of lifting human beings off the ground. The first practical success came in 1783, when the Montgolfier brothers constructed a hot-air balloon that carried two men up into the sky on a twenty-five-minute flight. Hot-air balloons were followed by other lighter-than-air devices filled with hydrogen; by the end of the nineteenth century, dirigibles could take off, land, and move along a course chosen by the pilot. Dirigibles, however, were slow, unwieldy, and expensive. In the late nineteenth century, inventors devised many designs and models of flying craft that were heavier than air, none of them successful. By the turn of the century, the competition was intense, as was the publicity that surrounded the inventors and their attempts to fly.

To the surprise of all, the first to fly in a heavier-than-air device were two bicycle makers from Ohio who had quietly been studying and performing experiments with almost no publicity. The great achievement of the Wright brothers that separated them from all other inventors was devising a system of controls so that the pilot could not only take off, fly, and land but also control the aircraft in all three dimensions. After three years building gliders that could be controlled in the air, they had a gasoline engine built that was small and light enough to power a glider and made a propeller that could efficiently translate the engine's power into motion through the air. On December 17, 1903, at Kitty Hawk, North Carolina, they flew their aircraft, the *Flyer*, four times, the last time covering a distance of 852 feet in fifty-nine seconds.[2] During the next few years, the Wright brothers built better aircraft, but in great secrecy. They tried to interest the U.S. Army into giving them a contract for airplanes sight unseen, but were rebuffed. From October 1905 to May 1908 they refused to fly at all, for fear of imitators. Yet their very secrecy only incited others.

In November 1906, Alberto Santos-Dumont, a wealthy Brazilian living in Paris who had already built several dirigibles, became the first person to fly in Europe; his airplane, the *14-bis*, took off under its own power without a headwind to help it, but it was not as fully controlled as the Wright brothers'. It was the Wrights' public demonstration in 1908 that showed others how to control an airplane in flight. Very quickly, inventors in Europe and North America improved upon Santos-Dumont's feat. That same year, the American Glenn Curtiss, a manufacturer of engines and motorcycles, built and flew his first airplane, the *June Bug*, and offered to build airplanes commercially. A year later, the Frenchman Louis Blériot crossed the English Channel in a plane built with Gabriel Voisin. By then, many entrepreneur-inventors were putting together a dazzling variety of heavier-than-air craft in Europe and North America. The most successful, like Blériot's, already had the shape associated with airplanes ever since: the engine and propeller in front, wings behind the engine, movable ailerons for control, a tail with a rudder and elevators in the rear, and wheels for takeoff and landing. In their ability to fly and maneuver in the air, such craft left the Wright brothers' design, resembling several kites strung together, far behind.

The limiting factor after that was the lack of power, as existing engines were weak and heavy and prone to overheating. Planes were so underpowered they could barely climb with two men aboard. In 1909 Louis and Laurent Séguin began manufacturing a seven-cylinder rotary engine they called Gnôme that produced fifty horsepower and weighed only 165 pounds. Because the crankshaft was bolted to the airplane fuselage while the rest of the engine spun around with the propeller, this engine stayed cool, although it had a tendency to make the airplane respond in strange ways to changes in direction. Once a pilot was trained to control it, a plane equipped with a Gnôme engine was capable of flying faster and maneuvering better than all previous models. By 1914 Gnôme engines produced up to a hundred horsepower and could propel planes at almost ninety miles per hour. Until well into World War I, such rotary engines dominated the skies of Europe.

Despite the tremendous advances made by 1914, airplanes were still slow, unreliable, and dangerous contraptions made of wood and cloth, with the pilot sitting precariously out in the open. Yet their military potential was obvious even before the Great War broke out in August of that year. When the war began, France had 141 military planes plus another 176 in reserve or in flight school, Russia had 250 (many of them useless for lack of parts and maintenance), Germany had 245 plus 10 dirigibles, and Britain had 230 (of which half were fit to fly).[3] Almost from the day that conflict began, the belligerents used airplanes for reconnaissance, to find out where enemy forces were. Within a few weeks, the pilots began shooting at each other. The war accelerated improvements in aircraft, engine design, and fire power. By early 1915, planes equipped with machine guns—the first fighters—began chasing each other through the sky, alleviating the tedium of trench warfare with the spectacular heroics of aerial dogfights. Bombers were equipped to carry not only explosive and incendiary bombs, but also poison gas, another technology introduced during the war. By the end of the war, Germany had produced 48,000 planes, France 67,000, Britain 58,000, and the United States almost 12,000.[4]

When the war ended, the belligerents had many warplanes left. For a decade, the vast number of surplus aircraft hindered technological advance, and most manufacturers went out of business. During the 1930s, fast monoplanes made of metal with sleek streamlined bodies and retractable landing gears began replacing World War I–vintage planes. A sign of things to come was the British Supermarine S.6B that won the Schneider Trophy for the world's fastest airplane in 1929 and 1931; in the latter year it was powered by a 2,300-horsepower Rolls Royce engine and reached a speed of four hundred miles per hour. It was the prototype of the Spitfires that saved Britain during the Battle of Britain in 1940. Airplane design also advanced in another direction with the development of efficient passenger planes, mainly in the United States. Like the warplanes, these too were monoplanes with streamlined all-metal bodies and retractable landing gears: first the Boeing 247 of 1933 that could transport ten passengers in comfort at 183 miles per hour, then the Douglas of 1935 that, under the names DC-3, Dakota, and C-47, became the most widely used airplane of all time.

Early Colonial Air Campaigns

Soon after Louis Blériot crossed the English Channel in July 1909, military officers in Europe began to consider using aircraft in colonial wars. Major Baden-Powell called for the use of aircraft in "savage warfare" and noted that the "moral effect on an ignorant enemy would be great, and a few bombs would cause serious panics."[5] A few years later, a British official in West Africa wrote optimistically that aircraft "would be invaluable against the hill tribes, and the terror caused by them would probably do away with the need for bloodshed."[6]

To the Italians goes the honor—or the shame—of being the first to use aircraft in warfare. Italy invaded Libya, then a province of the Ottoman Empire, in the fall of 1911. On October 21, two Blériot, three Nieuport, two Farman, and three Austrian Etrichs planes arrived in Tripoli. Two days later Captain Carlo Piazza took off in a Blériot on a seventy-minute reconnaissance flight over Turkish lines. The next day, ground fire hit Captain Riccardo Moizo's plane while on a reconnaissance flight, but did not damage it. On November 1, Italian planes began dropping bombs on Turkish positions. These bombs were little more than two-kilogram grenades that the pilot dropped over the side of his airplane, yet the Turks were outraged and claimed that the Italians had bombed a hospital. Later the Italians brought in dirigibles that carried larger bombs. By and large, however, bombing had little effect on the course of the war. Airplanes were used much more effectively for observation, aerial photography, and artillery spotting.[7]

The planes used in Libya aroused a lot of attention from Western observers. Opinions differed as to their effect. G. F. Abbott, a British correspondent with the Turkish forces, wrote: "Next morning it appeared again but, instead of passing our camp, it just whirred over our heads in a wide curve from north-east to north-west and vanished. The Arabs, beyond emptying their rifles at the visitor, showed small emotion at the visit. I could detect no consternation or dismay, scarce even surprise, in their faces."[8] Another war correspondent, Francis McCullagh, wrote, "as a rule, the bombs discharged from it bury themselves harmlessly in the sand. The Arabs have no barracks or permanent works which can be injured; and as they now scatter whenever they see an aeroplane approaching, practically no loss is ever inflicted on them by the grenades. The women and children in the villages are practically the only victims, and this fact excited the anger of the Arabs."[9]

Aircraft may not have had much effect in Libya, and they certainly did not give Italy an advantage in World War I, but they did have one long-term impact. They inspired Major Giulio Douhet, who participated in the campaign, to publish his famous book on air power, *Il dominio dell'aria* (The command of the air) in 1921, a book that has influenced military aviation strategists ever since.[10] In it, he advocated bombing an enemy's cities, industries, infrastructures, and government and military centers in order to terrorize the civilians, who would then convince their government to sue for peace. At its core, this doctrine was based on a psychological assumption about human nature, namely that civilians were weak and easily terrified and that a nation equipped with bombers could bypass its enemy's army and navy and win by intimidation alone.

In comparison to the Italian invasion of Libya, the use of aircraft in the U.S. punitive expedition to Mexico in 1916 was a slapdash affair, ill befitting the nation that had inaugurated the era of heavier-than-air flight. From 1908 to 1913, the U.S. government had spent only $453,000 on airplanes (Germany, for example, had spent $28 million), leaving it fourteenth in the world in military aviation.

On March 9, 1916, the Mexican revolutionary Francisco "Pancho" Villa crossed the border to Columbus, New Mexico, and killed seventeen Americans. A week later a force commanded by General John Pershing entered Mexico in pursuit of Villa. As part of this expedition, the First Aero Squadron under Captain Benjamin Foulois was sent to New Mexico with eight used Curtiss JN-2 and JN-3 "Jennies." These underpowered biplanes had a top speed of eighty miles per hour and a range of fifty miles; they were also very unstable, with a tendency to stall and go into a dive.[11]

No sooner had the planes been unloaded from the train and reassembled than General Pershing ordered them to fly to Casas Grandes in Mexico, one hundred miles away, just as night was falling. One plane crashed, another returned to Columbus with engine trouble, and the rest landed along the way and did not reach Casas Grandes until the next day.[12] In the following weeks, the Jennies were sometimes used for reconnaissance, but mainly to carry messages between the scattered units of Pershing's expeditionary force. They carried neither guns nor bombs. In the high altitude of northern Mexico, they had difficulty getting airborne or gaining altitude. By late April, only two were still in flying condition. When Captain Foulois requested ten new airplanes, Pershing approved but Secretary of War Newton Baker refused on the grounds that "All airplanes available for service are already with the Pershing Expedition." Eventually, they were replaced by Curtiss N8 and R2 airplanes. No better than the Jennies, they were poorly made of shoddy materials with many missing parts and often broke down in flight. Curtiss, then the most prominent American airplane manufacturer, was overwhelmed by the sudden demand.[13] Although these planes had little impact on the course of the expedition, they did give American aviators their first experience of flying in wartime and demonstrated the weakness of American aviation compared to that of the European powers then engaged in a bloody war. As a result of the publicity the planes received, the U.S. Congress authorized $13 million for military aviation.[14]

During World War I, the warplanes of the great powers were fully occupied on the battlefields of Europe. Once the war ended, however, troubles on the borderlands of empires and the availability of surplus warplanes led to a sharp escalation in the use of air power in colonial wars.

One such place was the mountainous border area between British India and Afghanistan that has never come under the control of any government. In May 1919 some Afghan warriors invaded India's Northwest Frontier, starting what is known as the Third Afghan War. In two earlier Afghan wars, in 1839–42 and again in 1878–

81, British India had faced enormous difficulties and had very little success. Understandably, the viceroy of India, Baron Chelmsford, was reluctant to send ground troops into that mountainous country in a costly and probably futile effort to punish the Afghan government. Instead, small airplanes were sent to drop a few bombs on frontier towns like Jalalabad and to strafe enemy warriors.

Meanwhile, a large Handley Page 0/100 twin-engine bomber was dispatched to India to bomb the Afghan capital of Kabul. Originally designed to reach Germany, that plane seemed eminently suited to the task, since it could carry several heavy bombs. The day before it was scheduled to make its first bombing run, however, a storm blew up, ripped it off its moorings, flipped it over, and damaged it beyond repair. To replace it, another and even bigger plane, a Handley Page V/1500 four-engine bomber, was sent out to India. This huge aircraft had been designed to fly from Britain to Berlin and back, but only three were built, too late to take part in World War I. After a few months of flying around India taking bigwigs on joy rides, it was dispatched to the frontier. On May 4, 1919, it flew over the Khyber Pass and dropped four 112-pound bombs and sixteen 20-pound bombs on Kabul. Though the bombs did little damage, they persuaded the Afghan government to sue for peace.

Thus ended the Afghan challenge to British India, but not that of the Northwest Frontier tribes, who continued to fight until 1920. During that sporadic guerrilla uprising, the tribesmen learned to hide behind boulders from which their rifle fire downed several of the planes. Though the British reestablished a semblance of control over the province—which is all any government has ever been able to claim—the value of aircraft was ambiguous and the subject of a dispute between Indian Army and Royal Air Force officers. Air Commodore Webb-Bowen, who commanded the air operations, had to admit that "the RAF acting alone will never overcome a courageous people."[15]

More interesting is the case of British Somaliland because it involved only Britain and not the Indian government and because it influenced later, and much more important, cases of colonial air power. Somaliland (the northern half of today's Somalia) was a desert with little economic value, but it had been claimed by Britain

Figure 8.1. De Havilland DH-9A bomber flying over a desert, early 1920s.

because of its strategic location across from Aden on the approaches to the Red Sea. Back from the coast, the land was effectively controlled by Sayyid Mohammed Abdullah Hassan, known to the British as the "Mad Mullah," who periodically harassed British outposts. Once World War I was over, the British government turned its attention to this gadfly of the empire. The first idea was to send a thousand British soldiers to chase him down, a costly plan. In May 1919, Lord Milner, secretary of state for colonies, turned to Sir Hugh Trenchard, chief of the Air Staff. Trenchard replied: "Why not leave the whole thing to us? This is exactly the kind of operation which the R.A.F. can tackle on its own." At a conference chaired by Winston Churchill, then secretary of state for both war and air, it was agreed to send twelve de Havilland DH-9 bombers, each capable of carrying a 460-pound bomb, along with two hundred RAF personnel. In great secrecy, the planes were sent to Berbera on the Somaliland coast. In January 1920, six of them bombed the Mullah's fort and camp. In several days of bombing, they killed the Mullah's advisors, seven of his sons, and other relatives. Six sons, five wives, four daughters, and two sisters were captured by ground troops. Sayyid Mohammed himself escaped with one son, a brother, and a few followers to Ethiopia, where he died a year later. Including the mopping-up, the entire operation lasted three weeks and cost £83,000, a mere fraction of what a land-based expedition would have cost.[16] This outcome was a great inspiration for later air campaigns.

Somaliland was but one of a number of campaigns in which airplanes were used against Africans and Asians. On several occasions during the 1920s, the British used planes to subdue rebellious peoples in the Sudan and in the borderlands between Aden and Yemen, while the French did likewise in Syria and Lebanon and the Italians in Libya.[17] Nowhere, however, was air power embraced more enthusiastically and used more effectively than in Iraq.

Great Britain in Iraq

The British Empire reached its maximum extent in the aftermath of World War I. Yet its size belied its weakness. The British government faced huge deficits and was deeply in debt to American banks. Its armed forces had to be demobilized and their budgets slashed. Meanwhile, nationalists in Ireland, India, and Egypt were challenging Britain's authority while it was still deeply involved in Persia, Russia, Turkey, and the Caucasus. In short, the British Empire faced its greatest crisis in over a century.

Of all the new territories Britain had acquired in the war, Mesopotamia was the most troublesome. In an effort to secure the oil fields around Abadan in southern Persia, Britain had occupied Basra in late 1914 as part of the war against the Ottoman Empire. Two years later, British and Indian forces on gunboats pushed up the Tigris and Euphrates rivers, only to be defeated by the Ottomans.[18] They renewed their offensive in early 1917 and entered Baghdad in March of that year.

By the end of the war, British-led forces occupied all of Mesopotamia as well as Palestine and parts of Persia, but very superficially. In early 1919, when the Abu Salih tribe rebelled against the British presence, it was bombed, shelled, and occupied.[19] The war diary of

the 31st Wing of the Royal Flying Corps reported in April 1919:
"Bombing still continues to be carried out. No sooner has one area
been subdued than another breaks out into revolt and has to be dealt
with by aeroplane. . . . thus the Army has been saved from marching
many weary miles over bad country and sustaining casualties."[20] The
flying corps was overstretched, however, with only sixteen airplanes
in the region, only six of which were in flying condition at any one
time.[21] The brunt of the repression was carried by the army, with
twenty-five thousand British and eighty thousand Indian troops in
Mesopotamia in 1919, at a heavy cost to the Treasury. What Winston Churchill called "a score of mud villages, sandwiched between
a swampy river and a blistering desert, inhabited by a few hundred
half-naked families" cost as much to rule as an Indian province with
millions of inhabitants.[22]

In June 1920, Iraq (as the British now called Mesopotamia) rose
up in revolt. Led by former Ottoman officers, up to 131,000 tribesmen, half of them armed with military rifles left behind by the Ottoman army, challenged British authority. In response, Britain sent a
hundred thousand troops, most of them Indian, to the region, along
with eight squadrons of warplanes and four armored car companies.
It was the largest military operation that Britain engaged in between
the two world wars. It took six months and cost over a hundred
million pounds to crush the uprising.[23] This was more than the British government could afford. Yet, with the rise of Indian nationalism, it could no longer pass the cost of maintaining its empire on to
the taxpayers of India, as it had done so often in the past. The army
general staff discussed pulling out of Iraq, but the honor and reputation of the empire, not to mention the discovery of oil fields around
Mosul in the north of the country, made such a decision as unpalat-
able as staying.

Sir Hugh Trenchard and his civilian counterpart, Winston
Churchill, offered a way out of this dilemma. Already in December
1919, Trenchard had written:

> Recent events have shown the value of aircraft in dealing with fron-
> tier troubles and it is perhaps not too much to hope that before long
> it may prove possible to regard the Royal Air Force units not as an

addition to the military garrison but as a substitute for part of it. One great advantage of aircraft in the class of warfare approximating to police work is their power of acting at once. Aircraft can visit the scene of incipient unrest within a comparatively few hours of the receipt of news. To organize a military expedition takes time, and delay may result in the trouble spreading. The cost is also much greater and many more lives are involved.[24]

In a memorandum titled "On the Power of the Air Force and the Application of That Power to Hold and Police Mesopotamia" presented to the cabinet in March 1920, the Air Staff argued: "Great as was the development of air power in the war on the western front, it was mainly concerned with aerial action against enemy aircraft and co-operation with other arms in actions in which land and sea forces were the predominant partner. In more distant theaters, however, such as Palestine, Mesopotamia and East Africa the war has proved that the air has capabilities of its own."[25] Furthermore,

> The Air Staff are convinced that strong and continuous action of this nature must in time inevitably compel the submission of the most recalcitrant tribes without the use of punitive measures by ground troops. . . . With certain stubborn races time is essential to prove to them the futility of resistance to aerial attack by a people who possess no aircraft, but it is held that the dislocation of living conditions and the material destruction caused by heavy and persistent aerial action must infallibly achieve the desired result.[26]

Trenchard's plan called for six squadrons of fighters and bombers and two of transports, along with 228 armored cars and 3,405 RAF personnel, supported by two battalions of British and two of Indian infantry, at an estimated cost of little over four million pounds per year.[27] Behind the generous offer by Trenchard and the Air Staff to take over responsibilities for policing the empire was a self-serving motive. Both the British Army and the Royal Navy were eager to dismantle the upstart Royal Air Force and absorb its personnel and equipment, so the RAF needed a justification for its continued existence as a separate service.[28] As an Air Staff paper explained, "The efficacy of the Royal Air Force as an independent arm should be put to the proof by the transference to it of primary responsibility for the maintenance of order in some areas of the Middle East, preferably Mesopotamia."[29]

In response to the uprising in Iraq, Churchill, who had moved from the War Office to the Colonial Office, called a conference in Cairo in March 1921. In view of Britain's parlous financial situation, his main concern was reducing the cost of occupying Iraq.[30] To do so, he proposed to shift the political responsibility for that country to the Colonial Office and the military responsibility to the RAF, supported by the Iraqi Levies, an auxiliary infantry of Assyrian Christians and Kurds led by British officers, and a yet-to-be-created Iraqi army.[31] Churchill's vision of an Iraq controlled by air was simple and straightforward:

> Bagdad itself will be the main centre, and here will be assembled 1,800 high-class armed white personnel of the Royal Air Force. This force is quite capable of protecting itself in its cantonment against a rising or local disorder in the town or surrounding country. It is capable also of feeding itself indefinitely by air. Radiating from this centre, the aeroplanes will give support to the political officers and the local levies in the various districts, and will act against rebellious movements when necessary.[32]

Air Control in Action

With the approval of the British cabinet, the RAF officially took over the military responsibility for Iraq in October 1922. Air Vice Marshal Sir John Salmond commanded eight squadrons of fighters and light bombers, nine battalions of British and Indian infantry, three armored car companies, and several thousand Iraqi levies.[33]

Several kinds of airplanes were used in Iraq during the 1920s: Bristol Fighters, wood-and-cloth biplanes armed with two machine guns and 112 pounds of bombs and capable of speeds up to 125 miles per hour; de Havilland DH-9 bombers able to carry up to 450 pounds of bombs and fly at up to 114 miles per hour; and Vickers Vernons, twin-engine transports capable of carrying twelve passengers at up

Figure 8.2. Westland Wapiti fighter-bombers were used in Iraq after 1927.
Westland Wapiti © Simon Glancey, 2003.

to 118 miles per hour and the very similar Vickers Victorias used as
bombers. These were delicate machines not built for desert conditions, with engines that overheated, propellers that warped, and
tires that were easily punctured by thorns. Yet they were cheap and
easy to maintain and slow enough to land anywhere. After 1927
some of the older planes were replaced with Westland Wapitis,
fighter-bombers that carried two machine guns and up to 500 pounds
of bombs.[34]

There was some debate over the kinds of bombs that would be
most effective. In August 1920 Churchill wrote Trenchard sug-
gesting the development of gas bombs, "especially mustard gas,
which would inflict punishment upon recalcitrant natives without
inflicting grave injury upon them."[35] Churchill himself proposed to
use gas bombs "which are not destructive of human life but which
inflict various degrees of minor annoyance." In an interesting case
of double-think, the Air Staff found that gas bombs were "non-
lethal, but were not innocuous. They may have an injurious effect on
the eyes, and possibly cause death."[36] Churchill's adviser on Middle
Eastern affairs, Colonel Meinertzhagen, advised against the use of
mustard gas, that horrifying weapon that had killed and maimed so
many soldiers on the Western Front: "If the people against whom
we use it consider it a barbarous method of warfare, doubtless they
will retaliate with equally barbarous methods. . . . say what we may,
the gas is lethal. It might permanently damage eye-sight, and even

kill children and sickly persons." To this, Churchill replied: "I am ready to authorize the construction of such bombs at once. . . . In my view they are a scientific expedient for sparing life wh shd [*sic*] not be prevented by the prejudices of those who do not think clearly."[37] Two years later, the director of research of the RAF gave a name to the use of gas bombs: "I am of the opinion that for a given carrying weight of aeroplane high explosive bombs or shrapnel bombs are superior to any other form of frightfulness, except gas. If mustard gas can be accepted for this type of savage warfare, it should prove more efficient than any other known form of frightfulness."[38] Although the army had used gas shells in 1920 with "excellent moral effect," in the end, Colonial Office officials decided that poison gas bombs could not be used in a League of Nations Mandate, Winston Churchill notwithstanding.[39]

The use of air power, while driven by cost-benefit analysis and by Whitehall politics, elicited some elaborate justifications.[40] A British commander wrote in 1921: "[Sheikhs] . . . do not seem to resent . . . that women and children are accidentally killed by bombs."[41] Two years later, an RAF officer wrote: "A large percentage of the tribes fight for the mere pleasure of fighting. . . . We oppose the tribes with infantry, the arm that supplies them with the fight. Substitute aircraft and they are dealing with a weapon that they cannot counter."[42] In 1928, Air Vice Marshal T. Twidible Bowen wrote to Trenchard: "against ignorant and very superstitious savages it is desirable to achieve a great moral effect at the commencement of the operation by using aircraft in strength and continuously."[43] And in 1930 Trenchard told the House of Lords: "The natives of a lot of these tribes love fighting for fighting's sake. They have no objection to being killed."[44]

Given such attitudes, it is not surprising that the methods first used by the RAF were violent. As Wing Commander J. A. Chamier explained,

> the Air Force must, if called upon to administer punishment, do it with all its might and in the proper manner. . . . The attack with bombs and machine guns must be relentless and unremitting and carried on continuously by day and night, on houses, inhabitants,

crops and cattle. . . . This sounds brutal, I know, but it must be made brutal to start with. The threat alone in the future will prove efficacious if the lesson is once properly learnt.[45]

Some questioned the efficacy of this doctrine of "frightfulness." Secretary of War Laming Worthington-Evans, who had succeeded Churchill in the War Office, wrote, "The only means at the disposal of the Air Force, and the means in fact now used, are the bombing of women and children in the villages. . . . If the Arab population realize that the peaceful control of Mesopotamia ultimately depends on our intention of bombing women and children, I am very doubtful if we shall gain that acquiescence . . . to which the Secretary of State for Colonies looks forward."[46]

Likewise, Lieutenant General C. Deverell, chief of the Indian Air Staff, complained that airplanes were "aimed against the whole population, men, women, and children, with no distinction between combatants and non-combatants. . . . the weakest, the old men, women, and children, who are least able to endure it and who suffer the most."[47] In 1924, during the brief Labour Party interlude, Colonial Secretary James Thomas declared that bombing was unsportsmanlike: "within 45 minutes a full-sized village . . . can be practically wiped out and a third of its inhabitants killed or injured by four or five planes which offer them no real target and no opportunity for glory or avarice."[48]

The "humanitarian" argument was rather disingenuous, since ground forces used to crush rebellions were also known to produce casualties among non-combatants. Nonetheless, in response to these criticisms, the RAF began dropping leaflets on insurgent villages warning of an impending air raid. The idea, called "inverted blockade," was to persuade the inhabitants to flee before bombing their houses, livestock, and food and fuel supplies, thereby limiting the number of deaths. Salmond, the military commander in Iraq, explained that the new, non-lethal method of air control

can knock the roofs of huts about and prevent their repair, a considerable inconvenience in winter-time. It can seriously interfere with ploughing or harvesting—a vital matter; or burn up the stores of fuel laboriously piled up and garnered for the winter; by attack on livestock, which is the main form of capital and source of wealth to the less settled tribes, it can impose in effect a considerable fine, or seriously interfere with the actual food source of the tribe—and in the end the tribesman finds it is much the best to obey the Government.[49]

As the Air Ministry informed Parliament in 1924: "The compulsion exercised by the air arm rests more on the damage to morale and on the interruption to the normal life of the tribe than on actual casualties."[50]

Officially, Iraq was not a colony of Britain but a League of Nations Mandate. Behind this fiction was the reality of indirect rule, long practiced by Britain in the princely states of India and parts of Africa. Even after the suppression of the 1920 revolt, the situation remained precarious. At the Cairo Conference of 1921, the British decided to install Faysal I, son of Sharif Husayn of Mecca, as king of Iraq. Faysal was supported by Sunni dignitaries from the towns and by former officers of the Ottoman army who had joined the Arab Revolt during the war. Shiites and Kurds, who between them constituted three-quarters of the Iraqi population, were excluded from the council of ministers and other government offices.[51]

The imposition of a foreign ruler was not popular among Iraqis. In the spring of 1922 Shiite clergy protested against Faysal and the British presence. More serious was the disaffection among the Kurds who occupied the mountainous northern region of the country. At the time, Kurds were closely allied with the Turks across the border who claimed Kurdistan as part of Turkey. In the summer of 1921 the Kurds rose in revolt with Turkish support. By September, the Arab levies in the region were in retreat. A month later RAF bombing raids helped British troops defeat the uprising.[52] But British-Iraqi control of Kurdistan was never complete. Further revolts took place in 1924–25 and in 1930–32, each time requiring not only bombing raids but also the use of ground troops.[53]

Meanwhile, the Iraqi administration encountered new opposition from the inhabitants of the lower Euphrates, a region of swamps, irrigated lands, and canals that was very hard for ground troops and vehicles to penetrate. The people there refused to pay taxes or hand over their rifles. In this test of wills, the RAF was brought in to enforce the government's rule. Group Captain A. E. Borton, Air Officer Commanding in Iraq, reported: "The 8 machines engaged in the attack broke formation and attacked at different points of the encampment simultaneously, causing a stampede among the animals. The tribesmen and their families were put in confusion, many of whom ran into the lake making a good target for the machine guns."[54]

Even Churchill was shocked. In a telegram to Sir Percy Cox, the High Commissioner, he wrote: "Aerial action is a legitimate means of quelling disturbances or enforcing maintenance of order, but it should in no circumstances be employed in support of purely administrative measures such as collection of revenue." Cox responded that bombing was used against "a deliberate attempt to defy the national Government" and that it resulted in "a general improvement in behaviour and a noticeable decrease in highway robbery."[55]

A similar incident occurred in February 1923 when the authorities tried to collect taxes from an impoverished tribe in southern Iraq whose fields had been deprived of water by the diversion of a canal. In November, when the tribal leaders could not pay, the RAF was called in to bomb the area. Instead of producing revenue, the bombing caused an uprising. Two weeks of bombing killed 144 and wounded many others. More bombing in February 1924 killed a hundred men, women, and children, and much livestock. Finally, the tribal sheikhs agreed to pay their back taxes with money they borrowed at 60 percent interest per half year from moneylenders. But the people fled their villages rather than give up their rifles. An RAF intelligence officer admitted: "The primary cause of the recent outbreak was the growing irritation at demands for revenue which the tribes' poverty and fecklessness makes them unable to meet. That they have little or no money is reported form all sources, both official and unofficial."[56]

The use of bombing to enforce tax collection became a hot topic in the British press when several newspapers engaged in a "Quit Mesopotamia" campaign. The issue was also discussed in Parliament, where one newspaper reported that Lord Curzon, former viceroy of India and foreign secretary, "has interested himself in this question. I gather that Lord Curzon was not satisfied that there was any real difference between bombing for non-payment of taxes and bombing for non-appearance when summoned to explain non-payment of taxes." Despite the bad publicity, the RAF continued to punish Iraqi tax evaders under Conservative and Labour governments alike.[57]

In 1932 Britain's mandate expired and Iraq became an independent nation. What had British rule, especially air control, accomplished in fourteen years? From the British perspective, it had achieved its two main objectives. It had saved the RAF from its greedy rivals, the army and the navy. It had also kept Iraq and its oil in the British sphere at a much lower cost than any other method; as Churchill put it in 1923, "Our difficulties and our expenses have diminished with every month that has passed. Our influence has grown while our armies have departed."[58] British ground forces were reduced from thirty-two infantry battalions in 1921 to none in 1928; in consequence, the expense of maintaining the garrison dropped from £20 million in 1921–22 to just over £1.25 million in 1927–28. After that, the RAF and the Iraqi army alone maintained control over the country.[59] By a treaty signed in 1930, Britain kept two RAF bases in Iraq after independence. With them, Britain was able to overthrow a pro-Axis military regime in 1941 and later keep Iraq aligned with the West during the cold war. Not until 1958 did the RAF leave Iraq.

In exchange, British air control helped the Iraqi regime. In the early 1920s, it prevented a resurgent Turkish republic from annexing the oil-rich area around Mosul. It also saved King Faysal and the monarchy; as Leopold Amery, secretary of state for colonies, wrote in 1925: "If the writ of King Faisal runs effectively throughout his kingdom it is entirely due to British aeroplanes. If the aeroplanes were removed tomorrow the whole structure would inevitably fall to pieces."[60]

Air control was a cheap substitute for administration. It terrorized peasants into paying high taxes without providing any government services in return. It contrast to India, where British rule brought tangible benefits and earned the acquiescence of much of the population for over a century, British air control in Iraq consisted of pure violence committed by a foreign occupier in support of a foreign dynasty. Small wonder that when the British left in 1958, the royal family and its supporters were killed and Iraq came under the rule of ever more violent military regimes.

Air control had consequences far beyond the borders of Iraq. Confirming the results of the Somaliland experiment, it enshrined the idea that warplanes could give back to industrial countries the military advantage they had enjoyed in the nineteenth century and were in danger of losing as poorer societies acquired modern infantry weapons. As Air Minister Samuel Hoare wrote in 1925, it was "a striking testimony to the efficacy and economy of the air as the key force for exercising control in suitable areas of the Middle East, a subject that opens up a wide vista of future possibilities."[61] Like all technological advances, aviation gave its practitioners a permanent new power over nature. But its power over people proved to be ephemeral. The possibilities that Hoare saw before him extended but a short distance into the future.

Spain in the Rif

In 1912, when France established a protectorate over Morocco, it ceded the northern portion of that kingdom, called the Rif, to Spain. The Spanish claim to the territory did not translate into effective occupation, for the region was mountainous with many gorges and caves, and its inhabitants had a long history, going back to the Romans, of fighting one another and all intruders. Over the centuries, Spaniards had fought many campaigns in that territory, from which they had gained only two small coastal enclaves: Melilla, a Spanish town since 1497, and Ceuta, since 1580. After 1912 the French had some difficulty suppressing military opposition in their area, but the Spaniards faced a far more difficult task, not only because of the terrain and its inhabitants but also because of weaknesses in the Spanish army.

In a desultory campaign in 1913–14, Spanish troops pushed as far as Tetuan just south of Ceuta, Larache on the Atlantic coast, and Ksar-el-Kebir just inland from Larache. In support of this operation, they built three airstrips and bought a few small airplanes from French or Austrian manufacturers. During World War I, the

interior of the region remained quiescent under the control of Riffi
tribal leaders, in particular Muley Ahmed el Raisuni, whom the
Spaniards bribed with rifles, artillery, munitions, and supplies. In
September 1919, the Spanish government decided to undertake the
conquest of the Rif, which they expected would take two or three
years.[62] They greatly expanded their ground and naval forces, and
bought used planes from France and Britain, such as de Havilland
DH-4 fighter-bombers and DH-9 heavy bombers, Bristol F-2B fighters, Breguet-14 bombers, and Farman F-60 Goliath twin-engine
transports.[63]

In July 1921, twelve thousand Spanish troops under the command
of General Manuel Fernández Silvestre advanced south from Tetuan. At the village of Anual, they were surrounded and attacked
by eight to ten thousand Riffi warriors commanded by Mohammed
Abdel Krim, a tribal notable and businessman turned military leader.
The retreat of the Spaniards turned into a rout in which ten thousand Spanish troops, including Silvestre and his officers, were killed
and most of the rest were taken prisoner. According to official Spanish figures, the Riffi also captured 29,504 rifles, 392 machine guns,
and 129 artillery pieces.[64] After Anual, Spanish forces throughout the Rif retreated to their coastal enclaves. It was the worst
defeat suffered by European forces since the Italian disaster at Adua
in 1896.

To avenge the disaster, the Spanish dictator, Miguel Primo de
Rivera, prepared a new offensive for 1923–24. Spain sent 150,000
men to the Rif, but kept them in fortified enclaves. The Spanish
army was not trained to fight a colonial war but to prevent domestic
uprisings and provide employment for an inflated officers corps.[65]
Military service was very unpopular in Spain, and service in Morocco even more so. Recruits were poorly trained, poorly equipped,
and poorly paid. Most officers were incompetent. Instead of relying
on them for colonial campaigns, Spanish military leaders recruited
native troops; unlike the Gurkhas and other colonial soldiers employed by the British, however, these were local men of dubious
loyalty, many of whom deserted with their rifles.[66]

To supplement this weak infantry, the Spanish government
turned to air control like the British in Iraq. Besides their British-

made Bristols and de Havillands, they also bought French Breguets, Dutch Fokkers, and flying boats from Dornier in Germany and Savoia in Italy. They based three squadrons of twenty-four planes each at Melilla, two at Tetuan, and two at Larache.[67]

One difference distinguished the Spanish from the British version of air control: their use of poison gas. As early as August 1919, King Alfonso XIII had sent an envoy to Germany to acquire war materiel. In 1922 the Spanish army had purchased tear gas and chloropicrine gas shells from the French arms manufacturer Schneider.[68] A month after the disaster at Anual, the Spanish government initiated secret contacts with the Reichswehr. The German chemical manufacturer Hugo Stoltzenberg received a contract to manufacture chemical weapons in Spain. After a visit to Madrid, Stoltzenberg concluded that mustard gas would work well in the mountainous terrain of the Rif, as it would impregnate the crops and the sources of water.[69] By 1923 chemical bombs manufactured in Germany and in a factory near Madrid were being shipped to Melilla. Soon thereafter, Spanish planes began dropping mustard gas bombs on Riffi towns during market days. They also bombed livestock and burned crops with incendiary bombs, especially during the harvest season. Between May and September 1924, they dropped 24,104 bombs. In one twenty-five-day period in May 1925, they dropped 3,000 mustard-gas bombs, 8,000 150-kilogram TNT bombs, and 2,000 incendiary bombs on the town of Anjera alone.[70] The bombing caused many casualties and extreme suffering among the civilian population, but only increased the resistance of the Abdel Krim's soldiers, who avoided the towns during daylight.[71] Captain Ulrich Grauert and Lieutenant Hans Jeschonnek, two German officers sent to study the Spanish poison-gas campaign, were not impressed; in their opinion it was, "measured by German standards, completely unsystematically carried out."[72]

By mid-1925 it was clear that the bombing campaign was not bringing Abdel Krim to his knees. On the contrary, his followers had learned to keep mobile and to dig shelters and hide behind obstacles to escape the bombing. They had also acquired modern weapons in large quantities from international arms dealers as well as from Spanish soldiers they killed or captured. According to Spanish statistics, they had almost seventy thousand rifles, two hundred machine guns, and well over a hundred cannon, as well as the services of twenty German gunners and instructors and a Russian artillery colonel. One of their cannon, captured at Anual, shelled and damaged the Spanish cruiser Cataluña.[73]

Clearly outmatched, the Spanish government turned to France for help. In April 1925, Abdel Krim overplayed his hand by attacking French outposts along the border between the Rif and the French zone of Morocco. In July 1925, Primo de Rivera and Marshal Louis Lyautey, the French governor of Morocco, agreed to conduct joint operations against Abdel Krim.[74] Command of the French forces went to Marshal Henri Pétain, hero of World War I, who wrote: "The brutal fact is that we have been suddenly attacked by the most powerful and best-armed enemy that we have ever had to encounter in the course of our colonial operations."[75] In September, twelve thousand Spanish troops landed at Al Hoceima Bay, halfway between Melilla and Ceuta, supported by battleships, destroyers, and a hundred airplanes.[76] Meanwhile France had 160 airplanes in Morocco, including Breguet-14s capable of carrying up to 120 kilograms of bombs and larger Farman Goliaths that carried up to 480 kilograms of bombs. These planes, including a squadron of American volunteers, began an intensive bombing campaign against Riffi towns. They were also used to supply isolated garrisons, evacuate the wounded, and take aerial photographs.[77] By early 1926, Abdel Krim's twenty thousand soldiers were running out of ammunition and parts for their weapons. Against them were 325,000 French and 140,000 Spanish soldiers, along with eighteen squadrons of airplanes. After having waged the largest and most complex colonial resistance campaign to that time, Abdel Krim surrendered to the French in May 1926.[78]

Italy in Africa

If the once glorious Spanish Empire was reduced to fighting a desultory campaign for a small piece of Africa, Italy had far grander ambitions, especially after Benito Mussolini became its dictator and promised to revive the Roman Empire. At a less bombastic level, it was a matter of pride to many Italians to avenge the humiliating defeat of 1896 and to conquer a colonial empire that could at least compare with those of Belgium and the Netherlands, if not those of France and Great Britain. To achieve that end, any means were fair, even the most ruthless.

Like the Rif, Libya was relatively quiescent during World War I when the Italian army was fighting the Austrians in the north. In the 1920s, the Italians ensconced along the Libyan coast faced a rebellion by the Senussi of the desert. In campaigns in 1923–24 and in 1927–31, the Italian government sent tanks, artillery, armored cars, and almost a hundred planes to Libya. The planes bombed tribal encampments and livestock herds with phosgene and mustard gas, killing three-quarters of the nomads.[79] It was what military historians James Corum and Wray Johnson have called "one of the cruelest military campaigns in modern colonial history."[80]

Much worse was to come in Ethiopia, the scene of Italy's humiliation a generation earlier. To prepare for the invasion of that country, Mussolini sent 650,000 men and twenty million tons of materiel to Eritrea and Italian Somalia. The invasion began on October 3, 1935. After advancing rapidly to Adua, the Italians were stopped for two months by unexpected Ethiopian resistance. Like the Spaniards in the Rif, they were poorly trained and equipped for warfare in Africa.

Mussolini then appointed Marshal Pietro Badoglio to command the invasion. Badoglio advocated the use of air power to bomb the Ethiopians into submission. Before the invasion, he had written to Mussolini:

> Our advance to Adua must be preceded by the violent action of all our bombers on all the principal Ethiopian centers, from the border to Addis Ababa. Everything must be destroyed by explosive and incendiary bombs. Terror must be spread in the whole empire. I expect great results from this action, the only one in which the enemy, even if in these months he manages to acquire some planes, cannot oppose any appreciable resistance. I repeat: it is with aviation that we can crush the Ethiopian resistance.[81]

Once the war began, Mussolini's son Bruno, one of the pilots, wrote:

> We had set fire to the wooded hills, to the fields and little villages. . . .
> It was all most diverting. . . . the bombs hardly touched the earth
> before they burst out into white smoke and an enormous flame
> and the dry grasses began to burn. I thought of the animals: God,
> how they ran. . . . After the bomb racks were emptied I began throw-
> ing bombs by hand. . . . It was most amusing. . . . Surrounded by a
> circle of fire about five thousand Abyssinians came to a sticky end.
> It was like hell.[82]

Mussolini specifically directed his forces to use poison gas, in vio-
lation of the international convention of 1928 (the Kellogg-Briand
Pact) that Italy had signed. In late 1935 the Italian air force devised
a new method of delivering poison gas by spraying it from airplanes
over large areas. On March 29, 1936, Mussolini cabled Badoglio:
"Given the enemy's methods of fighting, I renew my authorization
to use gas of any sort and on any scale."[83] The gas used was "yprite"
or mustard gas, named after the infamous battle of Ypres in World
War I. So deadly was this gas to animals and plants as well as humans
that the Ethiopian army was forced to retreat. Appearing before the
League of Nations, the Ethiopian emperor Haile Selassie described
the Italian method of warfare.

> It is not against soldiers only that the Italian government has con-
> ducted this war. They have concentrated their attacks primarily on
> people living far from the battlefield, with the intention of terrorizing
> and exterminating them.
>
> Vaporizers for mustard gas were attached to their planes, so that
> they could disperse a fine, deadly poisonous gas over wide areas. From
> the end of January 1936, soldiers, women, children, cattle, rivers,
> lakes, and fields were drenched with this never ending rain of death.
> With the intention of destroying all living things, with the intention
> of thereby insuring the destruction of waterways and pastures, the
> Italian commanders had their airplanes circle ceaselessly back and
> forth. This was their foremost method of warfare.

This horrifying tactic was successful. Humans and animals were destroyed. All those touched by the rain of death fell, screaming in pain. All those who drank the poisoned water and ate the contaminated food succumbed to unbearable torture.[84]

By the spring of 1936, 450 planes based in Eritrea and Somalia, including Caprioni bombers and Fokker reconnaissance planes, were carrying out this campaign of aerial terror.[85] Even after the Italian army entered Addis Ababa on May 5, 1936, resistance continued in remote areas. The Italian reaction was characteristically violent; Mussolini cabled to Badoglio's second in command General Graziani: "I once again authorize Your Excellency to initiate and systematically conduct a policy of terror and extermination against the rebels and the populations that support them. Without the law of tenfold retaliation the plague will not be wiped out in a reasonable amount of time. I await confirmation."[86]

Conclusion

The interwar years marked the apogee of the new imperialism, as the victorious powers expanded their colonial empires at the expense of Germany and the Ottoman Empire and independent Ethiopia. Yet the price of empire had risen steeply. Not only were resistance movements, now inspired by modern ideas of nationalism, better organized, but the availability of modern infantry weapons left over from the Great War made it more difficult and costly to repress insurgencies. Faced with anti-colonial resistance movements, the colonialists turned to a new weapons system that only they possessed: warplanes, along with the machine guns, bombs, and poison gas that they carried. Throughout the 1920s and 1930s, aircraft kept alive the colonial dream of controlling vast empires at minimal cost.

As military aviation evolved, so did insurgents' ability to avoid air attacks or defend themselves against them. In Somaliland, the British took Sayyid Mohammed by surprise. In Iraq, villagers were easily intimidated because they had nowhere to hide, were disorganized, and did not form a coherent resistance movement. From the

Rif War onward, however, insurgent movements became organized for resistance and learned to avoid air strikes by moving at night and hiding in caves or forests during the day.

As the Rif, Libyan, and Ethiopian campaigns showed, it was still possible to defeat colonial resistance movements, but at a steeply increasing cost. Defeating Abdel Krim required hundreds of aircraft and half a million soldiers from two nations. Mussolini's conquest of Ethiopia matched the scale of Napoleon's march into Russia. Not only did the cost in money, manpower, and materiel rise sharply during the interwar period, so did the level of inhumanity. The British used the word "frightfulness" and debated using poison gas, but backed away for both moral and tactical reasons. The Spaniards used poison gas, but ineffectively. The Italians, who had been the first to use airplanes in war, were also the ones who perfected the use of gas to massacre large numbers of defenseless Ethiopians and poison their livestock and fields. European imperialism may have reached its apogee (or nadir) in 1936, but the days when it was easy and cheap were over.

Aviation was a rapidly evolving technology, driven by the enthusiasm of inventors, the boldness of entrepreneurs, the demands of the public for faster, more comfortable transportation, and, in the 1930s, the competitive nature of international politics. Planes grew bigger, stronger, and faster, and warplanes carried more lethal weapons. In the arms race between the colonial powers and the resistance movements they encountered, it seemed clear that the colonial forces were winning; this was the lesson of Iraq and Ethiopia. Yet, as in all arms races, this was but a temporary advantage. The demands of military aviation in an age of world wars created weapons suited to conflicts between equals, not aimed at repressing colonial insurgencies and resistance movements. As technological advance, defined as speed and firepower, became an obsession among military aviators and strategists, they took for granted that more advanced warplanes would translate into ever greater superiority over insurgents and resistance movements. As later events would show, such techno-hubris was a disastrous illusion.

Notes

1. Henri-Nicolas Frey, *L'aviation aux armées et aux colonies et autres questions militaires actuelles* (Paris: Berger-Levrault, 1911), pp. 85–86. It is interesting to note that Frey refers to "the so-called 'inferior' races" rather than "the inferior races," perhaps because their possession of rapid-firing rifles made him more respectful.

2. On the Wright brothers, see Tom Crouch, *The Bishop's Boys: A Life of Wilbur and Orville Wright* (New York: Norton, 1989), and Crouch, *First Flight: The Wright Brothers and the Invention of the Airplane* (Washington, D.C.: Department of the Interior, 2002), biographies emphasizing their experiments; and Peter L. Jakab, *Visions of a Flying Machine: The Wright Brothers and the Process of Invention* (Washington, D.C.: Smithsonian Institution Press, 1990), emphasizing the technological aspects.

3. John H. Morrow, Jr., *The Great War in the Air* (Washington, D.C.: Smithsonian Institution Press, 1993), pp. 35–47; Hilary St. George Saunders, *Per Ardua: The Rise of British Air Power, 1911–1939* (London: Oxford University Press, 1945), p. 29.

4. Robin Higham, *100 Years of Air Power and Aviation* (College Station: Texas A&M University Press, 2003), p. 37.

5. Quoted in David E. Omissi, *Air Power and Colonial Control: The Royal Air Force, 1919–1939* (Manchester: Manchester University Press, 1990), p. 5.

6. Omissi, *Air Power*, p. 6.

7. Angelo del Boca, *Italiani in Libia*, 2 vols. (Rome: Laterza, 1986–88), vol. 1, p. 108; Luigi Túccari, *I governi militari della Libia, 1911–1919*, 2 vols. (Rome: Stato Maggiore dell'Esercito, 1994), vol. 1, p. 72; John Wright, "Aeroplanes and Airships in Libya, 1911–1912," *Maghreb Review* 3, no. 10 (November–December 1978), pp. 20–21.

8. Wright, "Aeroplanes and Airships in Libya," p. 21.

9. Francis McCullagh, *Italy's War for the Desert, Being Some Experiences of a War Correspondent with the Italians in Tripoli* (London: Herbert and Daniel, 1912), pp. 122–23.

10. Giulio Douhet, *Il dominio dell'aria: Saggio sull'arte della guerra aerea* (Rome: Amministrazione della Guerra, 1921).

11. Roger G. Miller, *A Preliminary to War: The 1st Aero Squadron and the Mexican Punitive Expedition of 1916* (Washington, D.C.: Air Force History and Museums Program, 2003), pp. 1–2; Herbert M. Mason, *The Great Pursuit* (New York: Random House, 1970), pp. 104–8; John D. Eisenhower, *Intervention! The United States Involvement in the Mexican Revolution, 1913–1917* (New York: Norton, 1993), p. 239.

12. Miller, *Preliminary to War*, pp. 20–22; Mason, *The Great Pursuit*, pp. 109–10; Eisenhower, *Intervention*, pp. 255–56.

13. Quoted in Eisenhower, *Intervention*, p. 256. See also Mason, *The Great Pursuit*, pp. 108–18, 221–22, and Miller, *Preliminary to War*, pp. 10–11, 41–51.

14. James S. Corum and Wray R. Johnson, *Airpower in Small Wars: Fighting Insurgents and Terrorists* (Lawrence: University of Kansas Press, 2003), p. 20.

15. Omissi, *Air Power*, pp. 101–3.

16. David Killingray, "'A Swift Agent of Government': Air Power in British Colonial Africa, 1916–1939," *Journal of African History* 25, no. 4 (1984), pp. 429–35. See also Douglas Jardine, *The Mad Mullah of Somaliland* (London: H. Jenkins, 1923), pp. 263–78.

17. Anthony Towle, *Pilots and Rebels: The Use of Aircraft in Unconventional Warfare, 1918–1988* (London: Brassey's, 1989), pp. 27–34; Killingray, "'A Swift Agent of Government,'" pp. 430–39; Corum and Johnson, *Airpower in Small Wars*, pp. 77–83.

18. On the river war, see Bryan Perrett, *Gunboat! Small Ships at War* (London: Cassell, 2000), pp. 143–56.

19. Jafna L. Cox, "A Splendid Training Ground: The Importance of the Royal Air Force in Its Role in Iraq, 1919–32," *Journal of Imperial and Commonwealth History* 13 (1984), p. 157.

20. Peter Sluglett, *Britain in Iraq, 1914–1932* (London: Ithaca Press, 1976), p. 262.

21. Mark Jacobsen, "'Only by the Sword': British Counter-Insurgency in Iraq, 1920," *Small Wars and Insurgencies* 2, no. 2 (August 1991), p. 327.

22. Charles Townshend, *Britain's Civil Wars: Counterinsurgency in the Twentieth Century* (Boston: Faber and Faber, 1986), p. 94.

23. Charles Tripp, *A History of Iraq* (Cambridge: Cambridge University Press, 2000), pp. 41–44; Toby Dodge, *Inventing Iraq: The Failure of Nation-Building and a History Denied* (New York: Columbia University Press, 2003), p. 135; Robert Stacey, "Imperial Delusions: Cheap and Easy Peace in Mandatory Iraq," *World History Bulletin* 21, no. 1 (Spring 2005), pp. 27–28; Jacobsen, "'Only by the Sword,'" pp. 351–52; Sluglett, *Britain in Iraq*, p. 41; Omissi, *Air Power*, pp. 22–24; Cox, "A Splendid Training Ground," pp. 160–61; Townshend, *Britain's Civil Wars*, p. 95; Corum and Johnson, *Airpower in Small Wars*, pp. 55–56.

24. Cox, "A Splendid Training Ground," p. 159.

25. Priya Satia, "The Defense of Inhumanity: Air Control and the British Idea of Arabia," *American Historical Review* 111, no. 1 (February 2006), p. 26.

26. Townshend, *Britain's Civil Wars*, p. 146.

27. Stacey, "Imperial Delusions," p. 28; Omissi, *Air Power*, pp. 16–22; Cox, "A Splendid Training Ground," pp. 157–60.

28. On the politics of the Royal Air Force in this period, see Barry D. Power, *Strategy without Slide Rule: British Air Strategy, 1914–1939* (London: Croom Helm, 1976), chapter 6.

29. Omissi, *Air Power*, p. 25.

30. Christopher Catherwood, *Churchill's Folly: How Winston Churchill Created Modern Iraq* (New York: Carroll and Graf, 2004), pp. 133–34.

31. Dodge, *Inventing Iraq*, pp. 135–36; Sluglett, *Britain in Iraq*, pp. 259–63.

32. Martin Gilbert, *Winston S. Churchill*, 4 vols. (London: Heineman, 1975), vol. 4, p. 803.

33. Omissi, *Air Power*, p. 31; Satia, "The Defense of Inhumanity," p. 32.

34. Towle, *Pilots and Rebels*, pp. 55, 18–19, 244. On planes, pilots, and desert patrols in the early 1920s, see Lieut. Gen. John B. Glubb's memoir, *War in the Desert: An RAF Frontier Campaign* (London: Hoddet and Stoughton, 1969).

35. Gilbert, *Churchill*, vol. 4, p. 494.

36. Townshend, *Britain's Civil Wars*, pp. 147–48.

37. Gilbert, *Churchill*, vol. 4, p. 810.

38. Killingray, "'A Swift Agent of Government,'" p. 432n15.

39. Omissi, *Air Power*, p. 160; Townshend, *Britain's Civil Wars*, pp. 147–48.

40. Lieut. Col. David J. Dean, USAF, describes British air control doctrine in very different terms, as an integrated system of British political officers on the ground linked by radio with RAF intelligence officers who could give recalcitrant tribes clear warnings by airborne loudspeakers, followed by precision bombing to achieve political goals with minimal violence. This fine ideal was seldom applied in real situations. See David J. Dean, *The Air Force Role in Low Intensity Conflict* (Mobile, Ala.: Air University Press, 1986), pp. 24–25, and Dean, "Air Power in Small Wars: The British Air Control Experience," *Air University Review*, July–August 1983, http://www.airpower.maxwell.af.mil/airchronicles/aureview/1983/jul-aug/dean.html (accessed January 28, 2009).

41. Satia, "The Defense of Inhumanity," p. 38.

42. Corum and Johnson, *Airpower in Small Wars*, p. 82.

43. Killingray, "'A Swift Agent of Government,'" pp. 437–38.

44. Satia, "The Defense of Inhumanity," p. 37.

45. Corum and Johnson, *Airpower in Small Wars*, p. 58.

46. Satia, "The Defense of Inhumanity," p. 35; Townshend, *Britain's Civil Wars*, p. 97.

47. Thomas R. Mockaitis, *British Counter-Insurgency, 1919–1960* (London: Macmillan, 1990), p. 29.

48. Towle, *Pilots and Rebels*, p. 20.

49. Townshend, *Britain's Civil Wars*, pp. 148–49; see also Mockaitis, *British Counter-Insurgency*, pp. 29–30.

50. Sluglett, *Britain in Iraq*, p. 265.

51. Tripp, *History of Iraq*, pp. 31–33, 45–47.

52. Cox, "A Splendid Training Ground," pp. 168–69; Sluglett, *Britain in Iraq*, pp. 81–86, 117–22; Omissi, *Air Power*, p. 27–32; Dodge, *Inventing Iraq*, p. 137; Jacobsen, "'Only by the Sword,'" p. 358.

53. Cox, "A Splendid Training Ground," p. 170; Omissi, *Air Power*, pp. 30–35; Corum and Johnson, *Airpower in Small Wars*, p. 61.

54. Cox, "A Splendid Training Ground," p. 171.

55. Gilbert, *Churchill*, vol. 4, pp. 796–97; Sluglett, *Britain in Iraq*, p. 264.

56. Sluglett, *Britain in Iraq*, p. 267; Dodge, *Inventing Iraq*, pp. 153–55.

57. Sluglett, *Britain in Iraq*, p. 264; Cox, "A Splendid Training Ground," pp. 171–73.

58. Stacey, "Imperial Delusions," p. 29.

59. Brian Bond, *British Military Policy between the Two World Wars* (Oxford: Clarendon Press, 1980), p. 16; Cox, "A Splendid Training Ground," p. 175; Sluglett, *Britain in Iraq*, p. 127. After 1930 the Iraqi army, led by Sunni officers, many of whom had served in the Ottoman army, replaced the Iraqi Levies, who were too closely associated with the British.

60. Sluglett, *Britain in Iraq*, p. 91.

61. Omissi, *Air Power*, p. 35.

62. Sebastian Balfour, *Deadly Embrace: Morocco and the Road to the Spanish Civil War* (New York: Oxford University Press, 2002), p. 52–55.

63. José Warleta Carrillo, "Los comienzos bélicos de la aviación española," *Revue internationale d'histoire militaire* 56 (1984), pp. 239–62.

64. Shannon Fleming, *Primo de Rivera and Abd-el-Krim: The Struggle in Spanish Morocco, 1923–1927* (New York: Garland, 1991), pp. 65–70; David S. Woolman, *Rebels in the Rif: Abd el Krim and the Rif Rebellion* (Stanford: Stanford University Press, 1968), p. 82; Balfour, *Deadly Embrace*, pp. 64–75.

65. See Daniel R. Headrick, *Ejército y política en España (1866–1898)* (Madrid: Editorial Tecnos, 1981).

66. Balfour, *Deadly Embrace*, pp. 56–57.

67. Corum and Johnson, *Airpower in Small Wars*, 71–72; Balfour, *Deadly Embrace*, 149.

68. Rudibert Kunz and Rolf-Dieter Müller, *Giftgas gegen Abd el Krim: Deutschland, Spanien und der Gaskrieg in Spanish-Marokko, 1922–1927* (Freiburg-im-Breisgau: Verlag Rombach, 1990), pp. 58–59.

69. Ibid., pp. 74–90.

70. Balfour, *Deadly Embrace*, pp. 141–43; Fleming, *Primo de Rivera and Abd-el-Krim*, pp. 141–42.

71. Balfour, *Deadly Embrace*, pp. 124–56.

72. Kunz and Müller, *Giftgas gegen Abd el Krim*, p. 135.

73. Général A. Niessel, "Le rôle militaire de l'aviation au Maroc," *Revue de Paris* 33 (February 1926), p. 509; Fleming, *Primo de Rivera and Abd-el-Krim*, pp. 136, 224–26; Corum and Johnson, *Airpower in Small Wars*, pp. 69–70.

74. Fleming, *Primo de Rivera and Abd-el-Krim*, pp. 229–40.

75. Woolman, *Rebels in the Rif*, p. 194.

76. Ibid., pp. 190–92; Fleming, *Primo de Rivera and Abd-el-Krim*, pp. 285–99.

77. Walter B. Harris, *France, Spain and the Rif* (London: E. Arnold, 1927), pp. 300–301; Woolman, *Rebels in the Rif*, pp. 202–3; Corum and Johnson, *Airpower in Small Wars*, pp. 75–76; Niessel, "Le rôle militaire de l'aviation au Maroc," p. 523.

78. Woolman, *Rebels in the Rif*, pp. 196, 204–5; Corum and Johnson, *Airpower in Small Wars*, pp. 72–77.

79. Eric Salerno, *Genocidio in Libia: Le atrocitá nacoste dell'avventura coloniale italiana, 1911–1931* (Rome: Manifestolibri, 2005), p. 64; Balfour, *Deadly Embrace*, p. 128.

80. Corum and Johnson, *Airpower in Small Wars*, p. 81.

81. Giorgio Rochat, *Guerre italiane in Libia e in Etiopia: Studi militari, 1921–1939* (Treviso: Pagus, 1991), p. 124.

82. Sven Lindqvist, *A History of Bombing*, trans. Linda Haverty Rugg (New York: New Press, 2001), p. 70.

83. Angelo del Boca et al., *I gas di Mussolini: Il fascismo e la guerra d'Etiopia* (Rome: Riuniti, 1996), p. 152.

84. Lindqvist, *A History of Bombing*, p. 70; see also Rochat, *Guerre italiane*, pp. 143–76.

85. Dennis Mack Smith, *Mussolini's Roman Empire* (New York: Viking, 1976), pp. 67–73; Angelo del Boca, *The Ethiopian War 1935–1941*, trans. P. D. Cummins (Chicago: University of Chicago Press, 1969), pp. 54–78; Ferdinando Pedriali, "Le arme chimiche in Africa Orientale: Storia, tecnica, obiettivi, efficacia," in Angelo del Boca et al., *I gas di Mussolini: Il fascismo e la guerra d'Etiopia* (Rome: Riuniti, 1996), pp. 89–104; Rochat, *Guerre italiane*, pp. 124–29.

86. Del Boca, *I gas di Mussolini*, p. 162.

❊ Chapter 9 ❊

The Decline of Air Control, 1946–2007

World War II brought about a major leap forward in military aviation. By the end of the war, jet aircraft had made their appearance, as did fleets of huge bombers that could destroy entire cities. Progress in military aviation accelerated further after the war, fomented by the rivalry between the United States and the Soviet Union. Less than half a century after the Wright brothers' first flight, the great powers had aircraft capable of flying faster than the speed of sound and delivering bombs that could annihilate the entire planet.

At the same time, the war radically transformed the relations between the great powers and weaker, economically backward societies. From the immediate aftermath of World War II to the present, the world has witnessed a series of wars in which nationalist and insurrectionary movements have challenged the military might of even the most powerful nations. Hence the paradox: in the postwar world, increasing power over nature has coincided with diminishing power over peoples, yet the great powers continued to look to technology as a solution to their failings.

Until recently, historians of military aviation have emphasized the technological changes and the rivalries between the great powers. The aircraft that occupy center stage in these works are naturally the most advanced and powerful. Descriptions of aerial warfare concentrate on the major battles: the Battle of Britain, Pearl Harbor and Midway, and the bombing of Germany and Japan.[1] Meanwhile, classic works on "small wars" have generally ignored the role of aviation.[2] Only recently have historians, inspired by the American debacle in Vietnam, turned their attention to this important topic.[3]

The goal of this chapter is to investigate the coincidence of increasing air power with declining ability of powerful nations to

impose their will on weaker ones. Let us consider the most celebrated cases: France in Indochina and Algeria, the United States in
Vietnam, the Soviet Union in Afghanistan, and the United States
in Iraq.

France in Indochina

The Japanese army had occupied French Indochina during World
War II. When Japan surrendered on September 2, 1945, the Vietnamese nationalist Ho Chi Minh declared the independence of his
country.[4] A few weeks later, French troops arrived to reoccupy Vietnam, for France intended to keep Indochina in the French Union.
After protracted but futile negotiations, a full-scale war broke out
between the French and Ho Chi Minh's followers, the Vietminh.
Very quickly, the French took control of the cities, while the Vietminh held the countryside, especially at night.

The environment of Vietnam was hostile to both conventional
ground forces and air control. The Vietminh readily melted into
the rural population. Half the country was mountainous and covered with dense rain forests, and most of the rest was either wooded
or occupied by rice paddies. Torrential monsoon rains from
May through September made both flying and driving vehicles difficult. Vietnam was the ideal terrain for ambushes and other guerrilla
tactics.

While the Vietminh became a guerrilla army, the French retained
the tactics and materiel of a conventional European army, for the
greatest threat France faced at the time was a major war with the
Soviet Union, not a colonial campaign. The French government
could send only 150,000 men to Indochina, all of them volunteers
or foreign legionnaires, for it was politically impossible to send draft-
ees to reconquer a distant colony; hence it had to rely on air power.[5]
Having just recovered from World War II, however, France had
very few planes to spare. Its air force consisted of a motley collection
of aircraft inherited from other countries: British Spitfires and
Mosquitoes, C-47 Dakota transports (known in the United States
as DC-3), German Ju-52/3 three-engine bombers (some assembled

in France), and Japanese Aichi "Val" dive-bombers. Most were not designed for guerrilla warfare and all of them were worn-out and in need of constant maintenance in the hot, damp climate of Indochina.[6]

The United States, ambivalent about resurrecting the European colonial empires, refused to provide France with newer planes. In June 1950, however, the outbreak of the Korean War transformed the French efforts in Indochina from a colonial campaign into an anti-Communist crusade. The Unites States quickly supplied the French with Grumman F8F Bearcat and F6F Hellcat fighter-bombers, Bell P63 Kingcobra fighters, Douglas B26 Invader (or Marauder) bombers, and Fairchild C113 Packet (or Boxcar) transports. In 1952 the French obtained Sikorsky and Hiller helicopters, mainly to evacuate wounded soldiers. They also used French-built Morane 500 Crickets, small slow planes designed for reconnaissance.[7] From that point on, the French had enough planes but not enough personnel to use them effectively. Nonetheless, command of the air over Vietnam did help the French in many ways. Instead of sending truck convoys along ambush-prone roads, they used paratroopers and transport planes. Fighters provided close air support to ground forces engaged in battles. And villages thought to harbor Vietminh guerrillas were bombed with napalm.[8]

In response, the Vietminh devised ways of avoiding French aircraft. They dug underground tunnel complexes, they moved at night or under the cover of trees, they used netting and other forms of camouflage, and they engaged French troops at close range to avoid artillery barrages. In contrast to the French, the Vietminh were fighting for their homeland and were willing to take huge numbers of casualties. Not only did they become more sophisticated, they also acquired more and better equipment. When the Chinese Communists defeated the Nationalists in 1949, they were able to train and equip Vietminh troops in China, out of reach of French reprisals. In September 1950 the Vietminh overran the French outposts along the Chinese border, capturing 13 artillery pieces, 125 mortars, 940 machine guns, 1,200 submachine guns, 8,000 rifles, and 1,300 tons of ammunition, enough to equip a whole division.[9] The tide was turning against the French.

In November 1953 General Henri Navarre decided to build a base at Dien Bien Phu in the mountains of northwestern Vietnam to cut the supply route from China to the Vietminh. It was to be supplied entirely by air. Eleven thousand French troops were dug into positions, supported by heavy artillery. Seventy planes a day were expected to deliver 170 tons of ammunition and 32 tons of food to the garrison, taxing the French air force to the maximum.

Meanwhile, the Vietnamese commander, General Vo Nguyen Giap, had a hundred thousand porters disassemble and transport hundreds of heavy guns and mortars and tons of ammunition over jungle trails through the mountains surrounding Dien Bien Phu. When the siege began in March 1954, the Vietminh's artillery quickly shut down the French airfield and their anti-aircraft guns forced transport planes to drop their supplies from 8,500 feet up, thus allowing much to land in Vietminh hands. Drizzle and fog made flying difficult. Sixty-two planes were shot down, crashed on landing, or were destroyed on the ground, and another 167 were damaged. After a two-month siege, Giap's forces overran what was left of the French base, forcing France to sign an armistice and evacuate most of Indochina.[10] The air power that the French had counted on to defeat the Vietminh had failed.

France in Algeria

On the night of October 31, 1954, just a few months after the defeat at Dien Bien Phu, a revolutionary movement rose up against the French in the mountains of Algeria. It adopted the name of the French resistance against the Germans in World War II: Front de Libération Nationale (FLN). The French government at the time considered Algeria to be a part of France, not a colony like Indochina. A million Europeans of French, Italian, or Spanish origin lived there. Algeria also had vast oil and natural gas reserves. Stung by its defeat in Indochina and recognizing the importance of Algeria to the French economy, the government decided to crush the rebellion with all means available. By the mid-1950s, the means available to France, both financial and technical, were considerably greater than they had been in Indochina a few years before.

First came a huge buildup in French forces. When the uprising began, France had approximately 60,000 troops in Algeria. By late 1956 the number had risen to over 400,000, plus another 100,000 police and auxiliaries. With this many troops, the French army was able to blockade every road, station soldiers in every village, and relocate much of the rural population to towns and cities. They also used psychological warfare, kidnaping, and torture to obtain information about the rebels and their plans.[11]

The insurgents, at first, were few in number and poorly armed. The uprising began with fewer than three thousand men. Only half of them had World War II–vintage infantry weapons; the rest had old rifles and shotguns. Their tactics were those of guerrilla armies: ambushes, hit-and-run raids, reprisals against civilian collaborators. The situation changed dramatically in 1956, when Morocco and Tunisia became independent. At its height in 1957, the FLN had 15,000 regular soldiers in Algeria and 25,000 in Morocco and Tunisia, plus another 90,000 part-time auxiliaries. Through the porous borders between Algeria and its neighbors, the FLN began receiving great quantities of modern weapons from Arab countries and the Soviet bloc. Morocco provided funds to purchase weapons from international arms dealers. Tunisia and Egypt handed over weapons left behind by Rommel's Afrika Korps. From them and from Czechoslovakia, the FLN acquired machine guns, grenades, mortars, bazookas, land mines, and tons of ammunition.

To stop the flow of arms, the French reinforced the borders with electrified fences and mine fields. They also seized several Yugoslav ships headed for Morocco loaded with arms. FLN attempts to breach the border with Tunisia ended in failure. By early 1958, the FLN was rapidly losing the battle for the countryside and seemed on the verge of losing the war.[12]

Among the French tactics was the use of air power. Unlike Indochina with its forests, Algeria was well suited to airplanes, for it is a largely dry land. By the late 1950s, France had recovered economically and was able to manufacture many aircraft and purchase others from the United States, in particular over three hundred North

American T-6 Texans (also known as Harvards), slow and inexpensive trainers armed with bombs, rockets, machine guns, and napalm tanks. They also used twin-engine B-26 Marauder bombers, F6F Hellcats and F4U Corsair fighter-bombers, and a variety of other American-made planes. By November 1957, they had 686 airplanes operating in Algeria, a far larger air force than had ever been used in a colonial war. With these, they could bomb villages at will, for the FLN had no anti-aircraft weapons. In February 1958 they even bombed Sakiet, a village in Tunisia suspected of harboring insurgents, killing at least eighty civilians and injuring scores of others, including many children, and causing international condemnation.[13]

Even these slow airplanes were overdesigned for a guerrilla war, however, so the French began using helicopters. From a single helicopter in 1954 the number rose to eighty-two in 1957 and to four hundred in 1960, including the twin-rotor Vertol H-21 Shawnee (known as "the Flying Banana"), the Bell H-13 Sioux, the Sikorsky H-19 Chickasaw, and the French Sud Aviation Alouette. Helicopters were very effective for reconnaissance, troop transport, medical evacuation, and surprise raids on villages. Many of them were heavily armed with machine guns and rockets, thereby introducing a new weapon into the world's arsenal: the helicopter gunship.[14] With their command of the air, at the first sighting of insurgents, the French could bomb a village or encampment, then immediately land troops who advanced quickly under covering fire from gunships, then fly away as soon as the firefight was over. From 1958 to 1960, Air Force general Maurice Challe carried out the "Plan Challe" to blockade all mountain villages and systematically eliminate the guerrillas. Civilians were herded into concentration camps where they endured horrible conditions. Within a year, the FLN had lost a third of its weapons and most of its regional commanders. The results were what Challe called an "impressive military success."[15] By 1960, the Algerian countryside was largely pacified.[16]

Then the fighting moved to the cities, where French soldiers, especially the elite paratroopers, took over police duties, raiding houses and torturing suspects. The FLN responded by exploding bombs in public places and employing other terrorist tactics. The violence and brutality of the repression—reminiscent of the Nazi occupation of France—turned even Algerians who had been pro-French against the French, and made the war so unpopular in France that it led to the fall of the Third Republic and brought a serious risk of civil war to France itself. In 1962, after part of the French army had mutinied in support of the die-hard European settlers, President Charles de Gaulle agreed to the independence of Algeria. Military victory had brought political defeat in its wake.

Much had changed since the days when the Royal Air Force could control Iraq with a few little biplanes or Italy could poison Africans at will. In Indochina and Algeria, as in the Rif, insurgents learned to move at night and hide during the day and in the shelter of cities, leaving only civilian non-combatants exposed to air strikes. Instead of terrifying people into submission, air strikes angered the population and provided fresh recruits to the insurgents, allowing them to claim the mantle of defenders of the people against the brutal foreigners.

As for the lessons so painfully learned by the French in Indochina and Algeria, they were entirely lost on the superpowers of the day. To the Americans as to the Soviets, France was a minor power that had caved in to Hitler's Wehrmacht and that had lost war after colonial war. What was there to be learned from such a sorry performance? While some writers commented on the reasons for the French defeats, few thought to inquire into the reasons for the victories of the Vietminh and the FLN.

The United States in Vietnam

Few wars in history have been as minutely dissected and argued over as the Vietnam War. How could the richest, most powerful nation on earth be defeated by what Henry Kissinger contemptuously called a "third-class Communist peasant state" and Lyndon Johnson described as a "damn little pissant country"?[17] How could it be that the world's most advanced military technology was ineffective against a far smaller, poorer, and weaker people? Why did the awesome American power over nature—the ability to dominate the sky, to drop thousands of tons of bombs anywhere at will, to defoliate thousands of acres of forests, to kill hundreds of thousands of people—not translate into power over such people as the Viet Cong and North Vietnamese?

The Geneva Accord that ended the French Indochina War was but a cease-fire. The North Vietnamese government never ceased to plan for the reunification, by conquest if necessary, of the two halves of Vietnam. Viet Cong guerrillas continued to operate inside South Vietnam. The South Vietnamese government, wracked by one military coup after another, never meant to hold a free election, which would have demonstrated its unpopularity. Corruption and other inequities grew so bad that even staunch anti-Communists like the Buddhist and Catholic clergy turned against the government. Yet North Vietnam and the Viet Cong were Communist and, in the cold war atmosphere of the time, any Communist anywhere was considered an enemy of the United States, and every anti-Communist regime was an ally or protégé. The United States got involved in the war for ideological reasons, not out of any interest in Vietnam per se.

At first, American involvement was limited to providing the South Vietnamese government with propeller-driven North American T-28 Trojan trainers, Douglas A-1 Skyraider fighter-bombers, and Sikorsky H-34 helicopters, as well as infantry and artillery weapons. To help South Vietnam, President John F. Kennedy had the U.S. military provide the South Vietnamese army with instructors and technicians, rising in numbers from about a thousand at the end of 1961 to almost sixteen thousand two years later.[18]

In August 1964, two American destroyers reported that they had been attacked by North Vietnamese torpedo boats in the Gulf of Tonkin. In the heat of the crisis, President Lyndon Johnson obtained from Congress a resolution that authorized the United States to enter the war in defense of South Vietnam. Eight months later, the United States began sending combat troops into Vietnam. The number of American troops quickly rose to 463,000 in 1967, reaching a peak of 541,000 in 1969. A minority of these were combat troops; most were support troops who provided the supplies, guarded the bases, and maintained the equipment necessary in a high-tech war.

Figure 9.1. Bell UH-1D transport helicopter disgorging soldiers in Vietnam.
Cody Images/US Army.

Though ground troops were numerous, much of the American effort took place in the air. The United States and South Vietnam had thousands of airplanes, more than had ever before been assembled in one theater of war. By the end of 1965, four hundred planes operated out of Tan Son Nhut airbase alone, making it one of the busiest airports in the world. The variety of planes was as tremendous as their numbers. Among the more prominent were the jet fighter-bombers: the F-4 Phantom, the A-4 Skyhawk, the F-101 Voodoo, and the carrier-based A-6 Intruder. The C-130 Hercules and the venerable C-47 (the military version of the DC-3) served as transports and gunships. The C-123 Provider was used to spray defoliants. Then there were the helicopters: the Bell UH-1 Huey transport helicopter and its sister the AH-1 Cobra helicopter gunship, and the Sikorsky CH-54 Skycrane heavy transport, among others.[19]

As in Algeria, helicopters and slower gunships were very effective in protecting convoys from ambushes and in transporting and supporting ground troops on search-and-destroy missions. They also saved the lives of wounded soldiers by evacuating them quickly to distant hospitals. For the same reasons, they were also very vulnerable to ground fire; by 1971, the United States had lost 4,200 helicopters to North Vietnamese and Viet Cong anti-aircraft fire.[20] In addition to conventional explosive bombs such as had been used since World War I, the United States adopted several more modern weapons: white phosphorus that destroyed flesh, pineapple bombs that released thousands of metal spikes, and napalm, a gel that burned and could not be extinguished.

One of the reasons the French had lost their war was because the environment of Indochina was so hostile to conventional forces. There were few roads or open ground for tanks and trucks. Wetlands, rice paddies, and steep mountains impeded the movement of soldiers. And the forest cover made it almost impossible for aircraft to see the enemy. Faced with the same problems that had stymied the French, Americans decided to change the environment. Since the thick forests of Vietnam made ambush easy, giant bulldozers flattened trees alongside important roads. To open up the landscape to air reconnaissance, the U.S. armed forces sprayed the defoliant Agent Orange to strip trees of their leaves. They also sprayed the herbicides Agent White and Agent Blue to destroy crops growing in Viet Cong–held areas. Between 1962 and 1968, 19,000 flights had sprayed six million acres, destroying 35 percent of Vietnam's hardwood forests and half of its mangrove forests. Although the armed forces claimed that these chemicals, unlike those used by Spain and Italy between the wars, were harmless to humans, nonetheless they caused stillbirths and birth defects, while the innumerable craters left by explosives became breeding grounds for malarial mosquitoes.[21]

Environmental warfare affected the civilian population of Vietnam far more than the Viet Cong or North Vietnamese military. While the South Vietnamese farmers suffered terribly, the Viet Cong devised ways to maintain and even increase their strength in the face of American firepower. They avoided areas that had been defoliated and blended in among the civilian population. Like guerrillas everywhere, they moved stealthily at night and on foot, to the consternation of an American officer who remarked: "Mobility means vehicles and aircraft. . . . The Vietcong have no vehicles and no airplanes. How can they be mobile?"[22] Instead of traveling in vehicles, they dug thousands of miles of tunnels, with underground storage, hospitals, and kitchens. They had much better intelligence than the South Vietnamese or American military, for they lived among the local peasants who accepted them out of fear or genuine support.[23]

As long as the fighting took place in the forests and mountains of South Vietnam, the war bred rivalry and resentment between the U.S. Air Force and the U.S. Army. The army wanted its own planes and helicopters to provide close air support for its ground troops, while the air force considered all aircraft to be in its jurisdiction. Besides, it preferred to use the latest jets, had little interest in counterinsurgency tactics, and did not like being placed in the position of responding to the army's demands. Without a clear strategy from above, the two services jockeyed for control of the war. The U.S. Air Force director of plans put it bluntly in 1962: "It may be improper to say we are at war with the Army. However, we believe that if the Army efforts are successful, they may have a long term adverse effect in the U.S. military posture that could be more important than the battle presently being waged with the Viet Cong."[24]

Unable to defeat the Viet Cong in the field, President Johnson and his military advisors turned to what they thought was America's strength: bombing. In July 1965, Johnson told his advisers: "We can bring the enemy to his knees by using our Strategic Air Command and other air forces—blowing him out of the water tonight. [But] I don't think our citizens would want us to do it."[25]

Despite the Bombing Survey that showed that bombing had had much less effect on Germany in World War II than air strategists had expected, air force and navy pilots still believed in the doctrine of strategic air offensive formulated by Giulio Douhet, Hugh Trenchard, and "Billy" Mitchell before the war. Convinced that wars were won by overwhelming air superiority, they were uncomfortable with the idea of fighting an unconventional guerrilla war. The compromise between the generals' and admirals' insistence on massive retaliation and what Johnson believed "our citizens would want us to do" was an operation called Rolling Thunder. Since the Viet Cong operating in South Vietnam seemed immune to air attack, the United States targeted North Vietnam, hoping to persuade that country to stop aiding the Viet Cong and enter into peace negotiations. Beginning in March 1965, bombing raids began in the south of that country and gradually moved north. By 1967 American planes were bombing North Vietnamese bridges, transportation routes, petroleum facilities, and industrial plants near Hanoi and Haiphong and up to the Chinese border.[26] Yet for fear of bringing on direct Chinese or Soviet intervention, Johnson forbade the bombing of the capital, Hanoi, or the harbor of Haiphong filled with Soviet ships, and of the Red River dikes and the North Vietnamese supply routes and training camps in southern China.[27]

Most of the fighter-bombers used in Operation Rolling Thunder were heavy, slow-turning F-105 Thunderchiefs; others included Phantoms, Intruders, Skyraiders, and Skyhawks. In addition, gigantic B-52 Stratofortress bombers coming from bases in Guam, Okinawa, and Thailand carried out up to sixty sorties a day. In the three years and nine months that Operation Rolling Thunder lasted, American planes flew 304,000 tactical sorties and 2,380 B-52 sorties over North Vietnam and dropped 643,000 tons of bombs. Including those dropped on South Vietnam, the United States dropped three times more bombs on Vietnam than it had in all of World War II.[28]

The idea of measuring a military campaign by the number of planes, sorties, and bombs was itself an innovation of the Vietnam War. It resulted not from looking at the results on the ground but from the intense rivalry between the air force and the navy. As General David Shoup, former commander of the Marine Corps, explained in 1969:

> So by early 1965 the Navy carrier people and the Air Force initiated a contest of comparative strikes, sorties, tonnages dropped, "Killed by Air" claims, and target grabbing which continued up to the 1968 bombing pause. Much of the reporting on air action has consisted of misleading data or propaganda to serve Air Force and Navy purposes. In fact, it became increasingly apparent that the U.S. bombing effort in both North and South Vietnam has been one of the most wasteful

and expensive hoaxes ever to be put over on the American people. . . . Air power use in general has been to a large degree been a
contest for the operations planners, "fine experience" for young pilots, and opportunity for career advancement.[29]

For all its impressive statistics, Operation Rolling Thunder failed
to persuade the North Vietnamese to enter into negotiations. The
destruction of power plants made little difference in a country where
very few people had electricity. The North Vietnamese replaced
their few oil storage depots with drums and dispersed tanks. Damaged bridges, railroad tracks, and roads were quickly repaired by a
half-million workers, many of them Chinese. Nor did bombing stop
the flow of materiel from China and the USSR to North Vietnam
or from North Vietnam to the Viet Cong, for both North Vietnam
and the Viet Cong were immune to the restrained bombing of the
sort practiced by the Johnson administration.[30] To prove it, in January 1968 the Viet Cong launched the Tet Offensive, showing their
strength throughout South Vietnam, even in the streets of Saigon,
and briefly capturing the ancient citadel of Hue. Though the American forces quickly recovered their previous positions, the political
damage was done. American and South Vietnamese troops were demoralized and the war divided and angered the American people.
On November 1, 1968, Johnson, in disgrace, ordered an end to the
bombing of North Vietnam. A few days later, Richard Nixon, the
candidate who promised that he knew how to end the war, was
elected president.

Nixon's plan was to win the war from the air rather than on the
ground. The first goal was to interdict the flow of men and materiel
from North to South Vietnam. The Viet Cong relied mainly on
locally obtained food and supplies, but were thought to receive fifteen to thirty-four tons of supplies a day, or 4 to 8 percent of their
needs, from the north. Since the Demilitarized Zone (DMZ) between the two countries was heavily fortified, most of the supplies
reaching the Viet Cong came down a series of trails through Laos
and Cambodia known as the Ho Chi Minh Trail. When the North
Vietnamese army took over most of the fighting after the Tet Offensive, the flow increased to seventy-five tons a day.[31]

The plan to stop this flow of men and materiel, called Operation Igloo White, was first developed by Secretary of Defense Robert McNamara. Thousands of sensors shaped to look like twigs, plants, gravel, or animal droppings were dropped all along the trail to sense the sound of truck engines, movements, body heat, or the smell of urine. Each contained a tiny transmitter that sent the information to a plane overhead. It in turn relayed it to the Infiltration Surveillance Center in Thailand. There, two IBM 360-65 computers analyzed the information and displayed the location of North Vietnamese truck convoys on computer monitors. Within two to five minutes of the sensing, B-52 bombers or Phantom fighter-bombers flying nearby were guided to the location of the convoy and their bombs were released, all by computer.

Igloo White was very expensive, costing around one billion dollars a year between the end of 1969 and the end of 1972. In return for that money, the air force claimed to have knocked out a total of 35,000 trucks in four years, a tremendous success. Yet air reconnaissance was able to spot very few damaged trucks, for the North Vietnamese quickly learned how to fool the sensors with tape recordings of truck noises and other decoys that attracted the bombers to empty sections of the trail. Though the air force claimed to have reduced the flow of supplies by 80 percent, the bombardment did not prevent the North Vietnamese army from increasing its strength in the south. What Igloo White did accomplish was the loss of three to four hundred airplanes and the flight or deaths of tens of thousands of Laotians and Cambodians living near the Ho Chi Minh Trail.[32]

In February 1972 Nixon had visited China in what he called "a journey for peace." Having driven a wedge between that country and North Vietnam, he felt confident that he could resume bombing North Vietnam. When the North Vietnamese felt emboldened to launch a tank and artillery offensive, the United States retaliated with a new bombing campaign. Operation Linebacker ran from May 10 to October 23, 1972, and targeted the harbor of Haiphong as well as industrial installations there and in Hanoi. This time, American planes carried laser-guided "smart" bombs that hit their targets more accurately than before.[33] They also carried long-range navigation equipment that allowed them to operate at night and in bad weather.[34] The North Vietnamese responded with conventional air defenses. They had acquired 204 Soviet MiG fighters, of which 93 were the latest model MiG-21s, and had also built hundreds of anti-aircraft sites armed with Soviet surface-to-air missiles. Though they seldom engaged in air combats, they used their fighters to draw the American fighter-bombers away from their intended targets and force them to jettison their bombs in order to engage in aerial dogfights.[35]

As negotiations between the United States and North Vietnam seemed deadlocked, the United States unleashed the heaviest bombing campaign of the war, called Operation Linebacker II. The goal was to destroy North Vietnam's will to fight, thereby saving South Vietnam from defeat while allowing the United States to withdraw its forces. From December 18 to 29, 1973, in what became known as the "Christmas bombings," American planes repeatedly targeted Hanoi and Haiphong. In January 1973 North Vietnam signed a peace accord by which it agreed to cease its offensives and allow the United States to withdraw its troops from South Vietnam in exchange for an end to the bombing. It waited patiently for two years while the South Vietnamese government continued to disintegrate. Then it unleashed a final offensive and conquered the south.

Did the peace accord truly bring "peace with honor," as President Nixon proclaimed, or was it a defeat disguised as a diplomatic maneuver? Nixon reached a short-term goal—to extricate the United States from a war it could not win—by abandoning a long-term goal—to save South Vietnam from a Communist takeover. The United States lost the war largely because of the corruption of the South Vietnamese government and the determination of the North Vietnamese and most of the people of the south to rid their land of foreigners after a century of occupation by the French, the Japanese, and the Americans. The endless fighting sapped the will of the American people to endure more deaths for a cause that few still believed in.

But the United States also lost because its armed forces were wedded to a military culture that was not suited to a counterinsurgency war. That culture was founded on the belief that wars are won by machines, and the more advanced and sophisticated the machines, the swifter the victory. This was especially true of the two high-technology services, the air force and the navy. In their eagerness to roll out impressive statistics—so many thousands of planes, so many hundreds of thousands of sorties, so many millions of tons of bombs—they demonstrated what one historian has called "a modern vision of air power that focuses on the lethality of its weaponry rather than on that weaponry's effectiveness as a political instrument."[36] That the weapons were lethal, there is no doubt; but their lethality was effective against the environment and the civilian population, not the North Vietnamese government or army. Looking back in his later years, Richard Nixon explained the American failure in these words: "Our armed forces were experts at mobilizing huge resources, orchestrating logistic support and deploying enormous firepower. In Vietnam these skills led them to fight the war their way, rather than developing new skills required to defeat the new kind of enemy they faced. They made the mistake of fighting an unconventional war with conventional tactics."[37]

In the end, the United States lost the war because it could not change the way it believed wars should be fought and because, for domestic political reasons, it could not send to war the millions of soldiers (or commit the mass atrocities) that defeating such a determined foe would have required.

The Soviet Union in Afghanistan

Afghanistan, as many invaders have learned to their dismay, is a most difficult country to conquer. Much of the land is rugged and mountainous, with bitterly cold winters. Inhabited valleys are narrow and mountain passes are blocked by snow much of the year. The Afghan tribes have often been at war with one another and most men have guns and experience fighting or hunting.

In December 1979 Soviet transport planes began landing troops in Kabul, ostensibly to show support for the Communist government of Afghanistan but actually as part of a well-planned campaign to conquer that country. Soon thereafter, seven motorized rifle divisions and an airborne division crossed the border and quickly seized the cities and the main roads. Within two years, the Soviet army in Afghanistan had grown to nearly 110,000 troops, supported by another 30,000 to 40,000 reserves in the neighboring Soviet republics of Turkmenistan, Uzbekistan, and Tajikistan.[38]

The Red Army waited until March 1980, when much of the snow had melted, to launch a major ground offensive. By then, many Afghan men, fired by their ardent belief in Islam and by their hatred of foreign invaders, had taken to the hills and prepared to fight back. If Leonid Brezhnev and the Soviet military leaders had expected the Afghan army to do most of the ground fighting, they were mistaken. So many Afghan soldiers deserted taking their weapons with them that the Soviet troops had to disarm the others and undertake the war themselves.[39] This proved a challenge, for the Red Army had not fought a war since 1945 and most of its troops were poorly trained.

At first, the Soviets tried the tactics that had served them so well in the Great Patriotic War against Hitler's Germany. They carefully planned offensives in advance, insisted on strict discipline and centralized control of all operations, and deployed scores of heavy tanks and armored personnel carriers. In the narrow valleys between towering mountains, such troops movements were vulnerable to ambushes, land mines, and anti-tank rockets. North Vietnamese military were brought in to advise the Soviets on counterinsurgency tactics, but the Soviet commander General Sokolov rejected their advice, preferring the traditional Soviet method of emptying rebel-held territories of civilians by killing them or forcing them to flee.[40]

The Afghan resistance fighters, called mujahideen, never formed a unified force like the Viet Cong and the North Vietnamese army. Instead, many groups, often tribal, formed armed bands. They built training camps in Pakistan, where up to two million Afghans had fled, and received help from Saudi Arabia and other anti-Communist Muslim countries, funneled through Pakistan. They used the same guerrilla tactics as the mountaineers of the Caucasus and of the Rif and Algeria. At first they were poorly armed with old military rifles and AK-47 semi-automatic rifles purchased in the bazaars of Pakistan. By 1982 they had obtained grenade launchers

Figure 9.2. An Afghan mujahideen demonstrates a handheld
surface-to-air missile, 1988.

and anti-tank mines. They traveled with caravans of mules, camels,
or trucks, mainly at night but sometimes in full daylight, even near
Soviet bases.[41]

To counter the threat of mujahideen attacks on their bases and
ground forces, the Soviets brought in massive air power. They built
a dozen air bases, the largest of which was Bagram in the north. To
avoid dangerous roads, they ferried troops and supplies in An-12 and
An-26 turboprop transports. MiG-21s and MiG-23s jet fighters, Su-
24 fighter-bombers, and Il-28 medium bombers were used in bombing raids. But the most important aircraft were helicopters. By
1981 the Soviet forces had five to seven hundred helicopters: Mi-6
and Mi-8 troop transports and especially the much-feared Mi-24,
known in the West as "Hind" and to the Soviets as "flying tank" or
"crocodile." This helicopter gunship was armor-plated and sported
cannon, rockets, and machine guns, and could carry up to eight fully
armed soldiers.[42]

By 1982 the Soviets had learned to use their air power more effectively. Rather than requiring headquarters' approval for every action,
they experimented with coordinating air and ground operations and
letting local commanders make decisions. Helicopters were very effective in preventing ambushes along roads. They landed troops on

mountain ridges overlooking valleys where ground troops operated. They transported commandos called "Spetsnaz" to ambush mujahideen convoys. They carpet-bombed mujahideen-held areas to depopulate them and dropped thousands of anti-personnel mines on fields and trails to blow up underfoot.[43] The American Central Intelligence Agency accused the Soviets of using a gas called "Yellow Rain" that caused itching and burning, but the humanitarian agency Doctors Without Borders operating in mujahideen areas could not corroborate this allegation.[44]

Yet in Afghanistan as in Vietnam, domination of the air and massive firepower failed to bring victory. Helicopters had half the carrying capacity at high altitudes that they had at sea level. They were adversely affected by the heat, dust storms, and high winds that were frequent in Afghanistan. Maintenance was difficult and often poorly done. Helicopters and slow lumbering transports were vulnerable to ground fire, especially on takeoff and landing. At first the mujahideen used heavy machine guns. Soon, however, they acquired Soviet-made SAM-7 surface-to-air missiles, allegedly from Egypt or China, but more likely from the Palestine Liberation Organization when the Israelis forced it to leave Beirut. These missiles forced planes and helicopters to fly at higher altitudes, where their bombs and rockets were less accurate. By 1986, the Soviets had lost some five hundred aircraft.

After 1986, Great Britain began supplying the mujahideen with hundreds of Blowpipe missiles. Then the United States—believing that any enemy of its enemy was its friend—supplied 2,700 Stinger heat-seeking anti-aircraft missiles to Pakistan, some (but not all) of which reached the mujahideen. Weighing only thirty pounds, Stingers were easily transported and could bring down a plane up to 15,700 feet away. At first the United States claimed that these missiles shot down one plane a day. Once the Red Air Force had learned to avoid them by flying high and launching flares, the number of planes lost dropped dramatically, but so did the accuracy of their bombing.[45]

The Stinger missiles did not in and of themselves defeat the Soviet Union, but they helped prolong a war between highly motivated mujahideen and heavily armed but poorly led and motivated Soviet troops. Although technically the Soviets retained command of the air, the war and the loss of so many aircraft caused a drain on the Soviet economy and undermined the Soviet political system at a time when it was beset by many other political and economic woes. In April 1988 Mikhail Gorbachev, leader of the Communist Party, ordered the withdrawal of Soviet forces from Afghanistan. By February 1989, the withdrawal was complete and the USSR began to fall apart. Once again, determined resistance fighters armed with portable infantry weapons defeated a great power equipped with heavy weaponry designed for a very different war.

U.S. Military Aviation after Vietnam

From the debacle of Vietnam, the United States armed forces drew a number of important lessons. One was that the number of American soldiers killed—some fifty thousand—was higher than the American people could accept; in future wars, the armed forces would have to use more machines and fewer combat troops. The air force and navy drew more specific lessons. In Vietnam, six million tons of bombs— three times as many as the United States had dropped in World War II—did not cow the Vietcong or North Vietnamese army into submission because so many bombs fell on inconsequential targets or missed altogether. In the process, the United States lost far too many planes and pilots. After 1975, the armed forces refocused their thinking on the confrontation in Europe with the Soviet Union, in particular the need to stop a potential Soviet invasion of Western Europe without setting off a global nuclear war. The means to achieve this objective was first to destroy the enemy's command and control systems and air defenses, then to damage the enemy's ground forces.[46]

What made such an ambitious plan seem feasible was the technological and economic circumstances in which the armed forces found themselves after the Vietnam War. Aided by government subsidies, the electronics and computer industries were in the midst of a frenetic development, leading the American economy and that of its allies into a "third industrial revolution."[47] The giant aerospace

defense contractors—McDonnell-Douglas, Lockheed, Boeing, Nor-
throp, and others—were at the forefront of this revolution. The administration of Ronald Reagan (1981–88) was especially eager to
fund military expenditures at a time when the Soviet Union, America's main rival, was in serious economic decline, followed in 1989–
90 by its collapse and fragmentation.

Throughout the cold war, aviation had played a prominent role
in the arms race with the Soviet Union. The result was the rise of
what Dwight Eisenhower called the "military-industrial complex,"
an alliance of military officers, arms (especially aerospace) manufacturers, and congressmen with bases or manufacturing plants in their
districts, supported by an elite of academic scientists and engineers.[48]
Even after the demise of the Soviet Union and the end of the cold
war, the complex did not disappear, for it was by then too deeply
entrenched in the American economy and political system. No
longer driven by fear of an enemy but by the challenge of the technology itself, the military-industrial complex continued to produce
ever more powerful and sophisticated warplanes and munitions. This
was a classic case of technological momentum driven by groups with
a vested interest in perpetuating the arms race, even without a
rival.[49] As a result, the United States moved far ahead of all other
nations in weapons systems, especially in aviation.

The first requirement of any air campaign is obtaining accurate
information on potential targets. Since the 1960s, satellites had provided high-resolution pictures of most of the world. To spot potential
enemy aircraft, the United States employed AWACS (Airborne
Warning and Control Systems) aircraft that carried powerful radar
that could see other aircraft up to 250 miles away.[50]

Once targets had been identified and located, the next task was
to destroy them from a safe distance. In all previous wars, hitting a
target meant bringing the bomber close to the target where air defenses were most dangerous. The correlation between precision and
risk is what had made bombing both dangerous and inaccurate in
all previous wars. The solution was precision-guided munitions such
as cruise missiles or bombs that could home in on their target while
the bomber stayed out of harm's way. Already toward the end of the
Vietnam War, the United States had deployed laser-guided bombs

called Paveway that could follow the light of a laser beam reflected off a target. While accurate to within six feet of the aiming point, laser-guided bombs required that the bomber or another plane keep the laser beam on the target for thirty seconds, a very long time in an air battle; furthermore, they could not be used through clouds, dust, or smoke. Other bombs introduced during the Vietnam War used a television camera to home in on an image of the target, or an infrared detector to strike hot targets. Laser-guided and electro-optical (television) guided bombs were estimated to be ten times more effective at hitting targets than "dumb" bombs. Even more sophisticated was a device introduced in the 1980s called LANT-IRN (Low Altitude Navigation and Targeting Infra Red for Night), a pod that guided the bomber toward the target, then released the bomb and guided it the rest of the way.[51]

For a bomber to locate a target and direct a laser beam at it was still dangerous if it was visible to the enemy's radar, hence vulnerable to anti-aircraft fire and surface-to-air missiles. In response to this threat, the U.S. Navy and Air Force acquired a missile called HARM (High-speed Anti-Radiation Missile) that could lock onto enemy radar in twenty seconds and destroy it. Even better were the Lockheed F-117 Nighthawk or stealth fighter-bombers, first flown in the 1980s. These strange-looking aircraft, all angles and flat surfaces covered with radar-absorbent material, were all but invisible to radar. Though they cost a hundred million dollars apiece, they proved invaluable in the Gulf War against Iraq in 1991.[52]

Numerous other planes also came out of the post-Vietnam development programs. Among the more radical were the F-15 Eagle and its two-seater version, the F-15 Strike Eagle, both large supersonic fighter-bombers capable of flying at night and in bad weather, and two lightweight fighters, the air force's F-16 Falcon and the navy's F/A-18 Hornet. These planes represented what one historian has called a "revolutionary discontinuity" in agility, maneuverability, power, and control. Equipped with engines so powerful they could accelerate the plane straight upward, they relied on computers and electronic controls to maneuver, a system called "fly-by-wire." The F-16E, for example, was controlled by 2.4 million lines of computer code.[53] For support of ground troops and to destroy tanks, the air

Figure 9.3. U.S. Air Force F-117 Nighthawk stealth fighter.

force developed the A-10 Thunderbolt II, familiarly known as the "Warthog" because of its ungainly shape, a slow but heavily armored plane able to survive machine-gun and even anti-aircraft fire and equipped with a thirty-millimeter cannon, the most powerful ever put on an airplane, as well as air-to-surface and air-to-air missiles. The army, meanwhile, acquired the AH-64 Apache attack helicopter equipped with laser-guided anti-tank missiles, the successor to the Vietnam-era Huey and Cobra helicopter gunships.[54] Most astonishing of all were the venerable B-52 Stratofortresses, some of which were older than their crew members; these huge planes could carry fifty-one 500-pound bombs or eighteen 2,000-pound bombs, much more than any other bomber. Once these 1950s-era bombers were equipped with the most modern electronics, they could launch smart bombs or cruise missiles from outside the range of enemy air defenses.[55] These and many other cutting-edge weapons systems were available to the U.S. armed forces, many of them stockpiled in Europe, when Iraq invaded Kuwait on August 2, 1990.

The Gulf War

The United States has fought two wars against Iraq—or is it one? The first one began in January 1991 and ended a few weeks later; the second began in March 2003 and is, as of this writing, still going on. The motives and politics for these wars will be debated for generations to come. The technologies involved, however, are already well understood. The question we need to address is how these two factors interacted.

When Iraq invaded Kuwait in 1990, the coalition that formed against it faced what appeared to be a formidable foe. Iraq had the largest army in the Middle East, with 800,000 troops, approximately 5,000 tanks, and over 3,500 artillery pieces.[56] Its air force was equally awe-inspiring, with 700 to 750 combat planes, mostly Vietnam-era MiG-21s, but also more modern MiG-23s, MiG-25s, the formidable MiG-29s, and French Mirage F-1s. It also had numerous Soviet bombers and fighter-bombers and helicopters.[57] Its air defenses used the most sophisticated Soviet and French technology; Baghdad was protected by seven times more anti-aircraft guns and surface-to-air missiles than had defended Hanoi at the time of Nixon's Linebacker II offensive.[58] But it was all an illusion, for the Iraqi army was poorly trained and had been used mainly to repress insurgencies and to fight the "human waves" of Iranian soldiers in the Iran-Iraq War (1980–88). Its air force was also weak, its pilots chosen for their political loyalty rather than their skill; fewer than half met Soviet standards and fewer than 20 percent met French standards.[59] The Iraqi dictator Saddam Hussein did not expect the United States (which had supported him in his war against Iran in the 1980s) to go to war over Kuwait. When war seemed imminent, he counted on the well-known American reluctance to endanger its soldiers and discounted the danger of an American air attack.

The United States and its coalition partners did not immediately rush into battle, but spent over five months preparing for war. By mid-January 1991, the coalition had amassed 2,400 aircraft and 1,400 helicopters in Saudi Arabia, Turkey, and the Persian Gulf. The U.S. Navy brought five aircraft carriers, each carrying seventy-five warplanes and accompanied by cruisers, destroyers, frigates, submarines, and support ships. The goals of this huge air armada were to isolate the Iraqi government from its armed forces by damaging telecommunications lines and electric power system, then destroy its air defenses, and finally weaken the Iraqi forces in Kuwait and in southern Iraq, thereby preparing the way for a land force of 600,000 troops and thousands of tanks, armored personnel carriers, and artillery pieces.[60]

The air campaign began on January 17. That night, F-117 stealth bombers, invisible to radar, destroyed Baghdad's communications center and air defense headquarters, as well as the presidential palace and the electrical power grid. With Iraqi air defenses in shambles, hundreds of other coalition combat aircraft delivered 1,700 direct hits in the next twenty-four hours without losing a single plane. B-52 bombers flew from Saudi Arabia, Spain, England, and the island of Diego Garcia in the Indian Ocean; some came from as far away as Louisiana, launched their cruise missiles, then returned to their base, a fourteen-thousand-mile nonstop trip.[61]

In forty-three days of bombing, the U.S. Air Force dropped roughly 61,000 tons of bombs, almost as many per month as in World War II and Vietnam. Only 6 to 10 percent were precision-guided munitions, but they caused 75 percent of the damage to Iraqi targets. Laser-guided bombs, in particular, hit their targets 98 percent of the time.[62] With the Iraqi air defense radars blinded by radar-seeking missiles, coalition warplanes flew over Iraq in relative safety; in 126,645 sorties, only 38 planes out of 2,500 were lost in combat, the lowest rate in aviation history.[63]

Confronted with such power, the Iraqi air force simply collapsed. The thirty-five to forty Iraqi planes that managed to take off were destroyed in combat. Two hundred more were destroyed on the ground or in their concrete bunkers. Another 120 to 140 fled to Iran, where their pilots were interned.[64] Having destroyed Iraq's air force and command and control system, the United States and its partners then turned on the Iraqi army in and near Kuwait. Using night-vision devices and infrared imaging, bombers could target Iraqi tanks and armored vehicles at night, forcing the Iraqis to bury them in sand. With all communications between Iraq and

Kuwait cut off, supplies could no longer reach the troops, who began to run out of food and water. By the time the ground war began on February 24, 1991, the Iraqi army had lost 60 percent of its tanks and artillery and 40 percent of its armored vehicles. It was no longer a fighting army.[65]

The ground war lasted a mere one hundred hours. Coalition tanks, self-propelled artillery, and armored personnel carriers ranged over the desert terrain almost at will. What Saddam Hussein called "the mother of all battles" turned into a rout, in which 150,000 Iraqi soldiers were killed or injured, and tens of thousands more were taken prisoner. Of the 148 Americans killed and 467 wounded in the war, 35 were killed and 73 wounded by "friendly fire," not by Iraqis.[66] This was war at its most asymmetrical, like the Spaniards at Cajamarca or the British at Omdurman. Air power enthusiasts rightly claim the victory for themselves; as Richard Hallion put it, "The Persian Gulf War will be studied by generations of military students, for it confirmed a major transformation in the nature of warfare: the dominance of air power. . . . Simply (if boldly) stated, air power won the Gulf war."[67]

There can be no doubt that superior air power won the battle for Kuwait, but did it win the war? If winning the war means destroying the enemy's armed forces, then the answer is yes. But if it means causing the enemy government to surrender, then surely not, for President George H. Bush's decision to order a cease-fire on February 28 stopped the war in mid-course and left Saddam Hussein in power. Many felt the goals of the war had been achieved, but others were frustrated to see victory snatched from their grasp. When George W. Bush (son of George H.) became president of the United States in 2001, he and his close advisors believed that the war with Iraq had not really ended, only temporarily interrupted.

The Iraq War

There is a vast and contentious literature on the reasons (real and imaginary) that led the second Bush administration to attack Iraq in 2003. Among them was the belief among neo-conservatives that the U.S. armed forces could achieve quickly and with few casualties the total victory that the president's father had denied them in 1991. This confidence came from the increase in American military power in the Middle East as well as from the weakening of the Iraqi military. If the Gulf War of 1990–91 was lopsided, the next round would be a victory march.

This confidence was strongest in the office of Secretary of Defense Donald Rumsfeld, who believed that thanks to advances in technology, the United States would need only one-third as many soldiers as had been brought to bear against Iraq in 1991. To Rumsfeld and his advisers, the traditional American military doctrine of overwhelming force was obsolete, replaced by information-age warfare. They put their faith in bombs that guided themselves with Global Positioning System (GPS) devices that could track their position by signals from satellites orbiting the earth, and then be guided to the target by an Inertial Measurement Unit (IMU). As one observer wrote, "the widespread availability of GPS-IMU weapons is increasing the confidence of Pentagon planners that smaller groups of soldiers bearing lighter equipment can achieve victory over more numerous enemy troops with the aid of devastatingly precise fire-power from the sky."[68]

The United States prepared for the next war both logistically and technologically. After the Gulf War, it built or expanded bases in Kuwait, Saudi Arabia, the United Arab Emirates, Bahrain, Qatar, Oman, Djibouti, and—after the bombing of the World Trade Center and the Pentagon in 2001—in Afghanistan, Kyrgyzstan, and Uzbekistan.[69] Many of the resources that had been stationed in Europe during the cold war were shifted to the Middle East.

The technology had also changed since the Gulf War. In addition to its old but still useful B-52 bombers and to the F-117s that had been the stars of the Gulf War, the U.S. Air Force could launch GPS-IMU bombs from B-1 Lancer supersonic bombers costing $200 million apiece and, most astonishing of all, from B-2 Spirit "stealth" bombers costing between $1 and 2.2 billion apiece. They could do so by day or night and in any weather, without requiring an aircraft to loiter nearby to guide them to their target. GPS guidance systems had been tried in the Gulf War, but only in expensive cruise missiles.

By 2003 the technology had improved to the point that such a guidance system could be fitted to any bomb. The result was a new generation of weapons: Joint Direct Attack Munitions (JDAM) that cost $20,000 to fit onto a "dumb" bomb and guide it up to eight miles; Joint Standoff Weapons (JSOW) with a range of fifteen to forty miles and costing $220,000 to $400,000; and Joint Air-to-Surface Standoff Missiles (JASSM) with a range of two hundred miles and a cost of $700,000. The days when Western nations had the advantage of less expensive and more cost-effective weapons than non-Western societies were long gone.[70]

With these, the air force and navy could achieve consistent accuracy without putting their planes or pilots in harm's way.[71] In order to find the targets, the U.S. armed forces used not only reconnaissance satellites and U-2 spy planes but also unmanned aerial vehicles like the Predator, a drone equipped with video cameras and satellite links that could stay in the air over enemy territory for thirty-three hours. With their help, the United States knew more about the disposition and movements of the Iraqi armed forces than the Iraqi commanders did.[72]

By 2003, the Iraqi armed forces were but the pathetic remains of what had seemed so formidable in 1990. For twelve years the northern and southern two-thirds of the country, declared a "no-fly" zone by the United Nations, had been patrolled by American and British planes that bombed any radar or anti-aircraft installation. Its equipment, damaged in the 1991 war, had never been rebuilt for lack of parts and maintenance. Its army was estimated at one-third the size it had been in 1991. Its few remaining combat planes had deteriorated and its pilots had received no training; when the war came, not a single one left the ground.[73] Thus, the outcome was even more foreordained than in 1991.

Yet the American attack during the night of March 21 to 22, 2003, was far more violent than in the Gulf War. The first day, six hundred cruise missiles and fifteen hundred combat planes hit a thousand targets. During the first thirty-three days, the United States and its lone ally, Great Britain, flew 1,576 sorties a day, two-thirds of them dropping precision weapons, with a loss of only two planes. They demolished buildings and heavy equipment. Simultaneously, their ground forces advanced as fast as the terrain and weather permitted, against almost insignificant Iraqi opposition. By the end of April, U.S. and British forces occupied most of Iraq. The new warfare—bringing to bear such overwhelming firepower that the enemy simply collapsed—became known as "Shock and Awe."[74] It had not been a war but a vast maneuver with live ammunition.[75] Only elusive targets that aerial reconnaissance could not pinpoint escaped for awhile; as one military writer put it, "The disadvantage of both weapons [i.e., JDAM and JSOW] is that in some applications . . . they require good intelligence on the exact location of the target."[76]

The Iraqi government and armed forces were abolished and their personnel dismissed. Saddam Hussein and his associates were captured and executed. And yet, as of this writing five years later, the war still rages on, pitting Shiites against Sunni and both against the U.S. occupation forces. How could the most advanced military technology the world has ever seen, deployed by the most powerful nation in the history of the world, fail so abysmally to control a country it had defeated so quickly? The American military achievement—overthrowing the Iraqi regime and destroying its armed forces—only aroused another, much more formidable enemy, namely Iraqi militias.[77] America's command of the air and its formidable firepower counted for little when Iraqis could easily obtain infantry weapons and homemade explosive devices. From a contest between military forces, in which the United States excelled, the war was transformed into a political struggle involving armed militias and terrorists, America's Achilles' heel. Williamson Murray and Major General Robert Scales, Jr., authors of a history of the war published soon after the invasion in 2003, intuited as much when they wrote:

> The conflict with Iraq engaged an enemy who had virtually no military capabilities left after an air war of attrition lasting over twelve years. Consequently, the conventional phase of the conflict was extremely lopsided and brief. . . . Nevertheless, unless advances in air power are coupled with intelligent thinking—by planners on

the ground—about the nature of one's opponent and of wars and their aftermath, past, present, and future, these improved technologies will ensure only that political and military defeats will come later, and at greater cost.[78]

Conclusion

In *Il dominio dell'aria*, Giulio Douhet asserted that bombing raids would terrorize a people into submission. His doctrine and variations thereof have had a powerful influence on aviation strategists ever since. It is this doctrine that led Field Marshal Herman Goering to order the bombing of Rotterdam and London in 1940. This doctrine led Allied leaders like Sir Arthur "Bomber" Harris and General Curtis LeMay to order the firebombing of German and Japanese cities in World War II. It also inspired Richard Nixon to order the bombing of Hanoi and Haiphong in 1972. Yet it has seldom worked, even when, as in World War II, entire populations were the enemy, not only their armed forces. More often, bombing reinforced the targeted populations' will to fight and their loyalty to their government. Douhet and his followers consistently misread the psychology of civilian populations.

In colonial and insurrectionary wars, the dominant or imperial power always assumes that insurgents do not represent the general population and can be targeted separately. Yet military technology does not allow such nice distinctions. Bombers are best at destroying immobile targets like buildings, or visible and slow-moving ones like tanks. Even when such bombing is extremely precise, as it has become in recent years, insurgents, who have little need for buildings or heavy equipment, can hide in caves or forests or melt into the civilian population. In an essay called "On Protracted War" written in 1938, Mao Zedong explained his view of asymmetrical warfare.

The theory that "weapons decide everything" . . . constitutes a mechanical approach to the question of war and a subjective and one-sided view. Our view is opposed to this; we see not only weapons but also people. Weapons are an important factor in war, but not the decisive factor; it is people, not things, that are decisive. The contest of strength is not only a contest of military and economic power, but also a contest of human power and morale.[79]

In asymmetrical wars, not only is bombing ineffective, it is often counterproductive. Here, too, Douhet's psychological assumption has proved false. Bombing, which is both violent and faceless, does nothing to win the "hearts and minds" of civilians (to use a Vietnam-era expression); even the shock and awe that bombing was expected to inspire among the Iraqi people in 2003 faded quickly. Instead, it inspires sympathy for the insurgents, who may be known to the civilians among whom they operate, or at least share the same language, ethnicity, and way of life. The boundary between insurgents and civilians is a porous one. For many civilians, especially unemployed young men, the temptation to join the insurgency is great. The more destructive the counterinsurgency, the more insurgents it creates. In short, the more military aviation has advanced technologically, the less effective it has become in asymmetrical conflicts. Rather than helping win wars, it helps lose them. The history of military aviation since World War II shows that ever-improving technology will, in the words of Murray and Scales, "ensure only that political and military defeats will come later, and at greater cost."

Notes

1. For examples of air power history, see Robin Higham, *100 Years of Air Power and Aviation* (College Station: Texas A&M University Press, 2003); Basil Collier, *A History of Air Power* (London: Weidenfeld and Nicholson, 1974); and James L. Stokesbury, *A Short History of Air Power* (New York: William Morrow, 1986).

2. The most famous such book, Colonel Charles E. Callwell's *Small Wars: Their Principles and Practice* (London: HMSO, 1896 and many subsequent editions) was written before aircraft existed. Another classic, Major General Charles W. Gwynn's *Imperial Policing* (London: Macmillan, 1934), mentions aviation, but downplays its role.

3. Among the most valuable such works are Anthony Towle, *Pilots and Rebels: The Use of Aircraft in Unconventional Warfare* (London: Brassey's, 1989); David E.

Omissi, *Air Power and Colonial Control: The Royal Air Force, 1919–1939* (Manchester: Manchester University Press, 1990); and James S. Corum and Wray R. Johnson, *Airpower in Small Wars: Fighting Insurgents and Terrorists* (Lawrence: University of Kansas Press, 2003).

4. The western half of Indochina, consisting of Laos and Cambodia, followed the same timetable as Vietnam, but were much more thinly populated and less thoroughly occupied and developed by the French.

5. Towle, *Pilots and Rebels*, pp. 106–16; Corum and Johnson, *Airpower in Small Wars*, pp. 139–46.

6. Lionel Max Chassin, *Aviation Indochine* (Paris: Amiot-Dumont, 1954), pp. 47–67. General Chassin was commander in chief of the French air force in Indochina from 1951 to 1953.

7. Chassin, *Aviation Indochine*, pp. 71–75, 176–86.

8. Towle, *Pilots and Rebels*, pp. 108–13, 247–48; Corum and Johnson, *Airpower in Small Wars*, pp. 145–56.

9. Towle, *Pilots and Rebels*, p. 107; Corum and Johnson, *Airpower in Small Wars*, pp. 150–60.

10. Chassin, *Aviation Indochine*, pp. 201–14; Towle, *Pilots and Rebels*, pp. 111–12; Corum and Johnson, *Airpower in Small Wars*, pp. 157–59.

11. Edgar O'Ballance, *The Algerian Insurrection, 1954–1962* (Hamden, Conn.: Archon Books, 1967), pp. 90, 141, 215; Towle, *Pilots and Rebels*, pp. 117–19; Corum and Johnson, *Airpower in Small Wars*, pp. 161–66.

12. Mohamed Lebaoui, *Vérités sur la Révolution algérienne* (Paris: Gallimard, 1970), pp. 127–38; Serge Bromberger, *Les rebelles algériens* (Paris: Plon, 1958), pp. 218–21, 251–55; Hartmut Elsenhans, *Frankreich's Algerienkrieg, 1954–1962: Entkolonisierungsversuch einer kapitalistischen Metropole: Zum Zusammenbruch der Kolonialreiche* (Munich: C. Hansen, 1974), pp. 381–83; O'Ballance, *The Algerian Insurrection*, pp. 39, 48–54, 88–89, 98, 138–40.

13. Martin Thomas, "Order before Reform: The Spread of French Military Operations in Algeria, 1954–1958," in David Killingray and David Omissi, eds., *Guardians of Empire: The Armed Forces of the Colonial Powers c. 1700–1964* (Manchester: Manchester University Press, 1999), p. 216.

14. Towle, *Pilots and Rebels*, pp. 117–25; Corum and Johnson, *Airpower in Small Wars*, pp. 166–70.

15. Corum and Johnson, *Airpower in Small Wars*, pp. 171–72.

16. General Michel Forget, *Guerre froide et guerre d'Algérie, 1954–1964: Témoignage sur une période agitée* (Paris: Economica, 2002), pp. 163–204; Alastair Horne, *A Savage War of Peace: Algeria, 1954–1962* (New York: New York Review Books, 2006), pp. 334–38.

17. Michael Adas, *Dominance by Design: Technological Imperatives and America's Civilizing Mission* (Cambridge, Mass.: Harvard University Press, 2006), p. 291; Max Boot, *The Savage Wars of Peace: Small Wars and the Rise of American Power* (New York: Basic Books, 2002), p. 292.

18. Ronald B. Frankum, Jr., *Like Rolling Thunder: The Air War in Vietnam, 1964–1975* (Lanham, Md.: Rowman and Littlefield, 2005), pp. 7–9.

19. Mark Clodfelter, *The Limits of Air Power: The American Bombing of North Vietnam* (New York: Free Press, 1989), p. 133; Frankum, *Like Rolling Thunder* passim; Towle, *Pilots and Rebels*, pp. 163–64.

20. Donald J. Mrozek, *Air Power and the Ground War in Vietnam* (Washington, D.C.: Pergamon-Brassey, 1989), pp. 114–28; Higham, *100 Years*, p. 248; Towle, *Pilots and Rebels*, pp. 159–65.

21. Raphael Littauer and Norman Uphoff, *The Air War in Indochina* (Boston: Beacon Press, 1972), pp. 91–96; Frankum, *Like Rolling Thunder*, pp. 88–92; Mrozek, *Air Power*, pp. 132–39; Towle, *Pilots and Rebels*, p. 168.

22. Quoted in Adas, *Dominance by Design*, p. 311.

23. Mrozek, *Air Power*, pp. 139–44; Towle, *Pilots and Rebels*, p. 165.

24. Mrozek, *Air Power*, p. 27. See also Corum and Johnson, *Airpower in Small Wars*, pp. 267–74.

25. Quoted in Adas, *Dominance by Design*, p. 291.

26. Robert A. Pape, *Bombing to Win: Air Power and Coercion in War* (Ithaca: Cornell University Press, 1996), pp. 175–84.

27. Boot, *Savage Wars of Peace*, p. 291.

28. Frankum, *Like Rolling Thunder*, pp. 20–21, 79; Clodfelter, *The Limits of Air Power*, p. 133; Higham, *100 Years*, pp. 245–52; Littauer and Uphoff, *The Air War in Indochina*, pp. 9–10. Per serviceman, the United States dropped twenty-six times more tons of munitions on Vietnam than in World War II; see Gabriel Kolko, *Vietnam: Anatomy of a War, 1940–1975* (London: Pantheon, 1986), p. 189.

29. General David Shoup, "The New American Militarism," *Atlantic Monthly* (April 1969), p. 55.

30. Pape, *Bombing to Win*, pp. 184–95; Clodfelter, *The Limits of Air Power*, pp. 131–36.

31. Pape, *Bombing to Win*, p. 192; Littauer and Uphoff, *The Air War in Indochina*, pp. 69–72; Higham, *100 Years*, p. 247.

32. Michael T. Klare, *War without End: American Planning for the Next Vietnams* (New York: Vintage, 1972), pp. 170–91; Paul Dickson, *The Electronic Battlefield* (Bloomington: Indiana University Press, 1976), pp. 83–95; Paul N. Edwards, *The Closed World: Computers and the Politics of Discourse in Cold War America* (Cambridge, Mass.: MIT Press, 1996), pp. 3–4; James W. Gibson, *The Perfect War: Technowar in Vietnam* (Boston: Atlantic Monthly Press, 1986), pp. 396–98.

33. On precision-guided munitions or "smart" bombs used in Vietnam, see Paul G. Gillespie, *Weapons of Choice: The Development of Precision Guided Munitions* (Tuscaloosa: University of Alabama Press, 2006), chapter 5: "Vietnam: Precision Munitions Come of Age."

34. Pape, *Bombing to Win*, pp. 175–201; Clodfelter, *The Limits of Air Power*, pp. 159–62; Mrozek, *Air Power*, pp. 101–3.

35. Higham, *100 Years*, p. 264; Clodfelter, *The Limits of Air Power*, pp. 131, 165; Frankum, *Like Rolling Thunder*, p. 152.

36. Clodfelter, *The Limits of Air Power*, p. 203.

37. Richard M. Nixon, *No More Vietnams* (New York: Arbor House, 1985), p. 56.

38. Edward Girardet, *Afghanistan: The Soviet War* (New York: St. Martin's, 1985), pp. 12, 32–33.

39. Edgar O'Ballance, *Afghan Wars, 1839–1992: What Britain Gave Up and the Soviet Union Lost*, 2nd ed. (New York: Brassey's, 2002), pp. 97–99.

40. Mark Galeotti, *Afghanistan, the Soviet Union's Last War* (London: Frank Cass, 1995), p. 199; O'Ballance, *Afghan Wars*, pp. 100–101.

41. Girardet, *Afghanistan*, pp. 32–38; Corum and Johnson, *Airpower in Small Wars*, 393–97.

42. Towle, *Pilots and Rebels*, pp. 194–201; Corum and Johnson, *Airpower in Small Wars*, pp. 391–92; O'Ballance, *Afghan Wars*, pp. 102–3, 122–23; Girardet, *Afghanistan*, pp. 30, 42–43.

43. Girardet, *Afghanistan*, pp. 37–43; Towle, *Pilots and Rebels*, pp. 197–98; Galeotti, *Afghanistan*, p. 196.

44. Girardet, *Afghanistan*, p. 33; O'Ballance, *Afghan Wars*, p. 114.

45. On the U.S. involvement in Afghanistan and the Stinger missiles, see George Crile, *Charlie Wilson's War: The Extraordinary Story of the Largest Covert Operation in History* (New York: Atlantic Monthly Press, 2003), chapter 28; Towle, *Pilots and Rebels*, pp. 192–205; Corum and Johnson, *Airpower in Small Wars*, pp. 393–97; and O'Ballance, *Afghan Wars*, pp. 145–63.

46. On the strategic doctrine, see Pape, *Bombing to Win*, pp. 211–12.

47. The first two were the classic Industrial Revolution of the late eighteenth and early nineteenth centuries with its cotton mills and steam engines, and the second industrial revolution of the late nineteenth and early twentieth centuries with its steel, chemical, and electrical industries and internal combustion engines.

48. Alex Roland, *The Military-Industrial Complex* (Washington, D.C.: AHA Publications, 2001).

49. I owe the concept of technological momentum to Thomas P. Hughes. See his *Networks of Power: Electrification in Western Society, 1880–1930* (Baltimore: Johns Hopkins University Press, 1983), p. 140.

50. Richard Hallion, *Storm over Iraq: Air Power and the Gulf War* (Washington, D.C.: Smithsonian Institution Press, 1992), pp. 308–12; Lon O. Nordeen, *Air Warfare in the Missile Age* (Washington, D.C.: Smithsonian Institution Press, 1985), p. 233.

51. David R. Mets, *The Long Search for a Surgical Strike: Precision Munitions and the Revolution in Military Affairs* (Maxwell Air Force Base, Ala.: Air University Press, 2001), pp. 28–29; Kenneth P. Werrell, *Chasing the Silver Bullet: U.S. Air Force Weapons Development from Vietnam to Desert Storm* (Washington, D.C.: Smithsonian Institution Press, 2003), p. 258; Williamson Murray and Maj. Gen. Robert H. Scales,

Jr., *The Iraq War: A Military History* (Cambridge, Mass.: Harvard University Press, 2003), p. 49; James F. Dunnigan and Austin Bay, *From Shield to Storm: High-Tech Weapons, Military Strategy, and Coalition Warfare in the Persian Gulf* (New York: Morrow, 1992), pp. 221–22; Hallion, *Storm over Iraq*, pp. 303–7; Nordeen, *Air Warfare in the Missile Age*, pp. 59, 230.

52. Hallion, *Storm over Iraq*, pp. 293–94; Werrell, *Chasing the Silver Bullet*, pp. 221–47; Dunnigan and Bay, *From Shield to Storm*, p. 204.

53. James P. Coyne, *Airpower in the Gulf* (Arlington, Va.: Air Force Association, 1992), pp. 74–75; Hallion, *Storm over Iraq*, pp. 276–92; Dunnigan and Bay, *From Shield to Storm*, p. 203.

54. Hallion, *Storm over Iraq*, pp. 23–24, 284–87; Coyne, *Airpower in the Gulf*, p. 78.

55. Dunnigan and Bay, *From Shield to Storm*, pp. 205–6.

56. These figures are from John Keegan, *The Iraq War* (New York: Knopf, 2004), p. 129. Other authors give different figures. Murray and Scales (*The Iraq War*, p. 82) write that Iraq had 5,100 tanks and cannon; according to Nordeen (*Air Warfare in the Missile Age*, p. 206), Iraq had 800,000 troops, 4,700 tanks, and 3,700 artillery pieces; and Pape (*Bombing to Win*, p. 252) puts the numbers at 300,000 men, 3,500 tanks, and 2,500 cannon.

57. Nordeen, *Air Warfare in the Missile Age*, p. 208; Hallion, *Storm over Iraq*, pp. 242–43; Pape, *Bombing to Win*, p. 227.

58. Werrell, *Chasing the Silver Bullet*, p. 223; Nordeen, *Air Warfare in the Missile Age*, p. 208; Hallion, *Storm over Iraq*, p. 169; Murray and Scales, *The Iraq War*, p. 4.

59. Pape, *Bombing to Win*, p. 227; Nordeen, *Air Warfare in the Missile Age*, p. 208; Murray and Scales, *The Iraq War*, pp. 81–82.

60. Chalmers Johnson, *The Sorrows of Empire: Militarism, Secrecy, and the End of the Republic* (New York: Henry Holt, 2004), pp. 219, 242; Nordeen, *Air Warfare in the Missile Age*, pp. 209–13, 231; Pape, *Bombing to Win*, pp. 220–22.

61. Coyne, *Airpower in the Gulf*, pp. 47–69; Hallion, *Storm over Iraq*, pp. 163–74.

62. Again, figures differ: According to Michael Puttré ("Satellite-Guided Munitions," *Scientific American*, February 2003, p. 70), 6 percent of the bombs were laser-guided. Gillespie (*Weapons of Choice*, pp. 137–38) says 8 percent; Hallion (*Storm over Iraq*, p. 188) says 9 percent; and Mets (*The Long Search*, pp. 35–36) says 10 percent. See also Werrell, *Chasing the Silver Bullet*, p. 258.

63. Norman Friedman, *Desert Victory: The War for Kuwait* (Annapolis: Naval Institute Press, 1991), pp. 147–68; Rod Alonso, "The Air War," in Bruce W. Watson et al., *Military Lessons of the Gulf War* (London: Greenhill, 1991), pp. 61–80; Michael Mazarr, Don M. Snider, and James A. Blackwell, Jr., *Desert Storm: The Gulf War and What We Learned* (Boulder, Colo.: Westview, 1993), pp. 93–124; Nordeen, *Air Warfare in the Missile Age*, p. 230; Pape, *Bombing to Win*, p. 228.

64. Hallion, *Storm over Iraq*, pp. 175–95; Dunnigan and Bay, *From Shield to Storm*, pp. 145–53; Coyne, *Airpower in the Gulf*, pp. 52–54.

65. Friedman, *Desert Victory*, pp. 169–96.

66. Werrell, *Chasing the Silver Bullet*, p. 249; Hallion, *Storm over Iraq*, pp. 231–47; Dunnigan and Bay, *From Shield to Storm*, p. 145; Murray and Scales, *The Iraq War*, pp. 4–7.

67. Hallion, *Storm over Iraq*, p. 1.

68. Puttré, "Satellite-Guided Munitions," p. 73. See also Todd S. Purdum, *A Time of Our Choosing: America's War in Iraq* (New York: Henry Holt, 2003), pp. 96–98.

69. Johnson, *Sorrows of Empire*, pp. 226–51.

70. On the declining costs of European firearms in the early modern period, see Philip T. Hoffman, "Why Is It That Europeans Ended Up Conquering the Rest of the Globe? Price, the Military Revolution, and Western Europe's Comparative Advantage in Violence," http://gpih.ucdavis.edu/files/Hoffman/pdf (accessed March 9, 2008).

71. Puttré, "Satellite-Guided Munitions," pp. 68–72; Mets, *The Long Search*, p. 43; Murray and Scales, *The Iraq War*, pp. 71–76, 155–61.

72. Purdum, *A Time of Our Choosing*, pp. 121–22; Murray and Scales, *The Iraq War*, p. 163.

73. Purdum, *A Time of Our Choosing*, pp. 99, 129; Murray and Scales, *The Iraq War*, pp. 82, 162–63; Nordeen, *Air Warfare in the Missile Age*, pp. 234–36.

74. The expression "shock and awe" was coined by former navy commander Harlan K. Ullman and James P. Wade in a 1996 paper for the National Defense University under the title "Shock and Awe: Achieving Rapid Dominance," http://www.dodccrp.org/files/Ullman_Shock.pdf (accessed January 28, 2009); Purdum, *A Time of Our Choosing*, p. 124.

75. Purdum, *A Time of Our Choosing*, pp. 5, 122; Murray and Scales, *The Iraq War*, pp. 110, 166–78; Keegan, *The Iraq War*, pp. 127, 142–43.

76. Mets, *The Long Search*, p. 43.

77. Perhaps the U.S. Army and Marine Corps realized this when they issued a new counterinsurgency manual—the first in more than two decades—in December 2006; see George Packer, "Knowing the Enemy," *New Yorker*, December 18, 2006, p. 62.

78. Murray and Scales, *The Iraq War*, p. 183.

79. Mao Zedong, "On Protracted War," in *Selected Works of Mao Tse-tung* (Peking: Foreign Languages Press, 1965), vol. 2, pp. 143–44.

✸ Conclusion ✸

Technology and Imperialism Redux

There are no laws in history. Nor is history merely a long string of factoids. By studying enough cases over a long enough span of time and in enough places, we see patterns emerge. What patterns can we discern in the history of Western imperialism over the past six hundred years? Clearly, technology matters. On several occasions, small groups were able to overcome the resistance of larger groups thanks to the weapons, the animals, and the equipment they used. The Portuguese in the Indian Ocean and the Spaniards in Mexico and Peru in the sixteenth century, and various European countries in Africa and Asia in the nineteenth, all translated technological gaps between themselves and their opponents into victorious empires.

But technologies are always environment-specific. While they provide their owners with power over nature, that power is limited to specific parts of nature. Thus, the Portuguese ships that dominated the Indian Ocean for a time were almost useless in the Red Sea; horses that served the Europeans so well in the Americas died of nagana in Angola and Mozambique. Environments can help or hinder conquerors; diseases helped the Spaniards in the Americas, but thwarted would-be invaders of tropical Africa for four hundred years. The organization and tactics that served the British well in India were ineffective in the mountains of Afghanistan.

Technology changes, and the environments that frustrate the imperialists of one era can succumb to the more advanced technology of a later era. In the early nineteenth century, steam-powered gunboats and ships brought Europeans into Burma, China, the Middle East, and parts of Africa. In the mid-nineteenth and early twentieth centuries, Europeans and Americans developed health measures that overcame the disease barrier of tropical Africa and firearms that defeated societies that had resisted Western invaders for centuries. Finally, in the early twentieth century, airplanes and bombs were able to overcome the resistance of peoples who had acquired the modern firearms that had given the West such an advantage a generation earlier.

These examples, then, confirm the statement made by Leon Kass that I quoted in the introduction: "What we really mean by 'Man's Power over Nature' is a power exercised by some men over other men, with a knowledge of nature as their instrument." And yet the "knowledge of nature" that provides "power over other men" is not the monopoly of any one culture. Although only a few non-Western societies can emulate the West's most advanced and elaborate technologies, almost all can acquire the simpler Western technologies of an earlier generation. When the Indians of Chile, Argentina, and the Great Plains of North America acquired horses, they became powerful enough to stop the European advance for several centuries. Likewise, modern firearms diffused from European-led armies in Africa to those of Ethiopia and the Rif and, after World War II, to Vietnam, Algeria, and Afghanistan.

Imperial expeditions were undertaken for a variety of motives, but sooner or later, the costs and benefits made themselves felt. The costs were both financial and human. Times when imperialists were successful were also times when the costs were low or outweighed by the benefits. Compared to the costs of warfare in Europe, the Spanish conquests in the Americas cost very little in treasure or manpower. Little Portugal, one of Europe's poorest kingdoms, could afford to maintain a navy in the Indian Ocean, thanks to the booty, protection payments, and trading profits it brought in. As diminishing returns set in, Western empire-building slowed down in the eighteenth and early nineteenth centuries, except in India, where the British succeeded in getting Indians to pay for their own conquest. In Algeria and the Caucasus, rising costs slowed the imperial expansion; in Afghanistan, they thwarted it. Western imperialism revived in the late nineteenth century as industrially manufactured weapons and medical advances lowered the cost of conquests in both money and lives.

Since the early twentieth century, the Western powers have dominated the sea and the air. For a while, air power proved effective in controlling the land, albeit at the price of ever-increasing barbarity. But after a generation, those who resisted the invaders and colonizers learned to adapt their defenses by operating at night, underground, in rough terrain, or in cities, where air power is either ineffective or exacts too high a price in lives and honor for the invaders to bear. Against well-trained resistance fighters, imperial forces could fly and bomb at will, but could seldom win. France, the United States, and the Soviet Union were defeated time and time again by smaller, poorer, and weaker peoples. In the face of such humiliations, some responded by seeking ever more sophisticated technologies.

In the early twenty-first century, not only is the United States far more powerful than any other nation on the planet, it also possesses far more power over nature than any nation ever did in the past. Technology—or rather the companies that produce the technology—sets policy. As Robert J. Stevens, the chief executive officer of the Lockheed Corporation—a company that receives 80 percent of its revenue from the U.S. government—explained: "We are deployed entirely in developing daunting technology" that requires "thinking through the policy dimensions of national security as well as technological dimensions." To ensure their closeness to the decision-makers, the *New York Times* reported in November 2004,

> [f]ormer Lockheed executives, lobbyists and lawyers hold crucial posts at the White House and the Pentagon, picking weapons and setting policies. . . . Men who have worked, lobbied and lawyered for Lockheed hold the posts of secretary of the Navy, secretary of transportation, director of the national nuclear weapons complex and director of the national spy satellite agency. . . . Former Lockheed executives serve on the Defense Policy Board and the Homeland Security Advisory Council, which help make military and intelligence policy and pick weapons for future battles.

As a result, Stevens claims, "With technology we've been able to make ourselves more secure and more humane. . . . I don't say this lightly. Our industry has contributed to a change in humankind."[1]

An article in *Scientific American* imagined a technological fix to the problem of terrorism.

> Suppose that U.S. intelligence finds indisputable evidence that a major terrorist is dining right now in a remote farmhouse in central Asia. Say also that the local political sensitivities prohibit calling in bombers for an air strike and that the meal is unlikely to last the two hours it would take a Tomahawk cruise missile to reach the site from its maximum range. How to respond? Pentagon weapons procurers hope to have an answer in an advanced turbine engine that can shrink a cruise missile's "time on target" to "tens of minutes". Such a system might catch the hypothetical terrorist chief before dessert.[2]

Leaving aside the notion of U.S. intelligence finding indisputable evidence, what does this project tell us? That the answer to terrorism is cruise missiles with advanced turbine engines? Or, in more general terms, that the proper response to an adversary is technology, and the more sophisticated the better? There have been times when such a response worked well, and other times when it failed. As of this writing, the United States is spending several billion dollars a week on maintaining its forces in Iraq, much of it on expensive equipment, with no victory or even honorable withdrawal in sight. Now that world attention is once more focused on a confrontation between a powerful, technologically advanced nation and a weaker, poorer one, it is time to revisit the history of such encounters, and learn its lessons.

Notes

1. Tim Weiner, "Lockheed and the Future of Warfare," *New York Times*, November 28, 2004, pp. 1, 4.

2. Steven Ashley, "Mach 3 Hunter-Killer: An Advanced Turbine Design for Versatile Missiles," *Scientific American* (September 2006), 26.

❋ FOR FURTHER READING ❋

The sources upon which this work is based are listed in the end-notes. Some of them, however, are particularly valuable and bear highlighting.

Only a few books have dealt with the general theme of this book. Carlo Cipolla's *Guns, Sails, and Empires: Technological Innovation and the Early Phases of European Expansion, 1400–1700,* published in 1965, was the first work to describe the relationship between techno-logical innovation and imperialism. It was the inspiration for my early work *The Tools of Empire: Technology and European Imperialism in the Nineteenth Century* (1981). Geoffrey Parker's *The Military Rev-olution: Military Innovation and the Rise of the West, 1500–1800,* 2nd ed. (1996), dealt in part with the theme of technology, but more so with the organization and financing of armed forces, both in Europe and overseas. Ronald Findlay and Kevin O'Rourke's *Power and Plenty: Trade, War, and the World Economy in the Second Millennium* (2007) discusses the economic and commercial aspects of Western imperialism. Michael Adas's *Machines as the Measure of Men: Sci-ence, Technology, and Ideologies of Western Dominance* (1989) de-scribes the ideologies of dominance, that is, the history of Western imperialism seen through the eyes of Western writers.

Much has been written on the theme of chapter 1, "The Discov-ery of the Oceans, to 1779." The very best introduction to this subject are the works of J. H. Parry, especially *The Discovery of the Sea* (1974). Also of interest is *When China Ruled the Seas: The Trea-sure Fleet of the Dragon Throne, 1405–1433* by Louise Levathes (1994) and *The Career and Legend of Vasco da Gama* by Sanjay Subrahmanyam (1997).

Naval warfare and politics in the Indian Ocean in the sixteenth and seventeenth centuries has also attracted scholarly attention. Two works by Charles R. Boxer are especially valuable: *The Portu-*

guese Seaborne Empire, 1415–1825 (1969) and *The Dutch Seaborne Empire, 1600–1800* (1965). Though dealing with the Mediterranean rather than the Indian Ocean, John F. Guilmartin's *Gunpowder and Galleys: Changing Technology and Mediterranean Warfare at Sea in the 16th Century* (2003) is particularly good on Ottoman naval technology. I am grateful to Giancarlo Casale for letting me read his Ph.D. dissertation, "The Ottoman Age of Exploration: Spices, Maps and Conquest in the Sixteenth-Century Indian Ocean" (Harvard University, 2004), in which he examines in detail the confrontations between Portuguese and Ottomans. In addition to Geoffrey Parker's *Military Revolution* mentioned earlier, two other books proved valuable in understanding this period of history: William H. McNeill's *The Pursuit of Power: Technology, Armed Forces, and Society since* A.D. 1000 (1982) and Kenneth W. Chase's *Firearms: A Global History to 1700* (2003).

The conquest of the Americas has been one of the most popular topics in history since the time of Columbus. Among the myriad books that have been written on the subject, I found a few especially useful. Ross Hassig's *Aztec Warfare: Imperial Expansion and Political Control* (1988) and *Mexico and the Spanish Conquest* (1994) elucidate the military aspects of the confrontation. The role of diseases is examined in William H. McNeill's *Plagues and Peoples* (1976) and two works by Alfred W. Crosby, *The Columbian Exchange: Biological and Cultural Consequences of 1492* (1972) and *Ecological Imperialism: The Biological Expansion of Europe, 900–1900* (1986). The centuries-long confrontation between whites and Indians in North America is the subject of Walter Prescott Webb's classic *The Great Plains* (1931), now dated but still fascinating to read.

Imperialism in Africa and Asia, the subject of chapter 4, also has a vast literature. On the role of diseases in delaying the European penetration of Africa, see Philip D. Curtin's *The Image of Africa: British Ideas and Actions, 1780–1850* (1964) and *The Rise and Fall of the Plantation Complex: Essays in Atlantic History* (1990). Bruce Lenman's *Britain's Colonial Wars, 1688–1783* (2001) and David B. Ralston's *Importing the European Army: The Introduction of European Military Techniques and Institutions into the Extra-European World, 1600–1914* (1990) are excellent accounts of the British conquest of India. The best work in English on the French conquest and the Algerian resistance is Raphael Danziger's *Abd al-Qadir and the Algerians: Resistance to the French and Internal Consolidation* (1977). An excellent introduction to Russian imperialism in the Caucasus is Charles King's *The Ghost of Freedom: A History of the Caucasus* (2008).

In contrast to the subjects of the previous chapters, steamboat imperialism has received relatively little scholarly attention, despite its importance. Louis C. Hunter's *Steamboats on the Western Rivers: An Economic and Technological History* (1949) is a classic of American history, but devotes little space to the impact of steamboats on white-Indian relations. R. G. Robertson describes the impact of diseases on North American Indians in *Rotting Face: Smallpox and the American Indian* (2001). A fine book on the introduction of steamboats to India is Henry T. Bernstein's *Steamboats on the Ganges: An Exploration in the History of India's Modernization through Science and Technology* (1960). On the impact of steam navigation on communications between Britain and India, see Halford L. Hoskins, *British Routes to India* (1928), and a more recent book, Sarah Searight, *Steaming East: The Hundred Year Saga of the Struggle to Forge Rail and Steamship Links between Europe and India* (1991). British naval actions in China are the subject of Gerald S. Graham, *The China Station: War and Diplomacy, 1830–1960* (1978), and Arthur Waley, *The Opium War through Chinese Eyes* (1958). On the role of steamboats in Africa, see K. Onwuka Dike, *Trade and Politics in the Niger Delta, 1830–1885* (1956), and Paul Mmegha Mbaeyi, *British Military and Naval Forces in West African History, 1807–1874* (1978).

On the role of diseases in the European penetration of Africa, two works by Philip D. Curtin are especially valuable: *Death by Migration: Europe's Encounter with the Tropical World in the Nineteenth Century* (1989) and *Disease and Empire: The Health of European Troops in the Conquest of Africa* (1998). The role of botany in nineteenth-century imperialism is the subject of Lucile H. Brockway's *Science and Colonial Expansion: The Role of the British Botanic Gardens* (1972). On the role of disease and medicine in the Spanish-American War, see Vincent J. Cirillo, *Bullets and Bacilli: The Spanish-American War and Military Medicine* (2004); and on the construction of the Panama Canal, see David G. McCullough, *The Path between the Seas: The Creation of the Panama Canal, 1870–1914* (1977).

There is an enormous literature on weapons, much of it written by and for enthusiasts, yet some of it is essential to understanding the use of weapons in colonial warfare. William Wellington Greener's *The Gun and Its Development* (9th ed., 1910) is a valuable compendium of information, as is William Young Carman's *A History of Firearms from Earliest Times to 1914* (1955). On the wars against the Indians of North America, see Joseph G. Rosa, *Age of the Gunfighter: Men and Weapons on the Frontier, 1840–1900* (1995). Two histories of machine guns are worth reading: John Ellis's *The Social History of the Machine Gun* (1975) and Anthony Smith's *Machine Gun: The Story of the Men and the Weapon That Changed the Face of War* (2002) are very interesting, though they only refer occasionally to colonial warfare. Few military writers paid attention to colonial campaigns; one exception is Colonel Charles E. Callwell, whose book *Small Wars, Their Principles and Practice* (3rd ed., 1906) went through several editions.

Like the history of firearms, the history of military aviation before 1945 is overwhelmingly biased toward "big" wars between major powers. The role of aviation in colonial and asymmetrical wars has received considerably less attention. Nonetheless, some works are noteworthy. James S. Corum and Wray R. Johnson's *Airpower in Small Wars: Fighting Insurgents and Terrorists* (2003) covers most of the twentieth century, as does Anthony Towle's *Pilots and Rebels: The Use of Aircraft in Unconventional Warfare, 1918–1988*. More specific conflicts are covered in David E. Omissi's *Air Power and Colonial Control: The Royal Air Force, 1919–1939* (1990), and in Sebastian Balfour's *Deadly Embrace: Morocco and the Road to the Spanish Civil War* (2002).

The contemporary period since 1945 has received a lot of attention, but much of it suffers from the shortsightedness inevitable in dealing with the recent past. Nonetheless, there are several interesting books that deal with air power in asymmetrical conflicts. Edgar O'Ballance has written about several such conflicts, in particular *The Algerian Insurrection, 1954–1962* (1967) and *Afghan Wars, 1839–1992*, 2nd ed. (2002). But it is the Vietnam War that has inspired the most analyses of air power, its successes and its failings; see Ronald B. Frankum's *Like Rolling Thunder: The Air War in Vietnam, 1964–1975* (2005), Mark Clodfelter's *The Limits of Air Power: The American Bombing of North Vietnam* (1989), and Donald J. Mrozek's *Air Power and the Ground War in Vietnam* (1989). Robert A. Pape's *Bombing to Win: Air Power and Coercion in War* (1996) also deals with the Vietnam conflict. Among the books that deal with the development of aviation technology after Vietnam, the best is Kenneth P. Werrell's *Chasing the Silver Bullet: The U.S. Air Force Weapons Development from Vietnam to Desert Storm* (2003). The Iraq war is covered in James P. Coyne's *Airpower in the Gulf* (1992) and Richard Hallion's *Storm over Iraq: Air Power and the Gulf War* (1992).